Assembly Language Using the Raspberry Pi

A Hardware Software Bridge

Robert Dunne

Assembly Language Using the Raspberry Pi: A Hardware Software Bridge

 Published in the United States of America by Gaul Communications, Downers Grove, Illinois.

ISBN 978-0-9701124-2-2 (hardcover)

ISBN 978-0-9701124-0-8 (digital)

18 17 16 15 14 13 12 11 10 9 8 7 6 5 4

To

the four ladies in my life

Carol

Shannon

Robyn

Kelly

Contents at a Glance

Contents

Preface

How do you use your Raspberry Pi? That's a question I ask my students as well as people I meet in the business world. Some use it for games and entertainment. Some use it as part of a workstation in a network of computers. Some don't even use it, but bought one simply because the price was very reasonable. I primarily use the Raspberry Pi as part of instruction in electronics and computer science. Education was the principal motive for the development of the Raspberry Pi, and you can learn about its creation through the Raspberry Pi Foundation, a registered charity in the United Kingdom that promotes computer literacy and enthusiasm.

Assembly language is the computer programming language closest to a CPU's (Central Processing Unit) "machine language." Programs written in assembly language are unique to a particular CPU design and are not portable from one CPU manufacturer or model to another. Programs written in higher level languages like Java, Python, C++, and Lisp are generally independent of the native hardware architecture of the CPU on which they will eventually run.

Audience for This Book

This book is part of a lecture and lab course in embedded systems. Embedded systems are internal computers that control almost everything today: automobile engines, television sets, furnaces, vacuum cleaners, cell phones, etc. The objective of this book is to introduce the novice to both CPU hardware and professional software development methods by working through sample assembly language programs. The intended audience is the following:

- Anyone wanting to learn assembly language, especially individuals interested in the Raspberry Pi, the ARM CPU, VFPv3 floating point coprocessor, and NEON coprocessor architectures in particular.
- Someone who already has assembly language experience, but now wants to become familiar with the ARM CPU, VFPv3 IEEE 754 floating point processor, and NEON vector processor resident in the Raspberry Pi.
- Electronics engineers who want to bridge the gap toward software development, and software engineers who want to bridge the gap toward CPU hardware operation.
- College students enrolled in embedded systems and computer architecture courses.

Expected background for someone reading this book to learn assembly language using the Raspberry Pi:

- Most of my electronics students have no programming experience. Some have had limited encounters with Basic, C++, or Java.
- Students should be able to start a Raspberry Pi and get it into Linux command line mode. I have included an appendix that should be adequate for most people needing some assistance. Note: Although I refer to the Raspberry Pi, almost any Linux or Unix system along with GNU utilities for an ARM processor will be adequate for most of the chapters in this book.
- My embedded systems students have already taken a course in digital electronics. In other words, they will be familiar with binary, hexadecimal, ASCII, and hardware concepts like registers. For those needing assistance or refreshing in these areas, I have included appendices.

With students having the above background, the approach taken in this book is to be very thorough in providing detailed programming examples and not very thorough in explaining binary, hexadecimal, and ASCII. I have taken this approach based upon several years of feedback from electronics students of embedded systems indicating where they want more detail and where they need less. Readers with a strong background in computer programming may find I develop coding examples more slowly than they might prefer, and others might say I didn't explain binary and hexadecimal adequately (although I do provide several appendices explaining a variety of background issues).

Why Did I Write a Book for Learning Assembly Language?

I've seen a lot of poorly written programs over the years, and I mean production code, not just classroom programming. The vast majority of it comes from people who have learned the elements of a programming language, but not how to develop software. It's comparable to an auto mechanic learning how to use wrenches, vacuum gauges, and oscilloscopes without knowing how an engine works. It's also like a soldier being taught to use a weapon but without being given the rules of engagement. In this book, I do introduce the "mechanics" of the ARM CPU and floating point processor within the Raspberry Pi, but I do it while subtly introducing software design patterns, object oriented design principles, and software development life cycles. This book is characterized by the following:

- Independent Study: Although this book is part of my embedded systems course, it does not rely on additional material presented in the course. It is expected that the student will learn assembly language by working through the examples in this book, and we'll focus on electronic interfaces and feedback control methods during class.

- Complete examples: Rather than provide just snippets of code, this book contains complete working programs.
- An emphasis on software design methods: Design patterns and program development life cycles are as important for assembly language application development as for higher level language development.
- Learn by doing: Prototype programs appear at the beginning of each chapter to get the readers' "feet wet" before diving into the principles to be learned.
- Flowcharts: Many students have stated they understand the concepts better when they can visually see the program flow.
- Review questions and exercises: Every chapter ends with a series of suggested exercises and review questions to augment the understanding of the material presented. Answers to some of the review questions are in the back of the book, but many questions are left unanswered and are available for classroom assignments.

Why the Raspberry Pi?

So, why did I specifically target this book at the Raspberry Pi when other development environments have been available to college students for years?

- ARM Processor: The ARM processor is not only a very interesting CPU architecture, but it is also one of the most popular CPUs currently in production.
- Floating Point Processor: I wanted my embedded systems students to be both knowledgeable and comfortable with measured analog data ranging from microvolts to hundreds of volts. This range of values is best represented in floating point numbers, and the Raspberry Pi SoC (System on Chip) design comes standard with a floating point processor. Several chapters are devoted to programming the VFPv3 and NEON coprocessors.
- Professional Quality: My students are in a curriculum to learn to produce industrial strength products. The hardware architecture (ARM CPU) and software (Linux and associated utilities) used with the Raspberry Pi are a common work environment for developing real-time embedded systems.
- Compile, Link, Execute: Although several sophisticated Integrated Design Environments (IDE) are available, I wanted the students to at least be grounded in the basics of how software is developed.
- Inexpensive: Although colleges can afford to purchase expensive microcomputer development environments, I wanted a system that students could also work on at home.

What Can Be Learned from this Book?

- Practical experience with the ARM instruction set
- Familiarity and experience with VFPv3 and NEON coprocessors
- Trade offs between floating point and integer programming
- Programming techniques like looping and searching
- Introduction to software design patterns
- Introduction to software documentation
- Acquaintance with software development life cycles

Book Organization

Each chapter presents a new programming technique and associated ARM assembly language instructions that builds upon the information and techniques developed in preceding chapters. Each chapter is similar to a cycle through the spiral model of software development. Within a chapter, the material is presented in sections in the following order:

1. Prototype: An assembly language program allowing the student to take a "test drive" through the concepts and techniques to be presented in the chapter.
2. Introductions: A brief description of each CPU instruction, assembler directive, Linux service and command line introduced in the chapter.
3. Principles: The concepts and techniques are explained in detail.
4. Coding and Debugging: Enhancements are made to the prototype, and the details of program execution are examined.
5. Maintenance: Every chapter ends with a series of suggested exercises and review questions to augment the understanding of the material presented.

Why Assembly Language and Why Now?

The structure and semantics of assembly language are not optimized for business, scientific, or engineering applications, but match the native hardware instructions of a particular CPU. Higher level languages developed in the 1950's, COBOL for business and FORTRAN for science, are much more appropriate for applications development than assembly language.

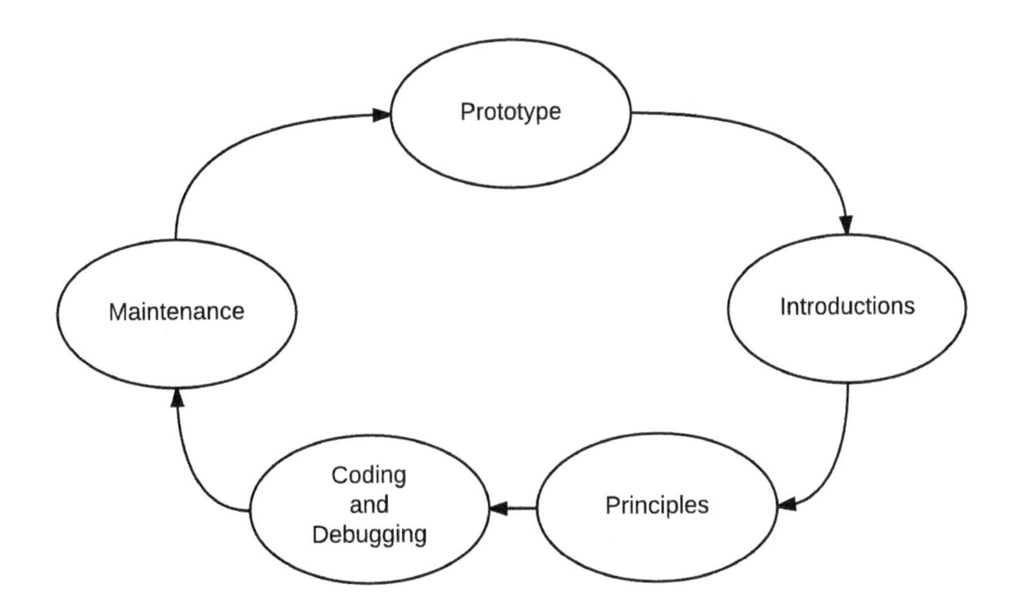

Figure P.0: Structure of every chapter in this book

With two strikes against it (locked to one CPU design and no applications relevance), why has assembly language ever been used?

- **1950s:** Before languages like FORTRAN and COBOL, assembly language was primarily all that existed. Some earlier computers didn't even have assembly language.
- **1960s:** A proliferation of application-relevant "languages" appeared that ranged from simplified "teaching" versions of FORTRAN (Basic, IITRAN, and many others) to artificial intelligence environments like LISP and object-oriented simulation languages like Simula. Assembly language was still used for some applications, especially those requiring high performance. Mainframe operating systems which were unique to each CPU were written almost entirely in assembly language at this time.
- **1970s:** Although mainframe and most minicomputer operating systems were still written in assembly language, portable minicomputer operating systems like UNIX were written in the C programming language. In the engineering and scientific world, many hybrid applications appeared that were a combination of assembly language with a higher level language like C or FORTRAN.
- **1980s:** Microcomputers were not only in desktop computers, but were integral to embedded systems ranging from automobile engines to data communications equipment. Due to the performance and memory limitations present at that time, many of these applications were developed in assembly language. Mainframes and minicomputers were using less and less assembly language for either applications or operating systems.

With the memory and performance issues long gone and the availability of a huge selection of application development languages, why would anyone use assembly language today? Even when it comes to performance, wouldn't we now use FPGAs (Field Programmable Gate Array) or CPLDs (Complex Programmable Logical Device) instead of CPUs? Do today's programmers know or even need to know any assembly language? The good programmers do. These programmers are well-rounded and not only understand application development but also have some knowledge of hardware capabilities.

Although the vast majority of applications and operating systems are now written in higher level languages, there are still some applications written in assembly language (like some embedded systems) and some operating systems are still in assembly language. Today, however, assembly language is primarily a bridge of understanding between programmers and computer engineers.

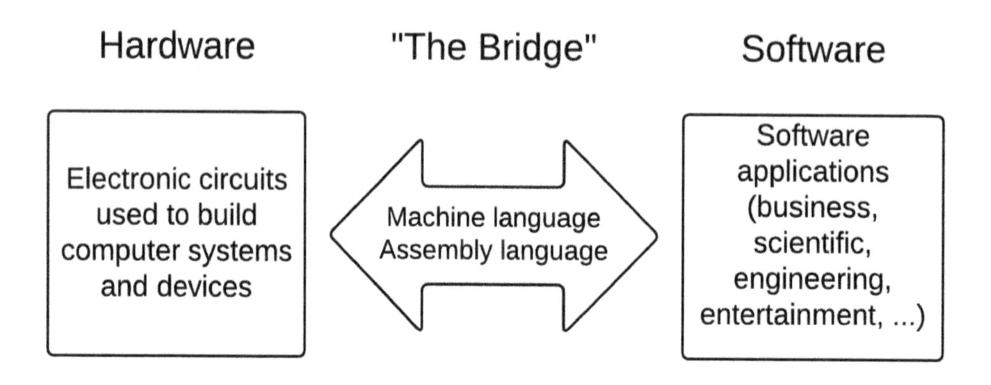

Figure P.1: Assembly languages bridges hardware and software.

What's Not in This Book That You May Expect to Find

Preliminary versions of this book have been "field tested" in the classroom by my electronics students, and I've been "advised" it was getting too many chapters and perhaps losing focus, so I decided not to include the following:

- Examples with assembly language with C: Although almost all of the embedded systems I have developed for production "in the real world" are a combination of assembly language with a higher level language like C, I decided to leave that subject to others or perhaps I'll include it in another book. I did, however, include a brief appendix on the subject.
- Electronics interfacing examples: As you might expect in electronics classes, I always have lab exercises that either drive motors or read very sensitive, even micro-volt data. Again, I felt I couldn't properly cover that material in a professional way without almost doubling the size of this book. Electronic interfaces, measurement, feedback, and control will be left for other separate

books dedicated to those subjects, but I do introduce some limited non-assembly language I/O in an appendix.

- Linux applications development: I use only enough Linux service calls to demonstrate the programming techniques and examples. File I/O, as well as task and thread management, are not included.
- Hardware reference: Although I go into great detail describing and using the ARM, VFPv3, and NEON processors, this is not a hardware reference book. Instruction timings and internal structural layouts are available on the Internet from hardware manufacturers and others.
- References: The Internet is full of references, with new ones appearing and old ones dropping every day. When I read a book or an article on the Internet, I use my favorite search engine to look up and cull any references I need. So I decided for this book, I'd only include references I felt were legally required or I wanted to point to a very specific reference out of an ocean of similar references. Besides I didn't intend for this book to become a scholarly document, but a set of practical examples to introduce assembly language programming.

About the Author

Robert Dunne has over 40 years of computer experience ranging from developing custom hardware interfaces on supercomputers to teaching technology courses in middle-school gifted-education programs. Starting out with degrees in physics and computer science, he was on staff at a national laboratory and a major engineering firm for ten years before becoming an entrepreneur in the development of embedded systems. Before working with the ARM architecture and Raspberry Pi, he had written well over 100,000 lines of assembler code developing systems and applications on nine unique CPU architectures encompassing mainframes, minicomputers, and microcomputers.

During the past ten years, he has taught three undergraduate courses per semester in digital electronics and embedded systems and is notorious for getting his students working on a lab project within the first 60 seconds of the very first class meeting.

Lab 0
Introduction and Computer Basics

This book is organized as a series of "laboratory" exercises consisting of one or more Raspberry Pi ARM, VFPv3, and NEON application programs. I call this first "chapter" Lab 0 for two reasons: 0) in computer science and electronics, we often start counting from zero (i.e., 0, 1, 2, 3, …) and 1) this first "lab" is different from the others because it's all background information with "zero" hands-on lab experience. The purpose of this chapter is to explain the format of the rest of the book and provide an introduction to CPU architecture and software development procedures.

Prototype

We begin every lab with a "test drive." The idea is to learn by example. The prototype program will contain ARM instructions and programming techniques that will be explained later in the "**Principles**" section of the lab.

```
mov     R2,#30      @ Maximum number of bytes to be entered.
mov     R0,#0       @ Code for stdin (standard input, i.e., keyboard)
mov     R7,#3       @ Linux service command code to read a string.
svc     0           @ Issue command to read string from stdin.
```

Listing 0.0: Example of assembly language source code

The following four steps will be performed to run each prototype:

1. **Edit (make the source code):** The source code such as that shown in Listing 0.0 has to be entered (or copied) into a text file. In my classes, I currently recommend using either the leafpad editor or GNU nano editor. They are both included with the Raspberry Pi Raspian Linux distribution and are very easy to use. If you're more comfortable with vi, vim, or emacs, please use it instead. I do. I just want the students to focus on the ARM architecture and assembly language and not be distracted by the complexity of other software. Lab 1 and

Appendix E will provide some help with editors if needed.

2. **Compile (make the object code):** Each "machine language" instruction executed by the ARM CPU is composed of several "groups of bits," which could be entered as binary numbers, but would be a lot of work. Each text line in the above sample source code corresponds to one ARM instruction. The "groups of bits" are represented by mnemonic names and decimal numbers. The "as" program, known as a compiler (or assembler in the case for assembly language), converts the text lines to the binary instructions (object code) needed by the ARM CPU.
3. **Link (make the executable program):** The "ld" linker program combines multiple object files into a single executable file.
4. **Execute (run the program):** This step is the objective of the previous three steps, but how do you know if your program is doing what you wanted it to do or is even doing anything at all? You'll need some type of I/O (Input/Output). For the first couple of labs, we'll use the echo command line to assist with a little output and the gdb (debugger) to "see" what's going on inside the program's memory and registers.

The following four GNU/Linux command lines will be used over and over in the labs. They are entered in response to the Linux command line prompt, which is initially "pi@raspberry ~$" which I will shorten to just ~$. If your Raspberry Pi is configured to boot straight into the Graphical User Interface (GUI), then you'll have to enter the command line mode from the LXTerminal desktop icon. We'll also be using the echo and gdb commands. Yes, there's a way to automate most of these commands, but for now it's good practice to know what's going on: edit, compile, link, execute.

```
~$ nano model.s
~$ as -o model.o model.s
~$ ld -o model model.o
~$ ./model
```

Listing 0.1: GNU and Linux commands to edit, compile, link, and execute

Introductions

Every lab will introduce new instructions, directives, services, and commands that were not used in preceding labs. This section will itemize them and give a brief description of each of the new items. A more complete explanation will occur in the "**Principles**" section.

1. **Instructions:** One line of text that is converted into one machine language instruction to be executed by the ARM CPU. Examples of these instructions: move, add, shift, multiply.
2. **Directives:** Commands to the "as" assembler providing options for generating the object code file. They typically begin with a period. For example the .global directive tells the assembler which statement labels are to be passed on to the linker.
3. **Services:** One of the functions of an operating system, such as Linux, is to provide services for application programs. These mostly include I/O services such as "read from a disk file" or "write to the display screen."
4. **Commands:** By commands, I generally mean the command lines such as those that invoke the editor (nano), the assembler (as), the linker (ld), the application program (./), and the debugger (gdb). In many of the labs, commands to the gdb debugger will also be introduced.

The vast majority of new instructions, directives, services, and commands introduced in each lab will be ARM instructions.

Principles

Although this book appears to be a lab manual, it's really a text book. In all my courses, I emphasize hands-on activities and make a point of getting into a lab exercise within 60 seconds of the start of the first class meeting. In other words, I put the lab before the lecture. The "**Principles**" section of each lab in this book is essentially the lecture that provides an explanation of what is being done. The **"Coding and Debugging"** section then goes back to the lab for enhancements and refinements.

The Hardware

When we refer to hardware, we're talking about the physical device. The hardware is easy to identify. You can see it and touch it. We will be looking at the hardware at the user level or highest level. We won't be looking into chip fabrication or the internal circuitry of the Raspberry Pi. In this book, the "hardware" refers to the CPU and floating point processor within the Raspberry Pi, not peripheral devices like sensors and motors.

When I ask my students during a lecture to tell me when the first computers were developed, I usually get answers like the 1980s or 1940s. Very few know that the basic architecture of almost all computers used in the past 50 years dates back to the 1830s with Charles Babbage's Analytical Engine.

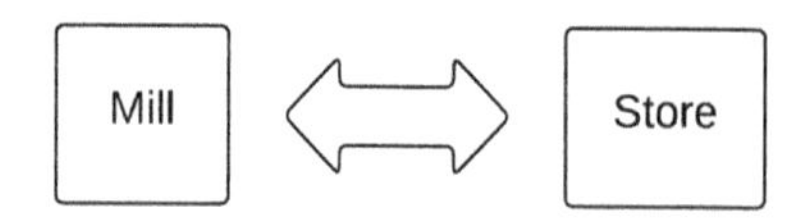

Figure 0.0: Babbage's Analytical Engine

Babbage's Analytical Engine consisted of two principal components:

- Mill: The hardware that did the work (arithmetic and logic operations)
- Store: Location for data storage of intermediate results

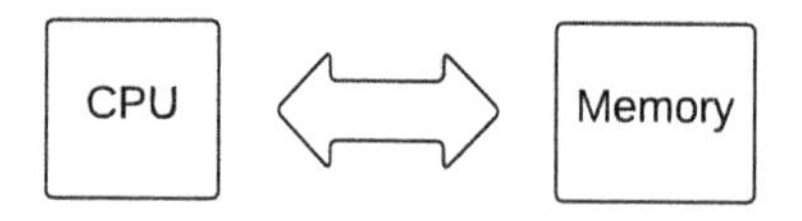

Figure 0.1: "Modern" computer hardware nomenclature

Babbage's mill and store correspond to today's computers:

- CPU (Central Processing Unit): The hardware that does the work (arithmetic and logic operations)
- Memory: Location for data storage of intermediate results

By no means am I implying there existed a positive progression of concepts and devices from Babbage's day to today. The computer pioneers in the 1940s recreated much of what was lost for nearly one hundred years. I personally observed the microcomputer software industry in the 1980s recreate the same mistakes and going down the same wrong paths as the mainframe developers did in the 1960s.

The Software

While the hardware is easy to identify, the software is somewhat nebulous. When I ask my electronics students to define software, I get a variety of circular definitions. "It's a program" and "the program is the software."

What's the program? I then ask, "What are you handed when you enter a school concert or play?"— A program, of course. Those programs consist of a sequence of songs or acts to be performed in a prescribed order. Computer programs are basically the same: a sequence of instructions to be performed in a prescribed order.

The Bridge

Today's software exists at multiple levels. Assembly language is the programming language that is closest to the physical "machine language" instructions executed by a CPU. I call assembly language the hardware/software bridge because it's a common ground. On one side we have electronic engineers and technicians while on the other side we have computer programmers and analysts. This assembly language bridge encourages software developers to get close to the hardware and hardware developers to approach the software controlling their electronic circuits.

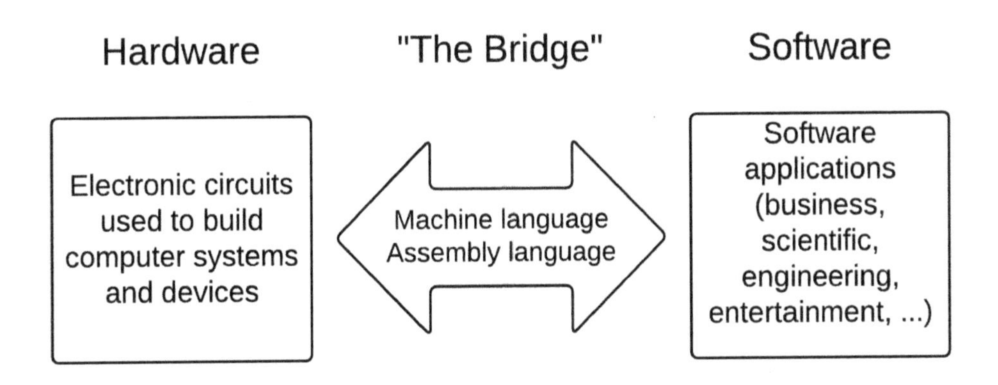

Figure 0.2: Assembly languages bridges hardware and software.

The Firmware

Since we've just mentioned the "hard" and "soft" ware, what's this "firmware" you've heard of that is obviously somewhere between the two? It is typically the software that is permanently resident in memory that is central to the operation of the hardware. This "firm" software runs when the computer powers up during the boot process. Much of the software that provides direct and detailed control of hardware devices (like disk controllers and automotive engines) is also in firmware.

The amount of firmware in a computer system can vary considerably depending upon the application. It could be as little as the amount of code needed to read a small portion of a boot device (like a disk) or as much as including all the software of an embedded system.

The Machine and Its Language

"Machine Language" generally refers to the binary codes that instruct the CPU what operation is to be performed and on what values.

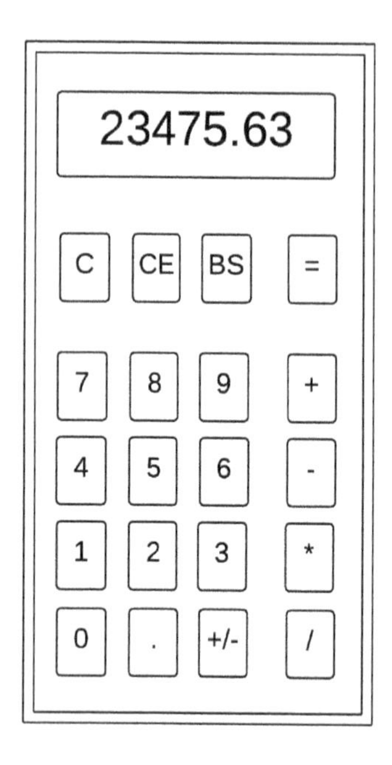

Figure 0.3: Simple calculator model

In order to understand machine language, think of a calculator which has the following features:

- **Display:** Current number being entered or current result of operations
- **Operations:** Clear (C), add(+), subtract(-), multiply(*), divide(/), display(=)
- **Data input:** Digits and decimal point, sign(+/-), clear entry (CE), back space (BS)

When we use a simple calculator, there is a symbiotic partnership to achieve the final calculation: The calculator does the work, and we provide the directions.

We enter the following sequence of instructions to perform the calculation 13×21+6:

1. **Clear**
2. **Add 13**
3. **Multiply 21**
4. **Add 6**
5. **Display**

Registers

CPUs contain a small number of fast-access memory locations called registers. Depending upon the particular CPU design, the number of registers varies from about five up to nearly one hundred. Some of the registers are accessible to user programs in assembly language, and some registers are only accessed by the CPU's electronics to perform its many tasks. The principal data register is generally referred to as the "accumulator." In our calculator example, the accumulator value is that "running total" or accumulated total that is in the display.

Most of the logical and arithmetic operations performed by CPUs are binary operations. Two numbers (called operands) are added, or two operands are multiplied, or one operand is subtracted from another. In our calculator example, the first operand is the value in the display (i.e., the accumulator), and the second operand is the number being entered.

Op-codes

Although a CPU could be constructed to work with the character string names of operations like clear, add, and multiply, it would be somewhat inefficient. Instead, the CPU designers assign an operation code (op-code) to represent each of the available CPU operations. For example, let the following six numbers be assigned to the following six calculator operations:

1. **Clear (C)**: Load zero into the accumulator.
2. **Add (+)**: Add the value of the operand (being entered) to the current accumulator contents.
3. **Subtract (-)**: Subtract the value of the operand (being entered) from the current accumulator contents.
4. **Multiply (*)**: Multiply the value of the operand (being entered) to the current accumulator contents.
5. **Divide (/)**: Divide the current contents of the accumulator by the operand (being entered).
6. **Display (=)**: Copy the current contents of the accumulator to the display line.

If we use the above numeric assignments to translate our previous sequence of instructions to calculate 13×21+6, we will get the following machine code:

Step number	Operation		Op-code with operand
1	Clear	translates to	1 : 0
2	Add 13	translates to	2 : 13
3	Multiply 21	translates to	4 : 21
4	Add 6	translates to	2 : 6
5	Display	translates to	6 : 0

Table 0.0: Translate "assembly code" into "machine code."

Note that in this simple model of a calculator being used as a computer, I've represented each instruction as a binomial: an op-code and operand pair. In Table 0.0's translation to machine code, binary operations like Add and Multiply are converted to form "op-code : operand" pairs. Unitary operations, such as Clear and Display, are converted to the form "op-code : 0" because there was no operand. So in this example, we would calculate 13×21+6 by entering "Clear, Add 13, Multiply 21, Add 6, Display" on a calculator, but the corresponding computer program (in machine language) would be the sequence "1 : 0, 2 : 13, 4 : 21, 2 : 6, 6 : 0."

Memory

What about the other half of Babbage's computer: the "store" ("memory" in today's terminology)? Babbage needed memory for the storage of intermediate results and so do we. In an arithmetic problem like 13×14+15×16, we can't just multiply 13 times 14, add 15, and then multiply by 16. The answer would be wrong because by convention, multiplication has precedence over addition: the 15 and 16 have to be multiplied before being added to the product of 13 and 14. There is an implied parenthesis in this calculation as follows: (13×14)+(15×16). With our non-memory calculator, we would have to write down the intermediate value of 13×14 and then reenter it after we calculate 15×16. Memory calculators do this "writing down" and reentering for us.

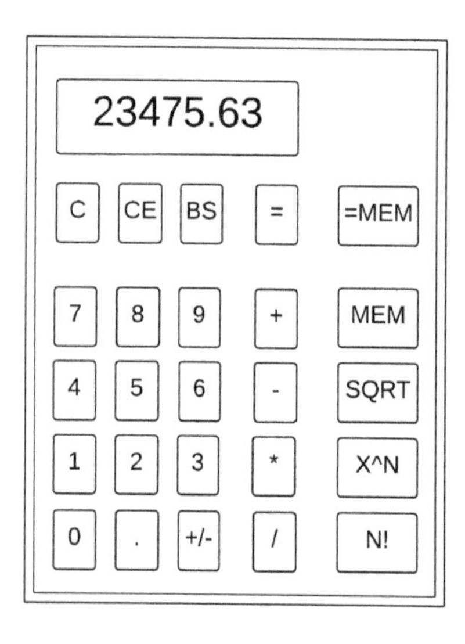

Figure 0.4: Memory Calculator with functions

In our calculator model shown in Figure 0.4, we have more operations, two of which are related to memory: Store (=MEM) and Load (MEM). Since we added two more operations to the calculator, we must also add two more op-codes: Store (op-code = 7) and Load (op-code = 8).

Op-codes 1 through 8 are defined as follows:

1. **Clear (C)**
2. **Add (+)**
3. **Subtract (-)**
4. **Multiply (*)**
5. **Divide (/)**
6. **Display (=)**
7. **Store (=MEM)**: Copy the contents of the accumulator into a memory location.
8. **Load (MEM)**: Copy the contents of a memory location into the accumulator.

Is this the best approach for using memory? The above Store command is fine, but the Load instruction is too limiting. In some simple memory calculators that store only one value in memory, there is an "Add memory" command, but what if you want to multiply using the value in memory or divide by it? What we really would like is to not just reload a saved value, but use it in any of the previously defined operations such as Add or Multiply.

Most CPU implementations include a "flag" in the instruction that indicates if the operand is immediate (value in the operand) or is from memory. Our binomial instruction format (op-code and operand) now becomes a trinomial (op-code, immediate-flag, and operand). If we use the above operations including the new immediate flag to write a little program to calculate 13×14+15×16, we will get the following code:

Step number	Operation		Op-code :i- flag : operand
1	Clear	translates to	1 : 1 : 0
2	Add 13	translates to	2 : 1 : 13
3	Multiply 14	translates to	4 : 1 : 14
4	Store Mem 0	translates to	7 : 0 : 0
5	Clear	translates to	1 : 1 : 0
6	Add 15	translates to	2 : 1 : 15
7	Multiply 16	translates to	4 : 1 : 16
8	Add Mem 0	translates to	2 : 0 : 0
9	Display	translates to	6 : 1 : 0

Table 0.1: Translate "assembly code" into "machine code."

The above system actually works fine, but can we improve on the performance? Storing intermediate results into memory always takes time. An accumulator is also memory, but it's very fast local memory inside the CPU, and we can also use it as an operand in our instructions. Although some CPUs only have one accumulator, the vast majority have several. The ARM has 16 user accessible general purpose registers, most of which can be used as accumulators for making calculations.

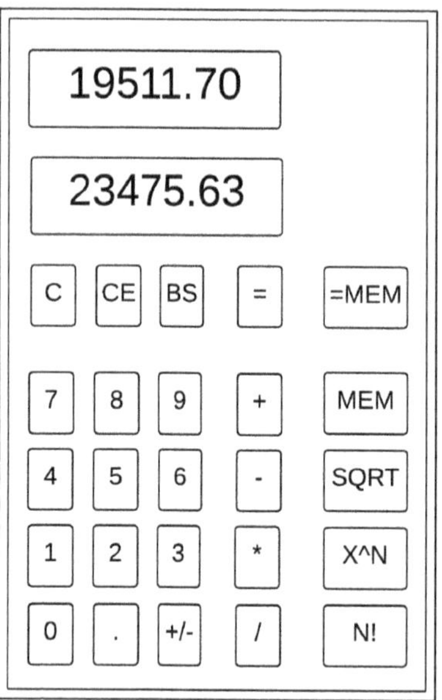

Figure 0.5: Two-accumulator Calculator

Let's expand our calculator model to have two accumulators and see how that changes our assembly language. First, our instructions are no longer trinomials consisting of three numbers, but now have a fourth component: we have to specify which accumulator is being used in the operation. In the previous one-accumulator calculator model, we only had one accumulator, so it was the only one that could be used and therefore did not need to be specified.

Step number	Operation		Op-code : acc : i-flag : operand
1	Clear A0	translates to	1 : 0 : 1 : 0
2	Add 13 to A0	translates to	2 : 0 : 1 : 13
3	Multiply A0 by 14	translates to	4 : 0 : 1 : 14
4	Clear A1	translates to	1 : 1 : 1 : 0
5	Add 15 to A1	translates to	2 : 1 : 1 : 15
6	Multiply A1 by 16	translates to	4 : 1 : 1 : 16
7	Add A1 to A0	translates to	2 : 0 : 0 : 1
8	Display A0	translates to	6 : 0 : 1 : 0

Table 0.2: Use two accumulators: A0 and A1

Any Other Instructions?

Computers are great for doing the same thing over and over again, but on different sets of input data. Sometimes, we write a program to perform the calculations differently, depending on the type and values of the data being processed. Making these decisions, as well as knowing when to exit these repetitive loops are done by branch instructions (a.k.a., jump instructions) and conditional test instructions. We'll examine these techniques in the next section on documentation, but we'll not implement them here in this calculator model.

Language Interpreter and Compiler

The "human language" source code of our programs must be translated to machine language in order to be executed by the CPU. This translation can be done all at once before any of the machine code is executed or it can be converted and executed line by line as it is needed. The two approaches are the following:

- Interpreter: Translate each line of source code to machine language line by line just before it is executed.
- Compiler: Convert the entire source code file to machine language all at one time.

Assembly language as well as C and Java are almost always compiled. Languages like Basic have traditionally been interpreted. There are merits to each approach which we won't go into here except saying that interpreter code is easier to write and debug, while compiled code offers much higher performance at execution time.

As in most things, there's always a slight modification in order. Java and Microsoft's .net programming languages like VB.net and C# don't compile all the way to machine language. They compile to an intermediate language that is very close to a generic machine language which is not tied to any particular CPU format. The code in this intermediate low-level language is executed by being run interpretively, but at a very high performance level. There are also some portability and security reasons for using this intermediate format. Assembly language is of course already at the machine code level and therefore goes straight to the native machine code.

Large Programs

Most of the programs appearing in the first few labs in this book are very short programs. In reality, most computer programs are fairly large consisting of tens of thousands of lines of code or much more. Keeping all this code in a single file would be ridiculous for many reasons (bulky, only one person working with it at a time, poor editor performance, etc.). These programs must be broken into modules of at most a couple hundred lines of code each. They can then be separately compiled into machine code, also know as object code. The object code files will then have to be linked together to form the working program.

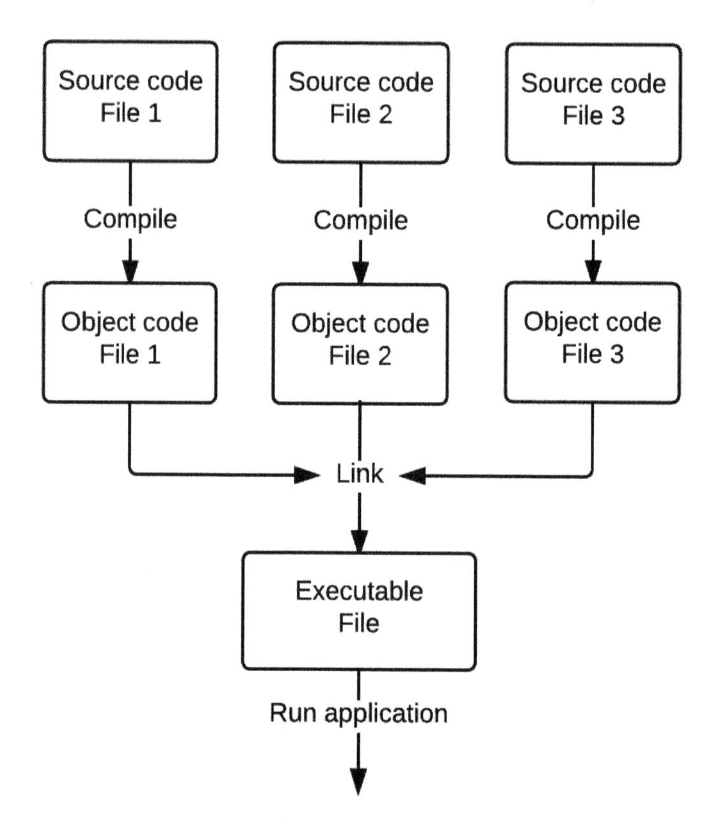

Figure 0.6: Dividing a program into modules

In figure 0.6 we see how a large program can be divided up into multiple source files and individually compiled into individual object files. These object files then have to be linked together to form a working program that can then be run.

Documentation

Program documentation is used during both construction and long term operation of a software application. It tells the development programmers what the application is supposed to do and tells maintenance programmers what the application is doing and how it's doing it.

Some computer languages are somewhat self-documenting. Assembly language is definitely not one of these. One of the first languages developed after assembly language was Cobol which has been present in business applications for over 50 years. Although professionally trained programmers were needed to write Cobol programs, almost anyone who could read English could read the Cobol program instructions and get a very good understanding of what was being done. Even today's commonly used languages like

C and Java contain structures like loops and objects which help identify what is being done in the program.

Documentation basically exists at three levels:

1. **Narrative**: A description in words, charts, tables, and examples explaining what the application does. To some degree it even recommends how the application should perform its assigned tasks.
2. **Diagrams**: There has been a variety of graphic modeling languages over the years beginning with traditional flowcharts through the Universal Modeling Language (UML). Their diagrams show program structure and flow.
3. **Internal documentation in the code itself**: All programs should have comments interspersed among lines of code saying what is being done, why it's being done, and how it's being done. For higher level languages like C and Java, internal documentation is important. In assembly language it is crucial.

I never really liked flowcharts. However, many students new to programming say this graphical approach helps them grasp the logic flow more readily. So I'll use them in the first several labs to explain some programming techniques and even use them to explain how an ARM instruction works. I've seen flowcharts used not only to document software, but also in the automotive, HVAC (heating, air conditioning), and other industries.

I'll be using three basic flowchart symbols (and then three alterations to one of them).

- **Process**: Identifies a task, such as adding three numbers.
- **Decision**: Shows alternate paths the program can take based upon current values in the data.
- **Terminator**: Identifies the beginning and ending points of a portion of the program

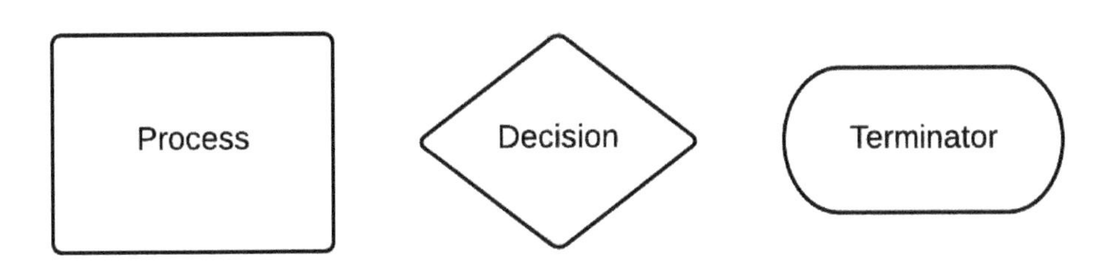

Figure 0.7: Basic flowchart symbols

Although I could use the process block throughout, I will also use three other symbols when the process is more specific: preparation, predefined process, and display.

- **Preparation**: A process like initializing a running total to zero (i.e., it is a process, but not the "main act")
- **Predefined Process**: A compound process like taking the square root (normally located external to the current program coding)
- **Display**: A process where the computer user receives a displayed message.

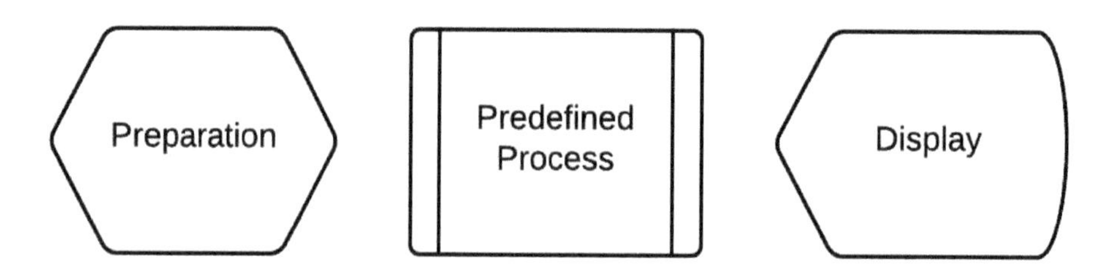

Figure 0.8: Specialized process symbols

Figure 0.9 illustrates a flowchart for a program that calculates the sum of a series of numbers using a calculator. The following steps are performed:

1. Turn on the calculator
2. Push clear to initialize the accumulator and set the display to zero.
3. Enter next number from keypad (This is actually a multi-step procedure where one or more numeric keys are pushed and may include a decimal point.)
4. Test whether the last number of the list has been added.
5. If not, push the "+" sign and then enter the next number in the list.
6. If the last number has been entered, then push the "=" key to display the final result.
7. Turn off the calculator.

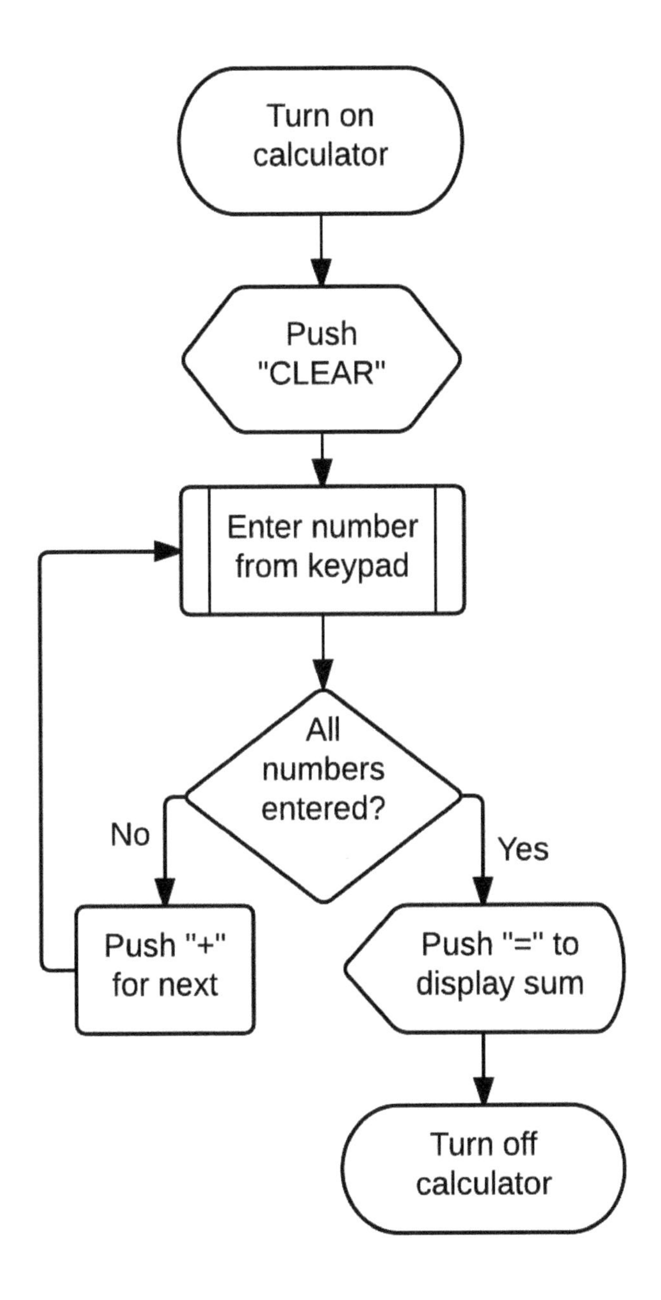

Figure 0.9: Program using predefined process

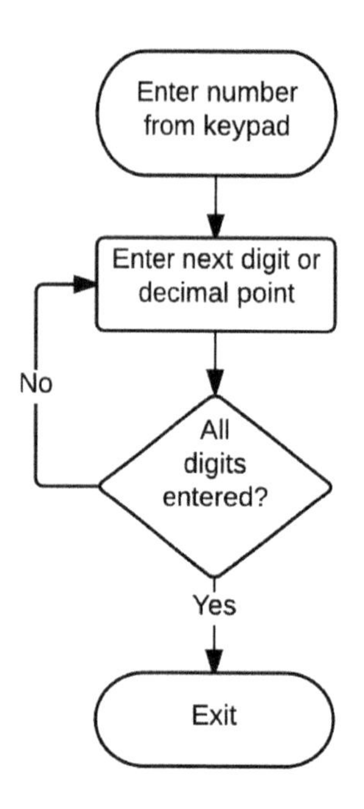

Figure 0.10: Predefined process "Enter number from keyboard"

Figure 0.10 illustrates the "predefined process" of "Enter number from keypad" called from the previous flowchart. Predefined processes, also know as subroutines, procedures, functions, and methods, are common to computer programming languages. Their use provides a structure leading to more reliable as well as more compact code. Most of the programming examples in this book involve building subroutines to perform specific tasks.

Figure 0.11: Connecting segments of a flowchart

The final flowchart symbol appearing in the first seven labs is the off-page link. It is used to divide a single flowchart into smaller segments. With a large piece of paper, off-page links would neither be necessary nor used very often. Because this book was designed to be compatible with the flowable text of eBooks where large images are not readable, I use the off-page link more often than I would have for a "paper-only" edition.

Although I think flowcharts are great as an introduction to how machine code instructions work and are to be used, I never really did like flowcharts for documenting program coding. It seemed like it took at least ten times longer to make the flowchart than to write the code. Apparently I'm not alone in that opinion because the traditional flowchart from the 1960s for program documentation has generally died out and been replaced by more practical techniques. The Universal Modeling Language (UML) has been available for many years for documenting software structure (usually for higher level languages) and development on many levels. Its "activity diagram" is the closest thing to traditional flowcharts, and it has some added real-time (i.e., embedded systems) features. Also, pseudo code which uses a somewhat arbitrary program-like verbal description of how the code works is more popular today.

Comments

Internal documentation is the description of the program appearing in the program itself and consists of two types:

1. **Global**: These comments describe what a section of code is doing. They normally consist of more than one line of text and are not on the same physical text lines as the actual computer instructions. Global comments are used in both assembly language as well as higher level languages.
2. **Local**: These comments share the text line with the actual program instructions. They are rarely needed in higher level languages, but are very important in assembly language to explain not what the code is doing but *why* the line of code is doing it.

The following excerpt from an assembly language program shows both global and local comments. There are almost as many ways to mark the beginning of a comment as there are programming languages. In this book, I will be using the @ (at-sign) which is commonly used with the GNU "as" assembler that comes standard with the Raspberry Pi. In other assemblers, both for the ARM as well as other CPUs, the semicolon is commonly used to indicate the beginning of a comment. The comment ends at the end of line.

```
@       Subroutine v_bin will display a 32-bit register in binary digits
@                   R0: contains the 32 bit value to be displayed in binary
@                   LR: Contains the return address
@                   Registers R0 through R7 will be used by v_bin and not saved

v_bin:  mov     R3,R0               @ R3 will hold a copy of input word to be displayed.
        mov     R6,#31              @ Number of bits to display (-1)

@       Loop through single bits and output each to the standard output (stdout)
        display.

nxtbit: mov     R4,#1               @ used to mask off 1 bit at a time for display
        and     R1,R4,R3,lsr R6     @ Select next 0 or 1 to be displayed.
```

Listing 0.2: Sample of local and global comments

Importance of comments: They're not necessary for a program to successfully run, and many programmers use very few comments. They're necessary for program maintenance, whether it be by a new programmer next week or by the original programmer a month or even several years later. When I was an undergraduate student and took a course in assembly language programming, my professor thought comments were so important that he subtracted one letter grade for each line of code that didn't have a local comment. That was a bit extreme, but I got the point. I confess that in my production code I don't comment every line, but I do comment much more than others. In this book, I will be commenting on almost every line to help set an example as well as explain what's going on in the code.

Coding and Debugging

In this section of each lab, we will be trying out particular coding sequences and programming techniques.

Coding

Almost all assembly language for all CPUs consists of four columns:

1. Labels: A name to be associated with the address of this instruction in memory.
2. Op-code: The operation being performed (add, sub, shift, ...)
3. Operands: Location of the data (usually a register combined with a constant, another register, or memory address)
4. Comments: Describes why the instruction is being used

Each column is separated by one or more blanks or tab characters. How wide is a column? Typical columns are 8 to 10 characters wide with the exception of the rightmost column which contains comments. The assembler doesn't care if it's one blank, two blanks, or more. We line up the columns of assembly code for the ease of reading by the programmers.

```
         .global   _start           @ Provide program starting address to linker
_start:  ldr       R1,=msgtxt       @ Set R1 pointing to message to be displayed
         mov       R2,#12           @ Number of bytes in message
         .data
msgtxt:  .ascii    "Hello world\n"  @ 12 character message (" " and "\n" each count as 1 char.)
         .end
```

Listing 0.3: Example of assembly language source code

In Case It Matters

One of the first questions programmers and hardware developers should ask when beginning a new computer language is, "Is it case sensitive?" In other words, are the commands and variable names, such as "Start, start, stArt, and staRT," all the same to the compiler or is each one unique? In programming languages like C, Java, and Python, as well as the Verilog hardware description language, each "start" in the above list is unique: "Start" is different from "start" which is different from "stArt" which is different from "staRT." In programming languages like Ada, Basic, Fortran, as well as the VHDL hardware description language, all of the above would be considered the same. Most assemblers, including the GNU "as" assembler that we're using, are not case sensitive.

So if case doesn't matter for assembly language, which should you use: add or ADD, mult or MULT, or any other combination? It's primarily your choice and style. From what I've observed, the majority of programmers and hardware developers today, including myself, use lowercase for programming and hardware development. In this book, I will provide the code listings and flowcharts using lower case, but in the text I will refer to them in upper case because they stand out better than just bolding or italicizing them (i.,e., the ADD instruction rather than the *add* instruction).

Debugging

The Late Rear Admiral Grace Murray Hopper was a computer pioneer who delighted many of us with her vivid stories of her experiences in the 1940s, 1950s and beyond. The origin of the term "bug" meaning a computer failure appeared in one of her favorite stories. In her case, that "first" bug really was a bug, a moth in particular, which got stuck in one of the computer's relays. The term stuck and expanded to encompass all types of computer failures, even those resulting from improper programming.

We will be using the GNU gdb debugger to investigate program execution:

- Examine register contents
- Examine memory contents
- Single step through programs (one instruction at a time)
- Set breakpoints to stop execution at particular instructions
- Modify register contents

```
(gdb) run
Starting program: /home/pi/model

Breakpoint 1, _start () at model.s:4
4       lsl         R0,R6       @ Multiply R0 by 4 (shift left 2 bit positions)
(gdb) i r
r0      0x11        17
r1      0x0         0
r2      0x0         0
r3      0x0         0
r4      0x0         0
r5      0x0         0
r6      0x2         2
r7      0x0         0
r8      0x0         0
r9      0x0         0
r10     0x0         0
r11     0x0         0
r12     0x0         0
sp      0x7efff860  0x7efff860
lr      0x0         0
pc      0x805c      0x805c <_start+8>
cpsr    0x10        16
(gdb)
```

Listing 0.4: Sample output from gdb debugger

Maintenance

Each lab in this book is somewhat like an iterative cycle of software development: We start with a prototype to examine. Then we study it to see how we can enhance it. Finally we provide maintenance support for the current version of the program until its next release. In this book, the Maintenance sections will provide review questions and suggested programming assignments to help the student solidify an understanding of the current material before going on to the next lab. Note: There are answers in the back of the book for the questions that begin with an asterisk (*).

Review Questions

1. What advantage does a second accumulator give a CPU? What's the disadvantage?
2. Most CPU instructions use two operands. If one operand is a register, what can the second operand be?
3. How are a compiler and an interpreter similar? How are they different?
4. * "By hand, without a calculator or computer," convert the following numbers expressed in decimal to binary format. See Appendix B if you need some background in binary.
    a. 21
    b. 63
    c. 16
    d. 129
    e. 13
5. * "By hand, without a calculator or computer," convert the following numbers expressed in binary to decimal format. See Appendix B if you need some background in binary.
    a. 1011
    b. 1100101
    c. 10110
    d. 100001
    e. 1111011

Programming Exercises

1. Write an assembly language program to calculate: 45×16÷7-46. Note: I'm not talking about using the real ARM instructions, but the simple ones made up for the calculator example (like the example in Table 0.0).

2. Hand assemble your "program" (from Exercise 1). That is, convert the op-code mnemonics to their numeric values (like example in Table 0.0).
3. Write an assembly language program for the memory calculator to generate: 31+45×37 (like the example in Table 0.1).
4. Rewrite the above exercise using the two-accumulator calculator without using memory locations.
5. Draw a flowchart for eating a bowl of soup. To start with, it should contain a process: "Dip spoon into soup" as well as a decision: "Is the bowl empty?" Other components will of course be necessary to complete the flowchart.
6. Add a "Branch If Zero" instruction to any of the calculators so that looping and conditional program flow can be achieved. Hint: You will need to assign an address to each instruction in your program, so you'll have a place to which to jump (or branch).

Lab 1
Compile, Link, Execute

Lab 1 generates and executes a simple assembly language program that outputs a single value to be displayed by a command line. It sets an example of the work flow (i.e., series of GNU/Linux command lines) used in many of the following labs.

Prototype

From a programmer's perspective, software development is a vicious cycle of modify the program, test the program, modify the program, test the program, modify the program, test the program until we are satisfied with the test results. As described in Lab 0, an assembly language program consists of lines of text which we will create and modify using a simple text editor. We then test the program by translating it into ARM machine code to be run from the command prompt. In this book, we use the standard Raspian Linux distribution, which includes the GNU assembler and linker that are run from the command line prompt.

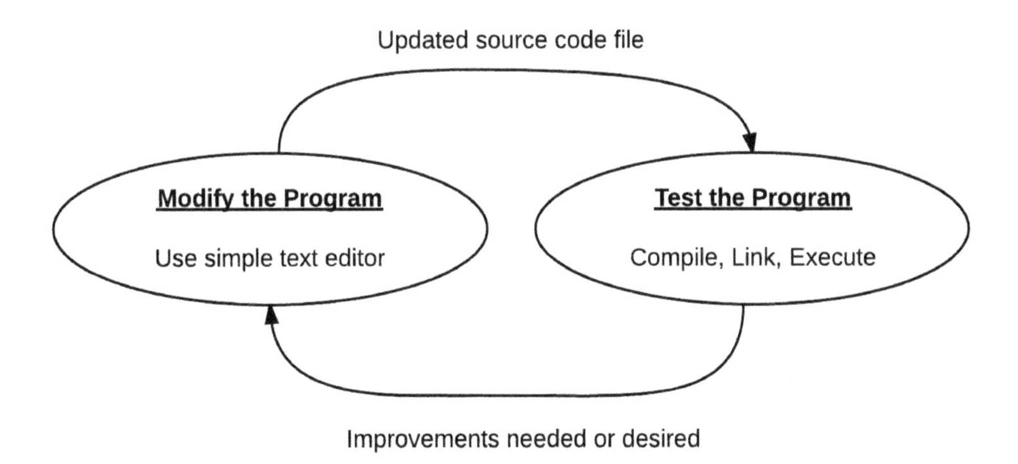

Figure 1.0: Program development: Cycle of modifying and testing a program

The compiling, linking, and testing must be done from the command prompt, but there are several ways to edit the source code program. Most of my students prefer to work using two open windows in the Linux GUI (Graphical User Interface) as shown in Figure 1.1.

1. Leafpad full screen editor
2. LXTerminal (Lightweight X Terminal emulator) for command lines

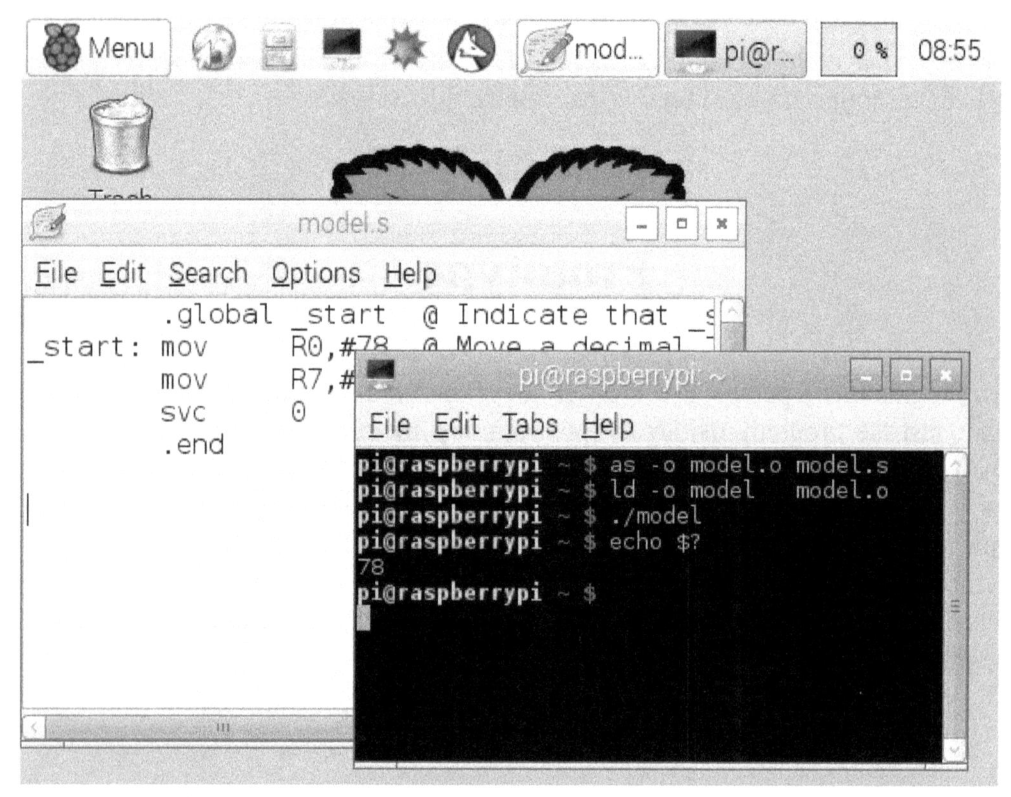

Figure 1.1: Two windows open: one for editing and one for testing

Each of these windows can be opened from the group of accessories under the pull-down menu as show in Figure 1.2. Existing source code files can also be opened by Leafpad by double clicking them from the file explorer.

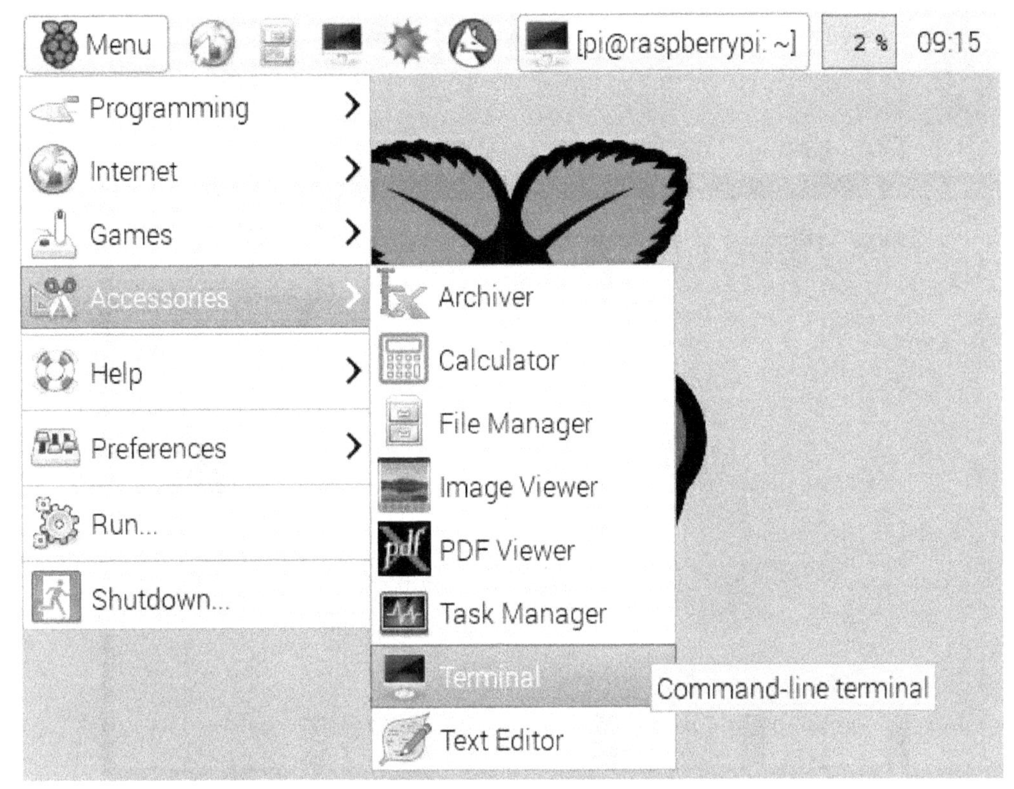

Figure 1.2: LXTerminal and Leafpad in pull-down menu

An alternate approach has been taken by many of my students where they work entirely from the command prompt. This can be done by either opening only one terminal window from the GUI or by configuring Linux to boot directly into command line mode (i.e., don't run "startx" to start the GUI). Instead of using the leafpad editor in its own window, they use an editor such as nano, vi, vim, or emacs that is opened from the command line. In this book, my edit-compile-link-execute work flow will show the nano editor being used, but please use whichever editor you prefer. Appendix E provides a few pointers on using Leafpad, nano, and vi for those who are new to editors or need a little refresher. A "word processor" program cannot be used to generate or update the assembler source code because it inserts many hidden commands for changing fonts, formatting pages, and including images that would upset the assembler.

A third approach involves remote access to the Raspberry Pi through a router. By using a PC-based application like PuTTY, files can be edited on the PC and then either uploaded or moved into nano or vi using a copy-paste procedure (control-C on the PC and shift-insert in nano). You may obtain the IP address needed by PuTTY by entering "hostname –I" from the command prompt using a keyboard already attached to the Raspberry Pi.

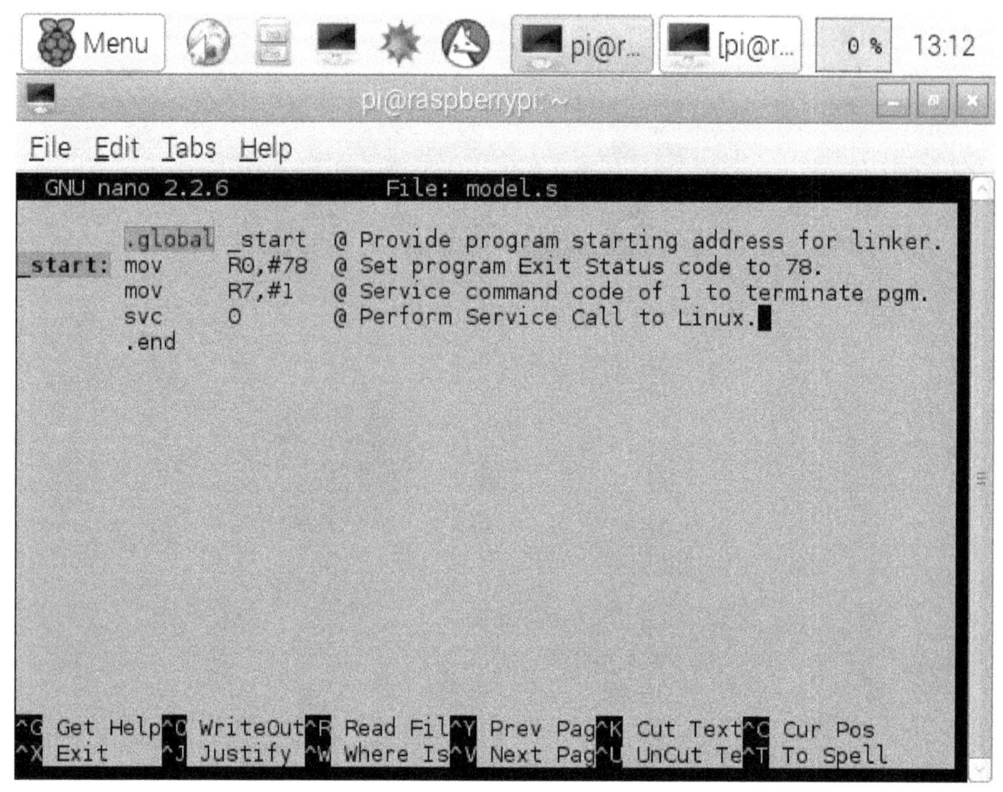

Figure 1.3: Nano editor with first program displayed.

Using the nano text editor or any other editor of your choosing, key the following very short program into a file named model.s. This is close to the shortest assembly language program you can write that does anything at all. A decimal value of 78 is loaded into register R0, and then a Linux service command is called to terminate the program.

```
        .global   _start    @ Indicate _start is global for linker
_start: mov       R0,#78    @ Move a decimal 78 value into register R0
        mov       R7,#1     @ Move a decimal 1 integer value into register R7
        svc       0         @ Perform Service Call to Linux
        .end
```

Listing 1.0: Program to load decimal 78 into exit status

Listing 1.1 provides a sequence of five GNU/Linux command lines that will be used for the next several program examples to generate and then run short computer programs written in ARM assembly language. Each command is entered in response to the command line prompt "pi@raspberrypi ~ $" which I've shortened to simply "~$." The first three commands of Listing 1.1 generate the executable program:

1. **nano:** The first command line uses the "nano" editor program to create or update the source text file containing the assembly language code. Of course, if you are editing using leafpad in its own windows, this command line is not present.
2. **as:** The GNU "as" (assembler) program reads the assembly language statements in the source text file and writes an object file containing machine code instructions.
3. **ld:** The "ld" (linker) program reads the object file, combines it with any needed utilities, and writes the executable file that is ready to run.

```
~$ nano model.s
~$ as -o model.o model.s
~$ ld -o model model.o
~$ ./model
~$ echo $?
```

Listing 1.1: Sequence of edit, compile, link, execute, display

Once you assemble this program using the "as" command, link it using the "ld" command, and execute it with the "./model" command, you can then display its output "status" value of 78 using the "echo $?" command. For a little variety, you may choose to change the 78 to some other value, but it does have an upper limit that isn't very large. Note: If you need a little background on use of nano or one of the other editors, please see Appendix E, and if the # (hash code, pound sign) character does not appear when you enter it from the keyboard, please see Appendix A.

Introductions

Since this is the first real lab, everything is new: instructions, directives, services, and also command lines.

	List of ARM instructions introduced in Lab 1		
1.0.2.	mov	R0,#78	@ Move "immediate" a decimal 78 value into register R0
1.0.4.	svc	0	@ Perform Service Call to Linux

	List of assembler directives introduced in Lab 1
.end	Identifies last line of text file (not required, but comforting to know nothing is missing)
.global	Inform the assembler of label to be passed on to the linker.

	List of Linux service calls introduced in Lab 1
1	Exit; [R7]=1 tells Linux to terminate the program.

	List of command lines introduced in Lab 1
nano	Text editor for updating program source code
as	Assembler program for converting source code to object code (machine language)
ld	Linker program combines one or more object file to make an executable file.
./	Execute user program in current directory
echo	Command line to display text message
ls	List names of files in a directory
cp	Copy one file to another
cat	List the contents of a file

Principles

How do we develop computer applications? How do applications work internally? How do we translate ideas into a sequence of computer instructions to be executed? These questions and more will be investigated by studying small applications running on the Raspberry Pi.

There are basically two types of computer applications:

- Procedural: The application developer knows what steps and what order is needed to solve the problem. This type of application covers the majority of all computer programs running today.
- Non-procedural: There's basically two types: 1) data manipulation languages and 2) computer applications that learn through multiple examples how to perform the task or solve a problem. The latter has traditionally been called AI (Artificial Intelligence), but today it is referred to as machine learning.

Procedural programming typically involves writing the program in a language (like Java, C, assembly, ...), converting it to machine language, and then running the machine language. Figure 1.4 illustrates the five GNU/Linux commands and three types of files that will be used in this and several of the following labs.

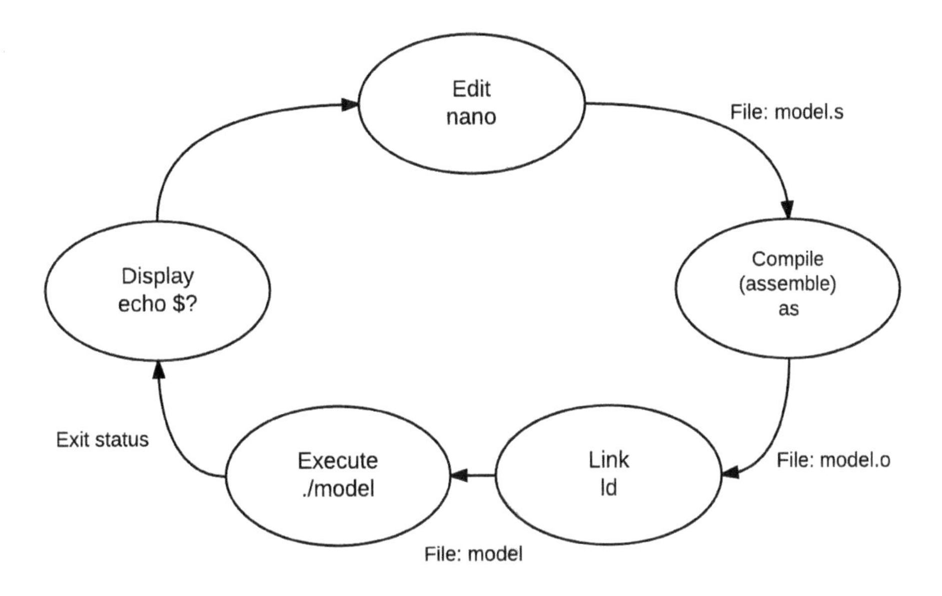

Figure 1.4: Work flow of testing a program update

File names appearing in Figure 1.4 and command line Listing 1.1:

- **model.s:** Source file containing assembly language
- **model.o:** Object file containing ARM machine code
- **model:** Executable file ready to run on a Linux/ARM based computer system.

In the event an error is detected by the "as" assembler command, then you'll have to go back and update the program source code using the nano editor. It is also possible that the linker ("ld" command) will catch some misspellings, and then you'll have to go back to the editor, correct the mistake, then reassemble the program and relink it again. Yes, there are ways to automate this process and make your life easier, but for now it's good practice to know what's going on. An easy way to provide some relief of re-entering command lines is to use the up-arrow key which will bring back command lines previously entered.

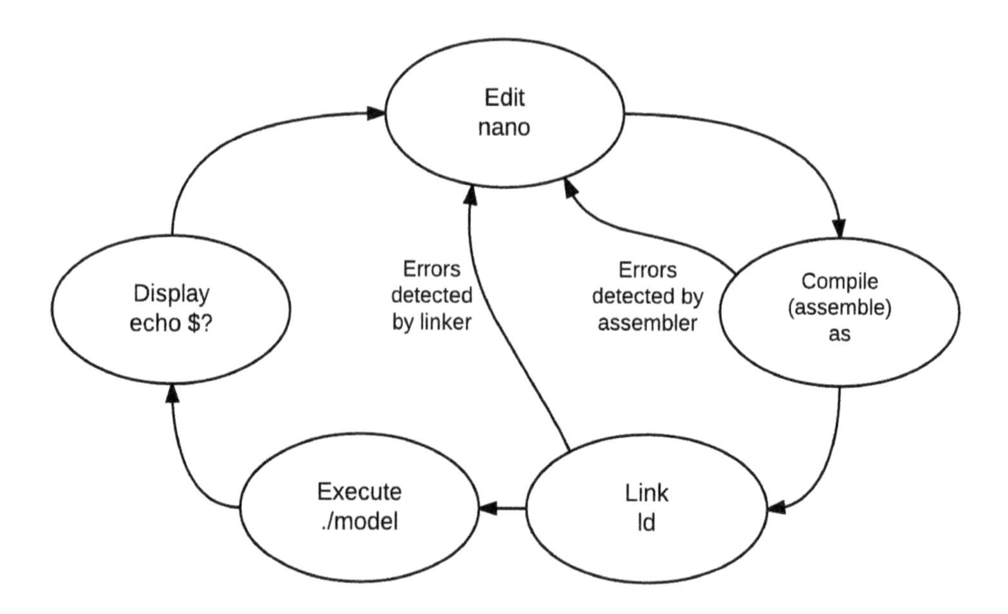

Figure 1.5: Errors found during assembly and linking

Actually any file names could have been chosen: *source1* or *PGM-2* or *mycode*. I chose *model* because I am subtly introducing you to using design patterns in software development. Design patterns are very important in the development of large programs and most important to the long-term program maintenance. My choice of file names will introduce you to the classic design pattern: Model View Controller (MVC). I'll have more to say about this later, but I thought I'd give you a heads-up on my choice of file names.

1. Compile: The information in a source file (usually text data) is used to generate an object file that is very close to the machine language of the CPU on which it is to run.
2. Link: The information from one or more object files is combined with standard system utilities (Input/Output routines, mathematics functions, etc.) to generate an executable file in the machine language of the desired CPU.
3. Execute: Actually run the generated program on the desired CPU.

In programs that are more than about 100 lines of code, you'll want to divide them into multiple source files as illustrated in Figure 1.6. Here the linker really is linking more than one object file in order to produce the complete working program. Here I've chosen three source files, but in reality hundreds of files could be involved to produce a large program. There are additional libraries and dynamic linking that can be used, but we'll leave those techniques for a different book.

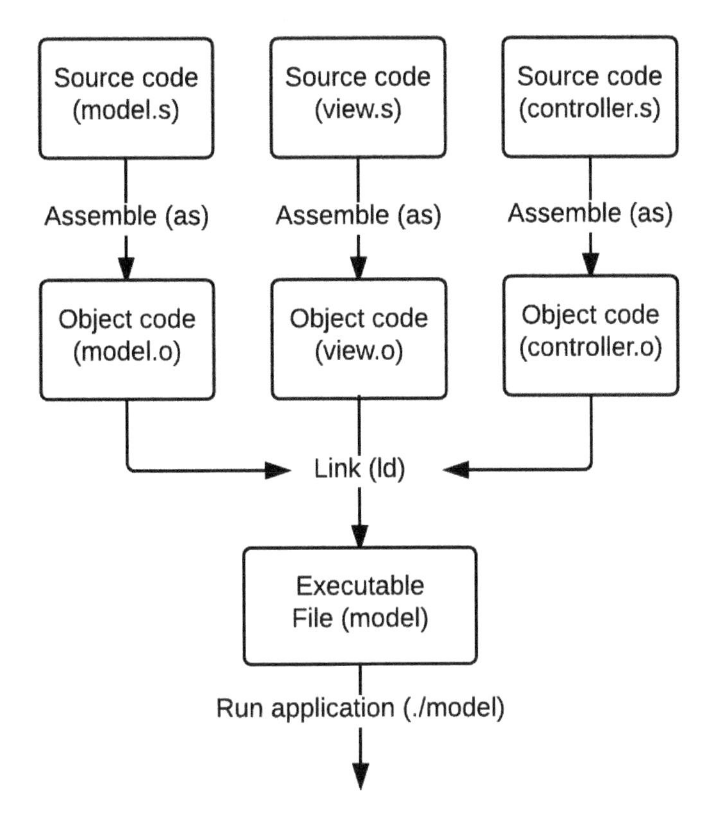

Figure 1.6: Build executable program from three source files.

Your Program Is Not Alone

How does a computer start, run, and stop? For the ARM processor and the Raspberry Pi in particular, the hardware knows where to start the first instruction after power-up or reset, and there is an ARM instruction that effectively halts its execution. Those are controlled by the Linux operating system in the environment in which we're working. Your program starts when Linux gives it control at the "_start" label; and when your program chooses to quit, it will return control back to Linux using a service call (svc 0 instruction). Actually, Linux never gives up full control to your program, but don't be concerned about that for now.

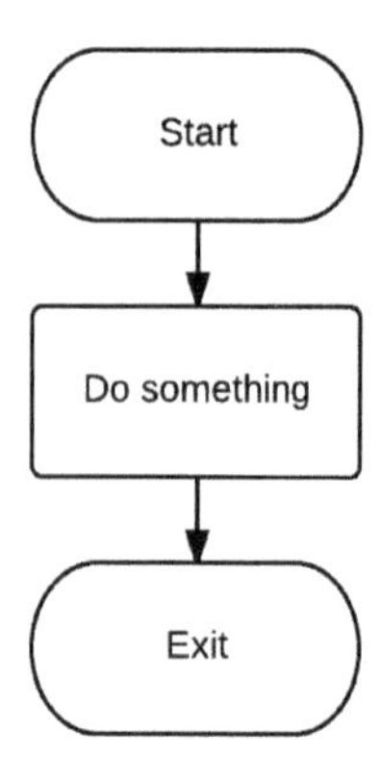

Figure 1.7: Simple program flow

The flowchart in figure 1.7 illustrates the major flow of a program: It gets CPU control, it does its intended job, and finally it returns control.

The only purpose of the first program (Listing 1.0) is to start and then quit. It "quits" by returning control to Linux using service code 7. Linux provides a variety of services to a running program such as reading and writing disk files, and communicating with keyboards, monitors, and networks. The contents of register R7 indicates what services is requested. For example, [R7]=3 tells Linux to read from a file or device, [R7]=4 tells Linux to write to a file or device, and [R7]=1 tells Linux that your program is ready to stop running and return full control back to Linux. Other registers provide Linux with addition information such as what data is to be written and which file is to be accessed.

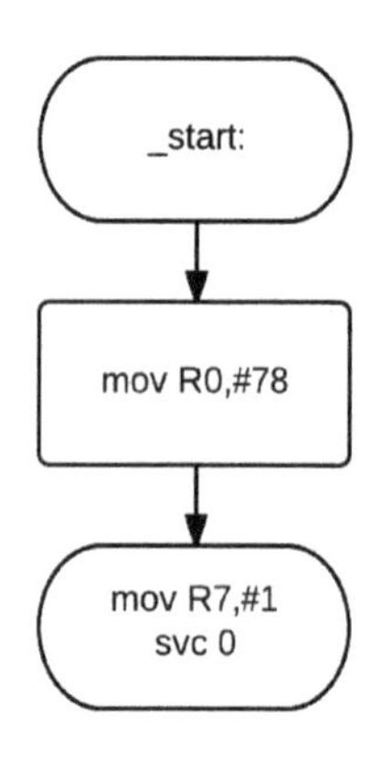

Figure 1.8: Flowchart of Listing 1.0

When an application terminates, it returns an "exit code" to Linux that can be tested in command line scripts. The value of the exit code will be placed into command line variable $? which can be displayed by the "echo $?" command line. Note: Although the exit code can range from 0 through 255, typical values for exit codes are 0 (implying a successful application) and 1 (implying the application somehow failed to perform as expected). Many of the other possible codes are loosely defined depending on application. For these first two labs, we will use the full 0 to 255 range, but in real applications, please use only 0, 1, or a few of the other specific "error" code numbers. Check the Internet for sysexits.h in Unix or Linux to get some of these codes.

Coding and Debugging

The beginning of a comment is marked by the @ in the GNU "as" assembler. In other assemblers for the ARM as well as the GNU assembler for other CPUs (such as the x86), the comment indicator is #, /, or semicolon. In other computer languages, comments are indicated by <!, /*, // - -, #, C, and even REM. The GNU "as" assembler also supports using /* to begin a comment with an associated */ to end the comment, but I won't be showing that in the examples.

Comments are a great way to locally document a program, but the comments provided in Listing 1.0 are worse than useless (except for helping someone learning to program in assembly language). Once a programmer knows the CPU architecture being used, having a comment repeat what the instruction does is ridiculous. Instead of saying *what* is being done, the comments should say *why* it is being done. The comments in Listing 1.2 are much better. Although the ".end" directive is not necessary, I find it comforting to know that I didn't lose part of the program in a truncated copy of cut and paste.

```
        .global  _start   @ Providing program starting address to linker
_start: mov      R0,#78   @ Set program Exit Status code to 78 (meaning ...)
        mov      R7,#1    @ Service command code 1 terminates program.
        svc      0        @ Issue Linux command to terminate program
        .end
```

Listing 1.2: Same program, but with better comments

It's good practice to enter the program yourself, but since there are over 50 program listings in this book, perhaps that's too much of a good thing. Appendix A shows how all the source code for this book can be easily downloaded over the Internet using git and GitHub. Listing 1.3 shows how the above model.s program can be copied using the cp command. The ls and cat commands are not necessary, but will list the contents of the directory and the copied model.s source code, respectively.

```
~$ ls RPi_Asm_Bridge
~$ cp RPi_Asm_Bridge/Listing_1_2.txt model.s
~$ cat model.s
```

Listing 1.3: Copy existing source code used in this book

Maintenance

Review Questions

1. What is the difference between source code and object code?
2. Why is there a linker? That is, why don't we go straight from source code to the executable file and skip this "middle man"?
3. * What is the main difference between a procedural computer program and a non-procedural one?
4. An assembler is an example of a type of compiler. What makes an assembler unique from other compilers?
5. * When updating a line of source code, should the comment on the line be updated as well?

Programming Exercises

1. Modify Listing 1.2 by removing the first line. Then compile, link, and execute as before. Where was the error caught? Where was it missed?
2. Modify Listing 1.2 by removing the # (hash code) from the second line of text (78 instead of #78). What happens when you compile, link, and execute?

Lab 2
Arithmetic Logic Unit

A major component of a CPU is the Arithmetic Logic Unit (ALU) where calculations take place. Lab 2 will be similar to Lab 1, but will actually perform some calculations rather than just load a value into a register and display it.

Prototype

We will begin with exactly the same command line work flow of edit (nano model.s or any other editor), assemble (as -o model.o model.s), link (ld -o model model.o), execute (./model), and display (echo $?) as we did in Lab 1. In the Coding and Debugging section we will replace the execute and display commands with a new debugger (gdb model) command line.

```
~$ nano model.s
~$ as -o model.o model.s
~$ ld -o model model.o
~$ ./model
~$ echo $?
```

Listing 2.0: Sequence of edit, compile, link, execute, display

Listing 2.1 demonstrates doing some arithmetic by adding two numbers together and getting the result to display in the exit status code.

```
        .global   _start     @ Provide program starting address to linker
_start: mov       R0,#17     @ Use 17 for test example (could be anything)
        mov       R6,#2      @ A second integer for test
        add       R0,R6      @ [R0] = [R0] + [R6], Add contents of R6 to R0
        mov       R7,#1      @ Service command code 1 terminates a program.
        svc       0          @ Issue Linux command to terminate program
        .end
```

Listing 2.1: Assembler source code for model.s

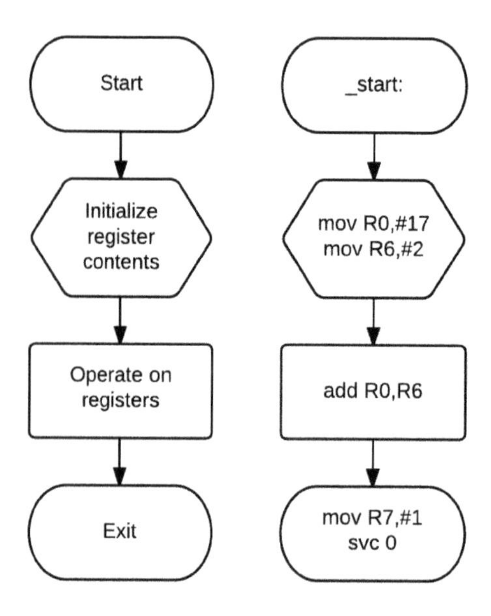

Figure 2.0: Prototype program for Lab 2

The program to be entered using the editor is illustrated in Figure 2.0 and provided in Listing 2.1. It has the following parts:

- Start: Special label "_start" indicates the location of the first instruction of a program. Recall from Lab 1 that the .global assembler directive enables the linker to "see" the _start label and know where to start the program.
- Preparation: Initialize contents of registers to be used in this example.
- Process: In this example, we're adding the contents of two registers.
- Terminate: Exit this program and return control to the Linux operating system.

After running the program in Listing 2.1 as is, try replacing the addition (ADD) instruction with subtraction (SUB) or multiplication (MUL). Also try different initial values for R0 and R6. Note that only the lower 8 bits of the answer are displayed by the echo $? command. For example, if register R0 is loaded with 11 and R6 with 30, then a "MUL R0,R6" instruction would result in 330 in R0, but the echo $? command will display 74 (which is 330-256). The decimal answer of 330 is actually 101001010 in binary. The decimal "answer" of 74 results because the high order (leftmost) bit representing 256 is lost and the remaining 01001010 (lower 8) bits give the decimal answer of 74. Refer to Appendix B if you need a review of binary numbers.

The program in Listing 2.2 is the same as that in Listing 2.1 except it uses a Boolean AND function instead of the arithmetic ADD instruction. It also has initial values entered in binary (0b1011 and 0b1100) which are more appropriate for logical operations.

```
        .global   _start        @ Provide program starting address to linker
_start: mov       R0,#0b1011    @ Use binary 1011 as first value for test
        mov       R6,#0b1100    @ Use binary 1100 as second value for test
        and       R0,R6         @ [R0] = [R0] & [R6] or 1011 & 1100 = 1000 (decimal 8)
        mov       R7,#1         @ Service command code 1 terminates this program.
        svc       0             @ Issue Linux command to terminate program
        .end
```

Listing 2.2: Logic (Boolean) example with binary inputs

In the above code, replace the Boolean "and" (AND) instruction with an inclusive "or" (ORR) or an exclusive "or" (EOR). Also try different initial values for R0 and R6.

Introductions

A variety of arithmetic, logical, and shifting instruction is introduced in this lab. The use of the debugger and its commands will also be new.

	List of ARM instructions introduced in Lab 2		
2.1.4.	add	R0,R6	@ [R0] = [R0] + [R6], Add contents of R6 to R0
2.2.4.	and	R0,R6	@ [R0] = [R0] & [R6] or 1011 & 1100 = 1000
2.3.3.	lsl	R0,#2	@ Shift R0 left 2 bits (i.e., multiply by 4)
2.4.4.	lsl	R0,R6	@ Shift R0 left by the value in R6

	List of gdb debugging commands introduced in Lab 2
b(reakpoint)	Define source code line numbers to suspend execution
i(nfo)	Dump all registers (r) or breakpoints (b)Dump all registers (r) or breakpoints (b)
l(ist)	Display source code lines
run	Start program execution and pause at next breakpoint
s(tep)	Single step (execute) the next instruction(s)
set	Load new value into a register

Principles

In Lab 0, we noted that the basic computer design consisting of a CPU with a memory goes back about 200 years to Babbage's design described as a mill with a store. When we look deeper into the construction of a CPU, we see that almost all of today's CPUs can be divided into three general constituents: ALU, registers, and instruction decoding/control logic.

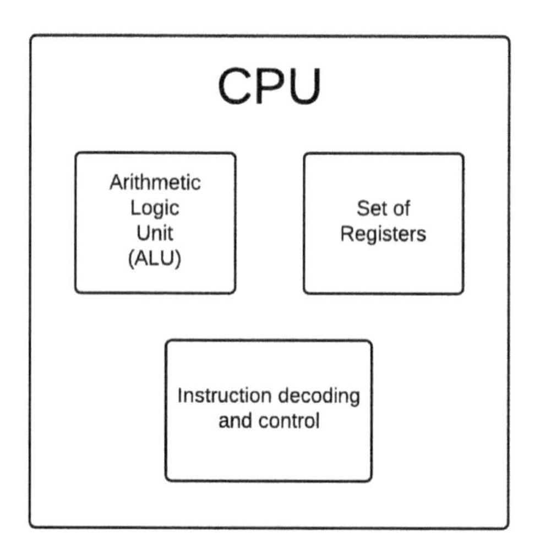

Figure 2.1: Major components of a Central Processing Unit

ARM Registers

The number of registers varies from CPU design to CPU design. The ARM has sixteen 32-bit general purpose registers, R0 through R15, which can be used as accumulators or as registers performing other functions like indexing into memory tables. Three of these registers, R13, R14, and R15, are also known as the LR link register, SP stack pointer, and PC program counter, respectively. Using these three registers as general purpose accumulators can have some undesirable results if not done very carefully .

ALU

The arithmetic portion of an ALU consists of the basic add, subtract, and multiply instructions. Many CPU designs also include a divide instruction, and many such as the ARM included in the Raspberry Pi do not. An ALU could of course also have a power or exponential instruction, but that is rarely the case. The Raspberry Pi does have a divide instruction and even a square root instruction, but those are not in the ARM processor but are in the floating point coprocessor that is included on the same "chip."

Logical Operations

The second category of operations performed by the ALU consists of the Boolean logic binary operations as depicted by the symbols and truth tables provide in Figure 2.2.

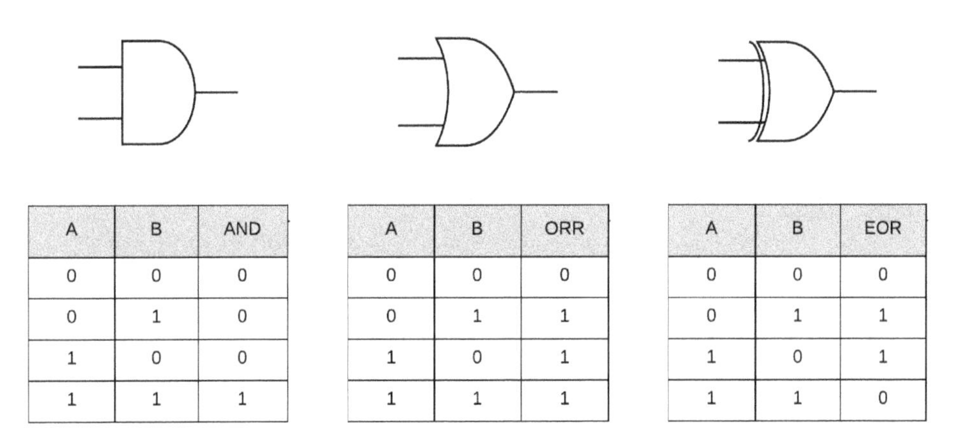

A	B	AND
0	0	0
0	1	0
1	0	0
1	1	1

A	B	ORR
0	0	0
0	1	1
1	0	1
1	1	1

A	B	EOR
0	0	0
0	1	1
1	0	1
1	1	0

Figure 2.2: Logical AND, Inclusive OR, and Exclusive OR

The logical instructions in almost all CPUs are "bitwise" logical operations:

- Thirty-two logical operations are performed in parallel. Figure 2.3 only shows a portion of the 32 pairs of corresponding bits being ANDed together.
- Figure 2.3 illustrates the AND instruction. The inclusive OR (ORR) and exclusive OR (EOR) are also bitwise instructions using 32 pairs of bits.

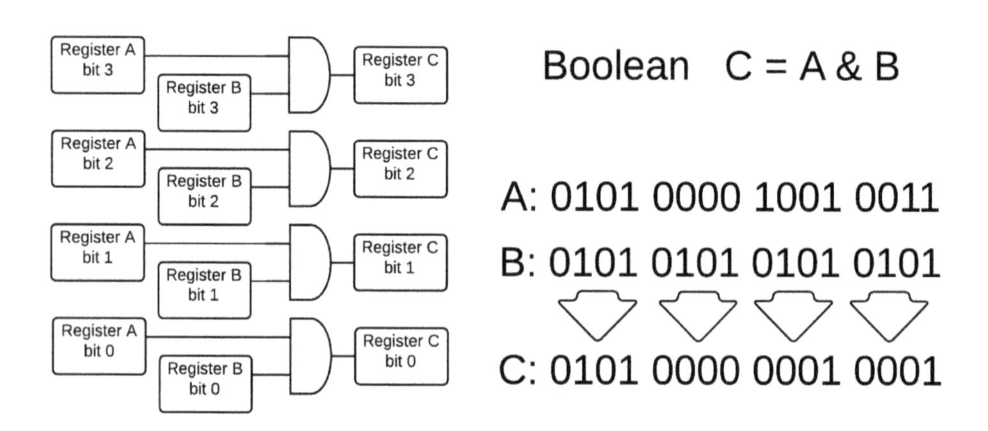

Figure 2.3: Examples of "bitwise" AND of two values

Bit Shift Operations

In addition to arithmetic and logic instructions, almost all CPU architectures include several instructions for shifting bits within a register. Most CPU architectures support three types of shifts:

- Logical: bits shifted out from either end of the register are discarded and new zero bits fill in
- Circular (also referred to as rotate): bits shifted out one end of the register come back in on the other side
- Arithmetic: similar to a logical shift except arithmetic shift brings in copies of the sign bit instead of zero.

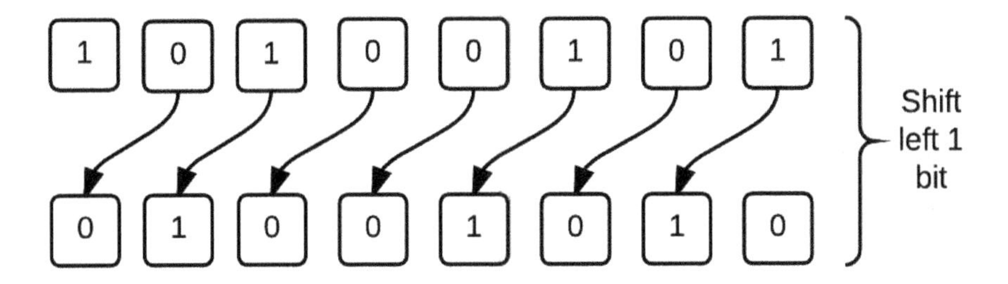

Figure 2.4: Logical left shift loses bit on left and brings in zero on right.

Logical shifts have many applications. Two common applications are for converting between serial and parallel and the second is for multiplying an integer by a power of two. For example, a one bit shift to the left is multiplying by two, while a two bit shift is multiplying by four. Some computers shift only one bit at a time. Not the ARM, because multiple bit shifts can be indicated either from the contents of a register or an immediate value in the instruction.

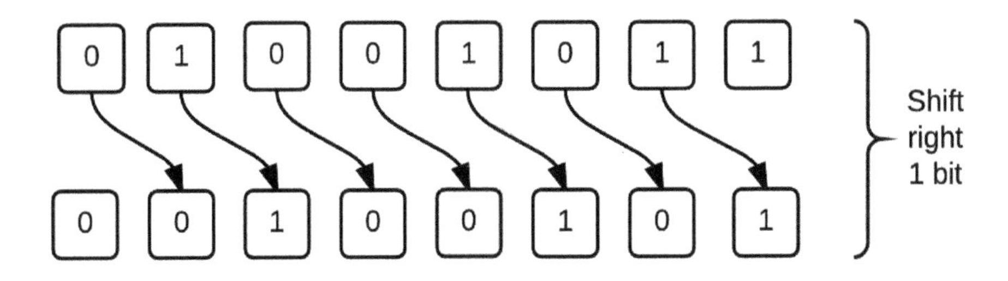

Figure 2.5: Logical right shift moves out bit on right and brings in zero on left.

A shift to the right is like dividing by a power of two, but be aware of two basic problems. Division can have a remainder which will get truncated, not rounded. Secondly, there are two problems with negative integers. A logical right shift will bring in a zero in bit 31, thereby converting a negative number to an inappropriate positive number. The arithmetic shift will solve the negative problem, but a rounding error is still present in the ARM architecture, so be careful using shifts to divide negative numbers. Please see Appendix B if you need an explanation why the high-order bit (bit 31) is a “1” for negative numbers

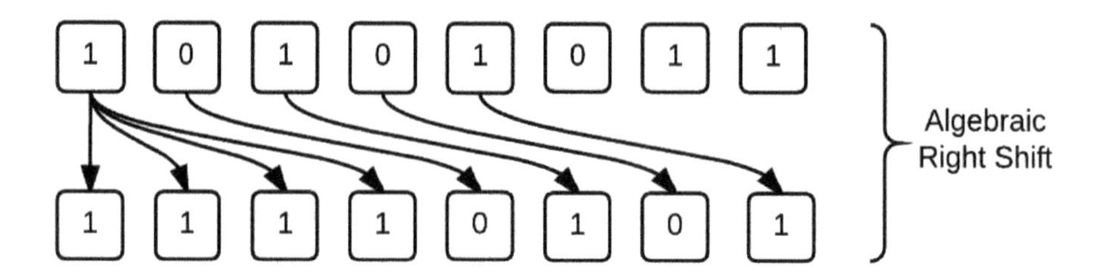

Figure 2.6: Algebraic shift copies the sign bit.

A circular shift, also referred to as a rotate, allows the bits to be shifted without losing anything out one end or the other. The ARM only provides a rotate to the right, but a rotate to the left can be done by a right rotate. For example, a left rotate of 5 bit positions is identical to rotating right 27 bit positions (32 minus 5). There is also an extended right rotate of one bit that copies bit 0 into a status bit known as "carry."

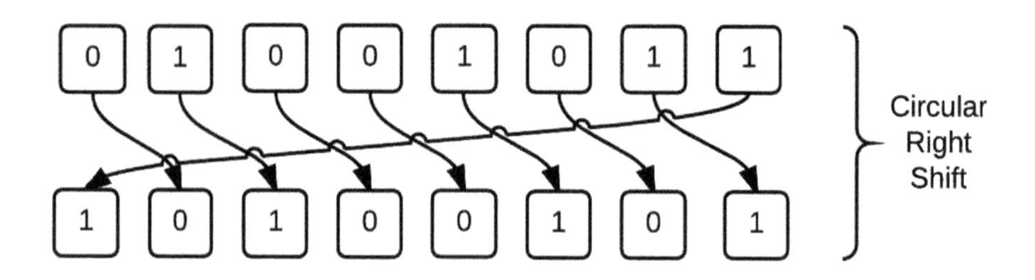

Figure 2.7: Rotate one bit position to the right — no bit lost

Listings 2.3 and 2.4 show the register R0 containing a decimal 17 being shifted 2 bits to the left using an immediate shift count or a shift count in a register, respectively. Both cases will result in R0 containing a value of decimal 68. Listing 2.3 demonstrates the Logical Shift Left (LSL) instruction, but I suggest you also try LSR and ROR (rotate) instructions.

```
        .global   _start    @ Provide program starting address to linker
_start: mov       R0,#17    @ Use 17 for test example (could be anything)
        lsl       R0,#2     @ Shift R0 left 2 bits (i.e., multiply by 4)
        mov       R7,#1     @ Service command code 1 terminates this program.
        svc       0         @ Issue Linux command to terminate program
        .end
```

Listing 2.3: Shifting a register's contents is like multiplying by a power of 2.

```
        .global   _start    @ Provide program starting address to linker
_start: mov       R0,#17    @ Use 17 for test example (could be anything)
        mov       R6,#2     @ A second integer for test
        lsl       R0,R6     @ Shift R0 left by the value in R6 (i.e., multiply by 4)
        mov       R7,#1     @ Service command code 1 terminates this program.
        svc       0         @ Issue Linux command to terminate program
        .end
```

Listing 2.4: Shifting a register by the value in another register

One final note about shift instructions on the ARM processor. Their assembly language coding might look similar to that of other CPUs, but the code generated has a surprise to be revealed in Lab 5.

Advanced RISC Machine (ARM)

Computers have never been fast enough. Application developers and users always want better performance, and electronics designers have generally been able to fulfill those expectations for decades. Of course, whenever one application is satisfied, another one that was previously "impossible" whets the appetite of application developers for continued performance enhancements. "Moore's Law" implies that computer performance will double every 18 months, mostly due to improvements in the packing density of transistors on integrated circuits. Oddly enough, this has proven to be the case for over three decades, far longer than many of us thought possible.

Not all of the performance improvements over the years have come from improvements in the structure of integrated circuits. Some have come from improvements in CPU design as well as running operations in parallel. As was presented in Lab 0, most instructions executed by a CPU have the following four components:

- Op code
- Location to receive result
- Location of first operand
- Value or location of second operand

This format is popular because applications generally need computers to evaluate "binary operations" such as C=A+B, where addition is just one of the many arithmetic and logical binary operations an ALU can perform. So where are these ABC variables located? Today, our computer systems have billions of bytes and words of storage in memory, but only about a couple dozen registers. Obviously, our variables are stored in main memory and only temporarily copied into the registers when we are using them in calculations. Figure 2.8 illustrates the general format of CPU instructions that was becoming popular around 1980. These instructions allowed the first operand (A in C=A+B example) to come from either a register or memory. The value for the second operand (B) could come from a register, memory, or be an immediate constant, such as C=A+6. The result of the calculation (C) could then be placed into a register or written into main memory.

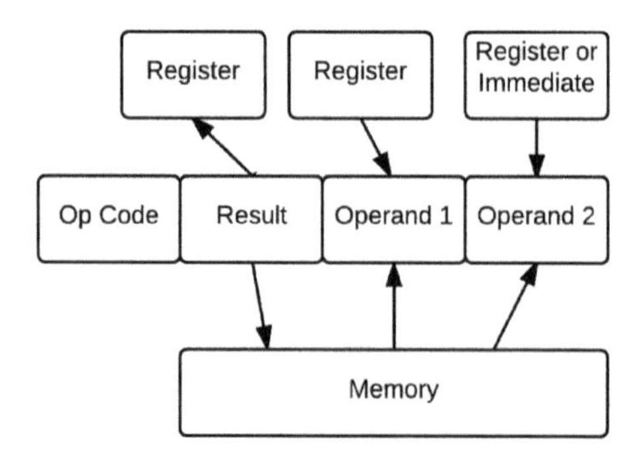

Figure 2.8: Instruction format in a CISC architecture

This instruction format actually enables C=A+B to be performed in a single CPU instruction and not even use any registers. The problem was one of execution speed, and it was being addressed in the early 1980s at the University of California, Berkeley, and Stanford University. The conclusion was to develop Reduced Instruction Set Computer (RISC) architectures. The primary difference between a RISC architecture and what was later referred to as a Complex Instruction Set Computer (CISC) architecture was that most instructions would only access data in registers, not in main memory. This difference accelerates program execution in two ways: the instructions are simpler in format thereby being easier to decode and of course getting data from registers is always faster than getting data from main memory.

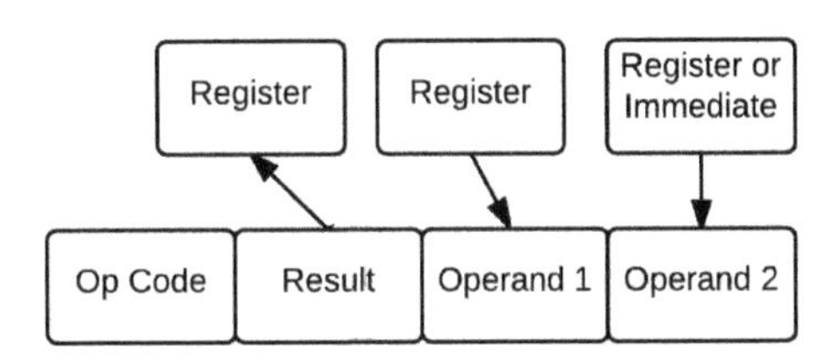

Figure 2.9: Instruction format in a RISC architecture

Obviously, variables in main memory will have to be accessed eventually, but by optimizing program segments to keep as much data in registers as long as possible does lead to faster program executions. The second obvious point is that there has to be at least a couple of instructions to access main memory: one to load registers and one to store register contents into memory. So, what does this have to do with the Raspberry Pi and the ARM processor in particular? Inspired by the research into RISC architectures, the British computer manufacturer, Acorn Computers, developed the Acorn RISC Machine (ARM), later renamed as the Advanced RISC Machine. With the exception of a few of memory load/store instructions, the ARM instruction format is very similar to that illustrated in Figure 2.9, but with an additional twist which we'll examine in detail in Lab 5.

Was this idea of a RISC architecture new in the mid 1980s? No, of course not. It was more a time of getting back to basics. Over a decade earlier, I was developing hardware/software applications for the supercomputer of that day: the Control Data CDC6600. Much of the speed of programs running on the CDC6600 came from CPU instructions that performed all ALU operations only in registers. Do all computers today have a RISC architecture? No. The one area where RISC architectures, and the ARM in particular, do have a monopoly is in portable devices like cell phones and tablets. RISC machines typically get much better throughput for the same amount of energy consumed, thereby getting much longer life from a battery charge.

Coding and Debugging

The output of logical operations in decimal as was done in the Prototype section was certainly not optimal. We will now start using the "gdb" debugger utility that comes standard with the GNU software available on the Raspberry Pi.

The debugger enables the following:

1. Set breakpoints
2. Examine register contents
3. Change register contents
4. Examine memory contents

Basically the same work flow is used: edit, compile, link, and execute, but we will need to include the -g option on the "as" command and replace the "./model" with the "gdb model" command line as shown in Listing 2.5.

```
~ $ nano model.s
~ $ as -g -o model.o model.s
~ $ ld -o model model.o
~ $ gdb model
```

Listing 2.5: Sequence of edit, compile, link, execute, debug

When gdb starts, it displays its copyright and license agreement (Listing 2.6), issues the "(gdb)" promp, and then waits for user input from the keyboard.

```
pi@raspberrypi ~ $ gdb model

GNU gdb (GDB) 7.4.1-debian
Copyright (C) 2012 Free Software Foundation, Inc.
License GPLv3+: GNU GPL version 3 or later <http://gnu.org/licenses/gpl.html>
This is free software: you are free to change and redistribute it.
There is NO WARRANTY, to the extent permitted by law. Type "show copying"
and "show warranty" for details.
This GDB was configured as "arm-linux-gnueabihf".
For bug reporting instructions, please see:
<http://www.gnu.org/software/gdb/bugs/>...
Reading symbols from /home/pi/model...done.
(gdb)
```

Listing 2.6: The gdb license statement

Listing 2.7 shows two commands being given to the debugger. First, the "list" command will display the program that is being run, including comments and line numbers. The "b 4" command is entered next to set a breakpoint at line 4 of the program. Note: Many of the commands can be shortened such as "l" for list and "b" for breakpoint, and there is a blank between the "b" and "4" in the command.

```
(gdb) list
1               .global   _start     @ Provide program starting address to linker
2       _start: mov       R0,#17     @ Use 17 for test example (could be anything)
3               mov       R6,#2      @ A second integer for test
4               lsl       R0,R6      @ Multiply R0 by 4 (shift left 2 bit positions)
5               mov       R7,#1      @ Service command code 1 terminates this
                                     program.
6               svc       0          @ Issue Linux command to terminate program
7               .end
(gdb) b 4
Breakpoint 1 at 0x805c: file model.s, line 4.
(gdb)
```

Listing 2.7: Setting breakpoints to pause execution

We now enter the "run" command to the debugger to start the program, and it runs until it reaches the first breakpoint, which is on line 4 and is just before the LSL instruction is executed. The "info registers" command ("i r" for short) is entered to provide the current contents of the ARM's registers. Notice that R0 contains 17 (hexadecimal 0x11) and that R6 contains 2.

```
(gdb) run
Starting program: /home/pi/model

Breakpoint 1, _start () at model.s:4
4       lsl         R0,R6        @ Multiply R0 by 4 (shift left 2 bit positions)
(gdb) i r
r0      0x11        17
r1      0x0         0
r2      0x0         0
r3      0x0         0
r4      0x0         0
r5      0x0         0
r6      0x2         2
r7      0x0         0
r8      0x0         0
r9      0x0         0
r10     0x0         0
r11     0x0         0
r12     0x0         0
sp      0x7efff860  0x7efff860
lr      0x0         0
pc      0x805c      0x805c <_start+8>
```

```
cpsr     0x10          16
(gdb)
```

Listing 2.8: Show current register contents

In Listing 2.9, we enter the single step "s" command to execute only the shift instruction, and the program is now suspended after just that one instruction is performed. Another "i r" command is entered which now shows R0 containing the value 68 as expected. From a hexadecimal view, we shifted 0x11 left two bit positions giving us 0x44.

```
(gdb) s
5        mov           R7,#1         @ Service command code 1 terminates this
                                     program.
(gdb) i r
r0       0x44          68
r1       0x0           0
r2       0x0           0
r3       0x0           0
r4       0x0           0
r5       0x0           0
r6       0x2           2
r7       0x0           0
r8       0x0           0
r9       0x0           0
r10      0x0           0
r11      0x0           0
r12      0x0           0
sp       0x7efff860    0x7efff860
lr       0x0           0
pc       0x8060        0x8060 <_start+12>
cpsr     0x10          16
(gdb)
```

Listing 2.9: Result of R0 shifted left two bits

We rerun the program in Listing 2.10, and when it suspends for the breakpoint at line 4, we use the "set" command to modify the contents of registers R0 and R6 to contain 5 and 4, respectively. Another "i r" command shows the changes have taken place. Another single step command is then taken. I'll leave the rest to the reader. The final value in R0 should be 5×2^4.

```
(gdb) run
Starting program: /home/pi/model

Breakpoint 1, _start () at model.s:4
4         lsl              R0,R6          @ Multiply R0 by 4 (shift left 2 bit positions)
(gdb) set $r0 = 5
(gdb) set $r6 = 4
(gdb) i r
r0        0x5              5
r1        0x0              0
r2        0x0              0
r3        0x0              0
r4        0x0              0
r5        0x0              0
r6        0x4              4
r7        0x0              0
r8        0x0              0
r9        0x0              0
r10       0x0              0
r11       0x0              0
r12       0x0              0
sp        0x7efff860       0x7efff860
lr        0x0              0
pc        0x8060           0x8060 <_start+12>
cpsr      0x10             16
(gdb) q
Quit anyway? (y,n) y

pi@raspberrypi ~ $
```

Listing 2.10: Modify register contents

The "q" command will exit the debugger and return control to the Linux command prompt. The debugger may ask "Quit anyway?" if the program has not reached the service call to exit.

Maintenance

Review Questions

1. What is meant by a "bitwise" logical operation?
2. Why don't many ALU designs include a divide instruction?
3. Which logical instruction can set a register contents to zero?
4. Which shift instructions can set a register to zero using a shift count of 32?
5. Using an Internet search, what are three examples of RISC CPUs in addition to the ARM?
6. Using an Internet search, what are three examples of CISC CPUs?
7. An important characteristic of RISC machines is that their arithmetic, logical, and shift instructions do not get their operands from memory, but only from registers or are immediate. Using an Internet search, what are two other common characteristics of most RISC CPUs?
8. * Which register will be changed after executing each of the following instructions, and what is its new value? Assume that each instruction begins with the following contents: R0 = 0, R1 = 1, R2 = 2, R3 = 3, R4 = 4, and R5 = 5.
    a. AND R2, R3
    b. ADD R4, R5
    c. LSL R1, #4
    d. EOR R3, #1
    e. MUL R4, R5
    f. SVC 0

Programming Exercises

1. Modify Listing 2.3 by initializing R0 with a negative number. Then using the compile, link, and debug command sequence as in Listing 2.5, compare the results (breakpoint at line 5) of an arithmetic right shift (ASR) to that of a logical right shift (LSR).
2. Modify Listing 2.3 to demonstrate how to set the contents of register R0 to zero using a shift instruction
3. Modify Listing 2.3 to demonstrate how to set the contents of register R0 to zero using a logical instruction

Lab 3
Subroutines and ASCII Output

As programs get larger and more complicated, it is extremely important that they are divided into components, and how the components are connected is critical. In Lab 3, we will begin this process of dividing up the application. The application is first divided between program and data. The program itself is then divided into multiple components called subroutines. Depending upon the programming language and application, subroutines are also know as functions, modules, procedures, and methods. In Lab 3, we will develop subroutine v_asc, the first of many subroutines that we will eventually save in a library which becomes a toolbox of utilities.

Prototype

Almost every CPU ever designed has an instruction that "calls" the operating system for services. We've already used the SVC (service call) instruction to terminate application programs. Now we will use it to write to the display monitor. In following labs, we will use it to read from the keyboard, and it can also be used to read and write disk files and Input/Output data lines as well.

```
        .global  _start        @ Provide program starting address to linker
_start: ldr      R1,=msgtxt    @ Set R1 pointing to message to be displayed
        mov      R2,#10        @ Number of bytes in message
        mov      R0,#1         @ Code for stdout (standard output, i.e., monitor)
        mov      R7,#4         @ Linux service command code to write string.
        svc      0             @ Issue command to display string on stdout
        mov      R0,#0         @ Exit Status code 0 for "normal completion"
        mov      R7,#1         @ Service command 1 will terminate this program
        svc      0             @ Issue Linux command to terminate program
        .data
msgtxt: .ascii   "Hey there\n" @ 10 character message (blank and \n each count
                               as 1 char.)
        .end
```

Listing 3.0: Output a string of ASCII characters

In Listing 3.0, a second Linux service request is being introduced to write a string of bytes. With R7 set to 4, the "svc 0" service call will write to the device specified by register R0. A value of R0=1 indicates that the string should be written to the standard output device "stdout," which is usually the computer's display monitor. The memory location of the string is loaded into R1 and its length into R2. We will now stop using the exit status code service request to display answers. The second "svc 0" which has R7=1 will terminate the program with a status code of 0 (from R0) implying a successful performance.

I've also introduced two more assembler directives: .data and .ascii. The .ascii directive tells the assembler to place a string of ASCII characters into memory. Special control characters can be indicated by a sequence beginning with a back slash (\n represents "line feed," hexadecimal value 0A). See Appendix D for some background on the ASCII character set if you like. The .data directive enables the assembler (with the help of the linker) to separate the areas of the computer memory dedicated to instructions from that of data values. In this example, it is not necessary, but it is setting a pattern for later programming examples.

Go ahead and run this sample program. You can use the same command sequences (Listing 2.0), but you will not need the echo command (unless, of course, you want to see the exit status of zero). Try different text messages, but be sure to change the value in register R2 to match the length of the message.

Introductions

In Lab 3, I will introduce the ARM instructions and assembly language directives that are typically used to divide a program into components. We will also begin using more appropriate techniques for displaying data through the Linux "write" service routine.

	List of ARM instructions introduced in Lab 3		
3.0.2.	ldr	R1,=msgtxt	@ Set R1 pointing to message to be displayed
3.1.4.	bl	v_asc	@ Call subroutine (Branch and Link) to display text
3.1.22.	bx	LR	@ Return (Branch eXchange) to the calling program

	List of assembler directives introduced in Lab 3
.ascii	Put string of ASCII characters into memory
.data	Inform linker to group following code with other data in memory
.text	Inform linker to group following code with other instructions in memory

	List of Linux service calls introduced in Lab 3
4	Write array of bytes to device

	List of gdb debugging commands introduced in Lab 3
c	Continue
x	Dump memory

Principles

One of the principal hallmarks of the industrial revolution was the use of interchangeable parts in the manufacturing process. In a similar manner, subroutines and operating system services are building blocks for developing large sophisticated software applications. Both subroutines and services are predefined program segments that can be used over and over again by calling them from the main application program. In most cases, the exact actions performed by these building blocks can be modified slightly based on a set of input parameters referred to as "arguments" and are usually contained in registers. After performing their assigned tasks, subroutines and services return program control to the instruction following the point from which they were called.

Linux Services

We've already been using one Linux service to quit our programs. This service (having [R7]=1) is unique in that it does not return control to the instruction after the service call. The new service that we will be using ([R7]=4) will perform the task of writing to the display monitor and then returning control to the instruction following the service call.

One of the main responsibilities of an operating system, such as Linux, is to provide services for application programs. A large portion of these services involves reading and writing peripheral devices (display monitor, keyboard, mouse, network, etc.) and disk files (real spinning disks as well as solid-state memory devices). The calling program must provide Linux with the details of what is to be performed:

1. What is to be done
2. Which device is to written or read
3. Where the data (buffer) is in the program's memory
4. How much data is to be written or read

In the case of Linux, as well as most operating systems, this information is provided in the CPU's registers. For some devices such as a disk or external memory, another parameter providing the location on the disk is many times required as well.

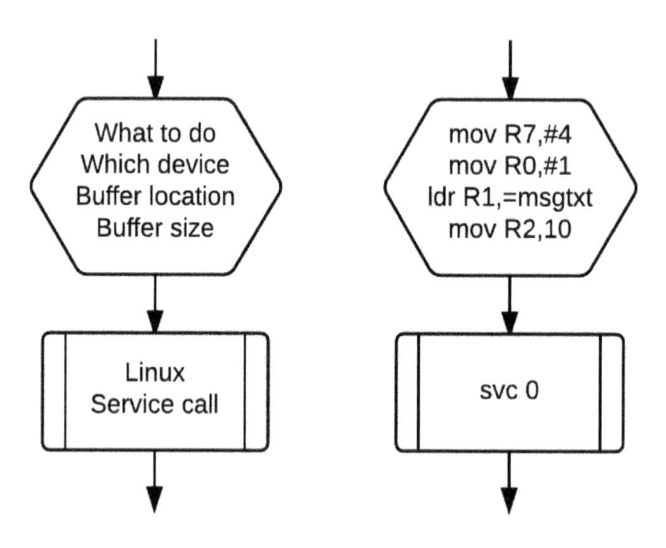

Figure 3.0: Linux service call to write to display monitor

A second responsibility of an operating system is to coordinate the sharing of a computer's resources among multiple tasks and users. This includes making sure each user receives its fair share of CPU time, memory, I/O device access, and other resources without interfering with other users. That's the main reason why in a Linux environment an assembly language program cannot read and write directly to devices.

Subroutines

A user application can and should be broken up into multiple building blocks. These subroutine building blocks are called in a manner similar to calling operating system services. In the ARM architecture the BL (branch and link) instruction calls subroutines in a manner similar to how the SVC (service call) instruction calls operating system services

A subroutine is a section of code that is "called" to perform a specific job. Examples of jobs a subroutine can perform:

- Display a number to the user
- Get keyboard input from the user
- Get input from a specific device such as a temperature sensor
- Change the speed of a motor
- Perform a particular type of analysis such as a least-squares fit of data

The advantages of using subroutines are many:

- Subroutines help organize the construction of the program.
- The code only takes up memory space once.
- It's less work to modify or correct one area of code rather than many areas.
- "Information hiding" occurs because one part of the program is unable to access data in another part and cannot accidentally change it.
- Division of programming assignments among different programmers is easier.

The disadvantages of subroutines are few.

- There is a slight performance degradation compared to "in-line code" due to the overhead of the call and return.
- It can lead to too much of a good thing: Too many tiny subroutines can lead to confusion.

Our first subroutine will be a very simple one. It will later be included in a library of subroutines which output various forms of data to the display monitor. Subroutine "v_asc" will display an ASCII message on the display monitor, a common occurrence in most computer programs. It's so simple, you might ask, "Why bother at all, just use the Linux SVC directly and reduce the additional overhead of a subroutine essentially calling the same type of subroutine." In a very small program with only a few calls to display, I would agree. However, for larger programs, a dedicated display subroutine provides a lot of flexibility from the maintenance perspective. Just for example, let's say we have developed a program with thousands of calls to display, and now the "marketplace" requires that we send our display messages to a different device (one that the simple Linux call cannot perform). Wouldn't it be more convenient to accommodate that change in one place in the subroutine's code rather than hunt it down in the large program and try to successfully change it thousands of times?

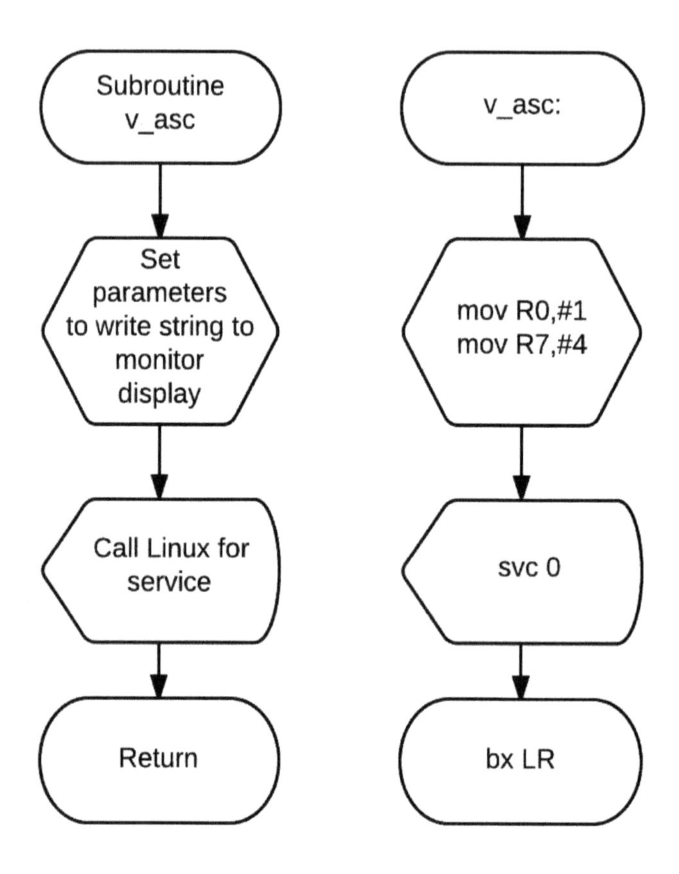

Figure 3.1: Subroutine v_asc displays messages

Figure 3.1 shows the structure of the v_asc subroutine that is called by the "main" program illustrated in Figure 3.2. These two components will replace the prototype program we just ran (Listing 3.0). When a program starts, Linux gives control to the main program identified by having the _start label. In general, a main program may call as many subroutines as it requires, and each of these subroutine can even call more subroutines, etc. There is even a special "recursive" type of subroutine that can call itself.

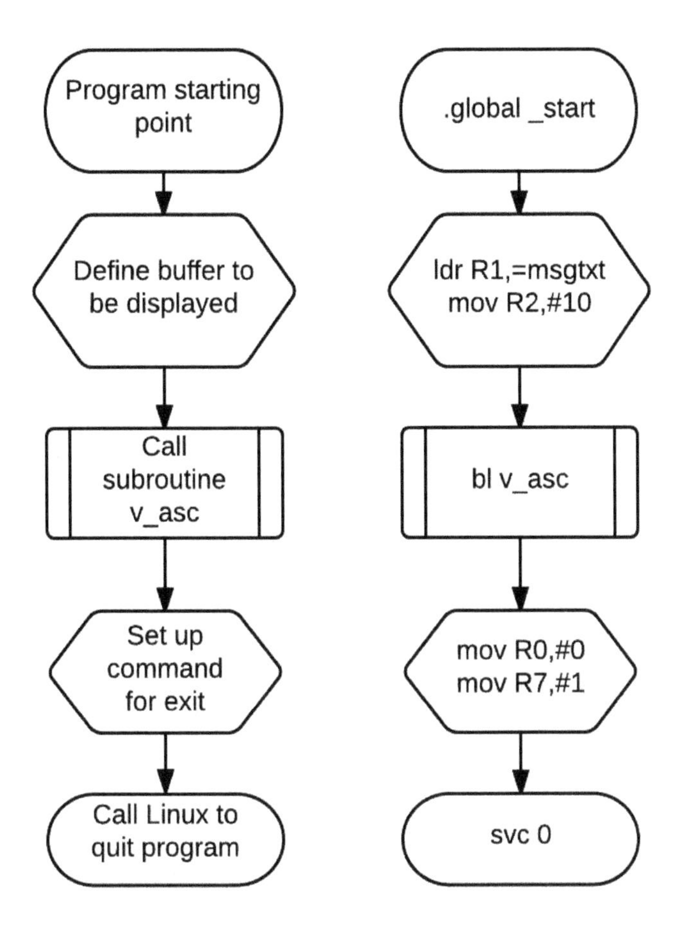

Figure 3.2: Main program calling subroutine v_asc

Functions, Methods, Procedures

The concept of a subroutine is also known by several other names. A "function" is a subroutine that not only performs a specified task, but also returns a value (usually in one of the registers). Some higher level languages such as C use the name "function," while the term "procedure" or "method" is the name used in other languages. No matter what the name, the concept is the same: a block of code that performs a predefined task (usually customized by data in registers), and it may return a value.

Interrupts

A third type of subroutine commonly used in embedded systems applications is the interrupt service routine. There are basically two types of interrupts on most CPUs:

- Software: The SVC instruction we've been using to call Linux kernel services in every program in this book. The SVC (service call) instruction is also known as the SWI (software interrupt).
- Hardware: A change in state of an I/O device essentially "calls" a device driver (subroutine) using an "instruction" that behaves almost identical to the SVC instruction.

Although interrupt handling is very important to embedded systems development, I will not be covering it in this book because its use is closely guarded by Linux and every other operating system supporting multiple users running on the same computer. However, once you become proficient in writing subroutines, you will be well on your way to writing good interrupt handlers.

Link Register (LR)

In the ARM architecture, a subroutine is typically called using the branch and link (BL) instruction which loads the address of the instruction following the BL into the LR link register. The idea is to provide the subroutine with a return address. Over the decades, this has been done in many different CPU architectures with instructions named BALR (Branch and Link Register), LMJ (Load Modifier and Jump), SLJ (Store Location and Jump), RJ (Return Jump), and of course CALL (not an acronym, just plain "call"). BALR and LMJ provided the return address in a register, while SLJ and RJ put it into the first word of the subroutine's memory, and CALL saves the return address onto the "stack," a concept to be introduced in Lab 4.

A subroutine must be careful to not accidentally lose its return address. If one subroutine calls another, the contents of the LR has to be saved beforehand otherwise the second BL will overwrite it. Also, the LR is a second name for register R14 which can be used as a general purpose accumulator.

Coding and Debugging

As Listing 3.1 illustrates, the operation of every subroutine should be locally documented in comments:

1. Name of the subroutine and what it does
2. Resources it needs (R1 points to the string to be displayed and R2 contains the length of the string).

3. How the subroutine is going to return to the program that called it (LR register contains the return address)
4. What is in the registers when the subroutine is complete and returns to the calling program

```
1.            .global    _start          @ Provide program starting address to linker
2. _start:    ldr        R1,=msgtxt      @ Set R1 pointing to message to be displayed
3.            mov        R2,#10          @ Number of bytes in message
4.            bl         v_asc           @ Call subroutine to view text string in ASCII.
5.            mov        R0,#0           @ Exit Status code of 0 for "normal completion"
6.            mov        R7,#1           @ Service command code 1 terminates program
7.            svc        0               @ Issue Linux command to terminate program
8.
9.            .data
10. msgtxt:   .ascii     "Hey there\n"   @ 10 character message (blank and \n each count as 1 char.)
11.           .text
12.
13. @         Subroutine v_asc will display a string of characters
14. @                    R1: Points to beginning of ASCII string
15. @                    R2: Contains length of string in bytes
16. @                    LR: Contains the return address
17. @                    Registers R0 and R7 will be used by v_asc and not saved
18.
19. v_asc:    mov        R0,#1           @ Code for stdout (standard output, i.e., monitor)
20.           mov        R7,#4           @ Linux service command code to write string.
21.           svc        0               @ Issue command to display string on stdout
22.           bx         LR              @ Return to the calling program
23.           .end
```

Listing 3.1: Setting up a subroutine to output ASCII string

Program Line Numbers

In Listing 3.1, I've included the line number for each line of the text file containing the program and data. These numbers are not actually in the text file itself, but usually provided by the editor, assembler, and debugger. Notice how I started from one rather than zero. I did this to match the values provided by the editor, assembler, and debugger.

Do I really start counting from zero? No. Of course not. So why did I do it in identifying the labs, listings, and figures in this book? Simply to get you accustomed to the idea that in computer software and hardware, there are many situations in which zero is the first value and not one.

Two additional instructions are introduced in Listing 3.1:

- Branch and link (BL): Calls a subroutine by saving the current program location (PC) in the link register (LR) and then branching (also known as jumping in many CPU architectures) to a another location in the program.
- Branch exchange (BX): Returns from a subroutine by branching to the program location specified in a register.

In labs 1 and 2, we used the exit status code to display the results of our calculations. From now on, we'll display our results using more appropriate techniques. We will also use the GNU gdb debugger to see intermediate values leading up to producing the final values to be displayed.

Since this lab is introducing the mechanics of subroutines and an ASCII display in particular, the two values we'll be examining are the following:

1. The contents of the link register (before and after the BL instruction)
2. The ASCII contents of the text message

Start the debugger like you did in Lab 2 (~$ gdb model), and then enter the l(ist) command to display the first part of the program.

```
(gdb) l
1               .global   _start        @ Provide program starting address to linker
2     _start:   ldr       R1,=msgtxt    @ Set R1 pointing to message to be displayed
3               mov       R2,#10        @ Number of bytes in message
4               bl        v_asc         @ Call subroutine to view text in ASCII.
5               mov       R0,#0         @ Exit Status code: "normal completion"
6               mov       R7,#1         @ Command code 1 terminates program
7               svc       0             @ Issue Linux command to terminate program
8
9               .data
10    msgtxt:   .ascii    "Hey there\n" @ 10 character message.
(gdb)l
11              .text
12
13    @         Subroutine v_asc will display a string of characters
14    @         R1: Points to beginning of ASCII string
15    @         R2: Contains length of string in bytes
16    @         LR: Contains the return address
17    @         Registers R0 and R7 will be used by v_asc and not saved
18
19    v_asc:    mov       R0,#1         @ Code for stdout (standard output)
20              mov       R7,#4         @ Linux service command to write string.
```

Listing 3.2: List source code line numbers using debugger

Next, set two breakpoints in the program using the "b" command to suspend program execution at two locations:

1. At line 4 so we can examine the registers before subroutine v_asc is called
2. At line line 19 so that we can examine the return address in the LR link register

Then start the program execution with the "run" command as illustrated in Listing 3.3. It will stop at the first breakpoint it encounters. At that point, enter the "i r" (info registers) command to the debugger.

```
(gdb) b 19
Breakpoint 1 at 0x8090: file model.s, line 19.
(gdb) b 4
Breakpoint 2 at 0x807c: file model.s, line 4.
(gdb)
(gdb) run
Starting program: /home/pi/model
Breakpoint 2, _start () at model.s:4
4        bl              v_asc          @ Call subroutine to view text string in ASCII.
(gdb) i r
r0       0x0             0
r1       0x100a0         65696
r2       0xa             10
r3       0x0             0
r4       0x0             0
r5       0x0             0
r6       0x0             0
r7       0x0             0
r8       0x0             0
r9       0x0             0
r10      0x0             0
r11      0x0             0
r12      0x0             0
sp       0x7efff860      0x7efff860
lr       0x0             0
pc       0x807c          0x807c <_start+8>
cpsr     0x10            16
(gdb)
```

Listing 3.3: Register contents before calling v_asc

Notice the following register contents shown in Listing 3.3:

1. The contents of the PC (program counter) is hex 807C before the call. This is the memory location of the BL instruction which is calling v_asc.
2. The contents of the LR link register is 0 before the call and is hex 8080 after the call. The 8080 address is (807C plus 4 equals 8080) the address of the instruction immediately after the BL subroutine call. We add 4 because in the ARM architecture each instruction is 4 bytes long. Note: The ARM CPU can be switched into a special 2-byte instruction format mode known as Thumb code (described in Lab 17 and Appendix N).

Continue the program execution by entering the "c" command as shown in Listing 3.4. The program will now be suspended at the next breakpoint which is at line 20 just after entering the subroutine.

```
(gdb) c
Continuing.

Breakpoint 1, v_asc () at model.s:20
20      mov         R7,#4        @ Linux service command code to write string.
(gdb) i r
r0      0x1          1
r1      0x100a0      65696
r2      0xa          10
r3      0x0          0
r4      0x0          0
r5      0x0          0
r6      0x0          0
r7      0x0          0
r8      0x0          0
r9      0x0          0
r10     0x0          0
r11     0x0          0
r12     0x0          0
sp      0x7efff860   0x7efff860
lr      0x8080       32896
pc      0x8090       0x8090 <v_asc+4>
cpsr    0x10         16
(gdb)
```

Listing 3.4: Register contents inside subroutine v_asc

While we're temporarily suspended inside the subroutine v_asc, let's take a look at the data to be displayed. Note that register R1 indicates the data to be displayed begins at address 0x100a0. The "x" command allows "dumping" data stored in memory. Use the "gdb help" command or see Appendix J for more options. The following are just a few of the options possible with the "x" command:

- x/10cb 0x100A0 — Display memory at hex address 100a0 as characters 10 bytes long
- x/10xb 0x100a0 — Display memory at hex address 100a0 in hexadecimal 10 bytes long
- x/1cs 0x100a0 — Display memory at hex address 100a0 as characters 1 string long (ends with byte of zero)

```
(gdb) x/10cb 0x100a0
0x100a0 <msgtxt>:   72 'H' 101 'e' 121 'y' 32 ' ' 116 't' 104 'h' 101 'e' 114 'r'
0x100a8 <msgtxt+8>: 101 'e' 10 '\n'
(gdb) x/10xb 0x100a0
0x100a0 <msgtxt>:   0x48 0x65 0x79 0x20 0x74 0x68 0x65 0x72
0x100a8 <msgtxt+8>: 0x65 0x0a
(gdb) x/1cs 0x100a0
0x100a0 <msgtxt>:   "Hey there\nA\025"
```

Listing 3.5: Examples using debug "x" command

After getting the above memory dumps using the gdb "x" commands, enter "c" to the (gdb) prompt to continue executing the program instructions. It will then display the message, subroutine v_asc will return to where it was called, and the program will "quit." You will still be in the gdb debugger, so you could restart the whole program again by entering "run" or you could exit to the Linux prompt by entering the "q" command.

Maintenance

Review Questions

1. What are two important functions provided by an operating system such as Linux?
2. Almost every CPU ever designed has a way to call subroutines which involves saving the current PC (Program Counter) in either memory or a register so that the subroutine knows where to return control. What is the disadvantage of saving the return address in a specific memory location, and what is the disadvantage of saving the return address in a register?
3. Using the BL instruction on the ARM CPU, what is the problem of one subroutine calling another?
4. A recursive subroutine is one that calls itself. A popular example of using a recursive subroutine is calculating a factorial where n! = n(n-1)(n-2)..1. What must be done with the return address in the link register to make recursive subroutines work?
5. * Which register(s) will be changed after executing each of the following instructions, and what are their new values? Assume that each instruction begins with the following contents: R0 = 0, R1 = 1, R2 = 2, LR = 36000, and PC = 32892.
    a. ORR R0, R3
    b. ASR R3, #1
    c. BL 34000
    d. BX LR
    e. ADD PC, #4

Programming Exercises

1. Write subroutine “v_byte” to output exactly one ASCII character that is contained in the low byte (bits 7-0) in register R0. Register R2 will not need to be specified because it is always going to be 1.
2. Write a subroutine similar to v_asc, but embed the output in brackets. For example if R1 contains the address of the word Computer, then [Computer] will be displayed. You may want your new subroutine to call v_asc and v_byte. Be careful to save the LR link register contents.

Lab 4
True and False

What is truth? How it's represented in a computer program is almost as confusing and inconsistent as it is in everyday life. In this lab, we'll examine the ARM's Current Program Status Register (CPSR), conditional program flow (a.k.a., branching or jumping), and some enhancements to the register load/store instructions. A memory storage technique known as the "stack" will be introduced and implemented. We will also develop subroutine v_bool, which displays True or False.

Prototype

Listing 4.0 contains the first version of a program that displays True or False depending on the contents of register R0. Here we designate False to mean R0 contains zero, and any non-zero value in R0 will mean True. For display purposes, this is consistent with many higher level languages.

```
1.          .global   _start       @ Provide program starting address to linker
2. _start:  mov       R0,#1        @ A value of 1 indicates "True"
3.          bl        v_bool       @ Call subroutine to display "True" or "False".
4.          mov       R0,#0        @ Exit Status code of 0 for "normal completion"
5.          mov       R7,#1        @ Service command 1 terminates this program
6.          svc       0            @ Issue Linux command to terminate program
7.
8. @        Subroutine v_bool will display "True" or "False" on the monitor
9. @                  R0: contains 0 implies false; non-zero implies true
10. @                 LR: Contains the return address
11. @                 Registers R0 through R7 will be used by v_bool and not saved
12.
13. v_bool: cmp       R0,#0        @ Set condition flags for True or False.
14.         ldrne     R1,=T_msg    @ Load pointer to "True" if z-flag clear.
15.         ldreq     R1,=F_msg    @ Load pointer to "False" if z-flag set.
16.         mov       R2,#6        @ Number of characters to be displayed at a
                                   time.
17.         mov       R0,#1        @ Code for stdout (standard output, i.e.,
                                   monitor)
18.         mov       R7,#4        @ Linux service command code to write string.
```

```
19.            svc     0             @ Call Linux command.
20.
21.            bx      LR            @ Return to the calling program
22.
23.            .data
24. T_msg:     .ascii  "True "       @ ASCII string to display if true
25. F_msg:     .ascii  "False "      @ ASCII string to display if false
26.            .end
```

Listing 4.0 Subroutine to display True or False

Go ahead and compile, link, and run the program as you did in the previous couple of labs (Listing 2.0). Also, recompile, link, and run the program with a zero value in R0 (line 2. _start: mov R0,#0) before calling subroutine v_bool.

```
~$ nano model.s
~$ as -o model.o model.s
~$ ld -o model model.o
~$ ./model
~$ echo $?
```

Copy of Listing 2.0: Edit, compile, link, and execute

Introductions

Lab 4 introduces new instructions (PUSH, POP, CMP) and "conditional" execution of the LDR (load address) instruction.

```
            List of ARM instructions introduced in Lab 4
4.0.13.     cmp     R0,#0       @ Set condition flags for True or False.
4.0.14.     ldrne   R1,=T_msg   @ Load pointer to "True" if z-flag clear.
4.17.13.    push    {R0-R7}     @ Save contents of registers R0 through R7.
4.17.22.    pop     {R0-R7}     @ Reload contents of registers R0 through R7.
```

Principles

Boolean algebra is a form of mathematics where the values of the variables have only two possible states: true or false. If we're working with a single bit, we can assign 0 to represent false and 1 to represent true or vice versa. However, with a 32-bit word or even an 8-bit byte, what are we going to do with all those extra bits? Let's examine a few of the misunderstandings of how to represent true and false.

By representing true as any non-zero value, we have the confusing example: [R1]=1 and [R2]=2 (both true), but the ARM instruction "AND R1,R2" generates a value of 0 indicating false which is obviously wrong. Another area of confusion is in the program exit codes. Some say a value of zero implies success (true) and a non-zero value implies failure (false). I prefer to think of an exit code value of zero implying "no errors" while a non-zero value indicates the type of error. Adding to the confusion is one of my favorite system design methodologies: "fuzzy logic" where there is a continuous range of degrees of "truth." The moral of this story: If you're not sure how true and false are represented in the system your working, please read the documentation.

Program Flow

A computer program is composed of a sequence of instructions to accomplish a task. Labels in assembly language programs identify special memory locations within the program. We have already used labels in the preceding labs for the first three of the following four reasons to use labels:

1. Identify the address of the start of the program
2. Locations of text messages and data tables
3. Subroutine entry point name and location
4. Identify location for alternate program flow

Why do we change the flow of a program?

1. Reuse a section of the program: Why recreate code that is already tested and working? Also, it reduces computer memory size requirements, but that's not as critical as it was years ago.
2. Difference in input data requires different processing.

In the previous lab, we "reused" a section of code in the form of a subroutine, and in the next several labs we will "reuse" code in the form of program loops. Here in Lab 4, we will use labels to identify the beginning of an alternate path through the program's logic.

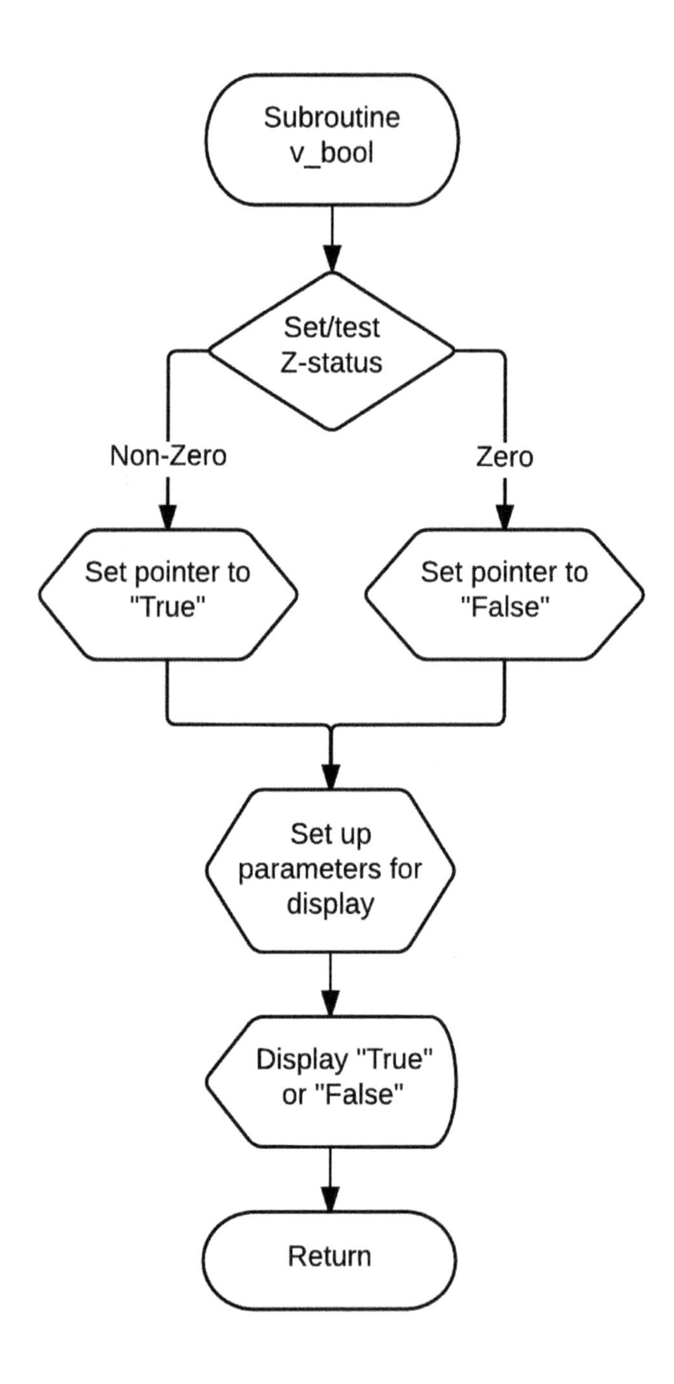

Figure 4.0: Flowchart of subroutine v_bool

The flowchart in Figure 4.0 illustrates what we want to do in subroutine v_bool: display "False" if zero or display "True" if non-zero. How do we do that in the code? How do we take alternate routes or paths in the flow of the instructions being executed?

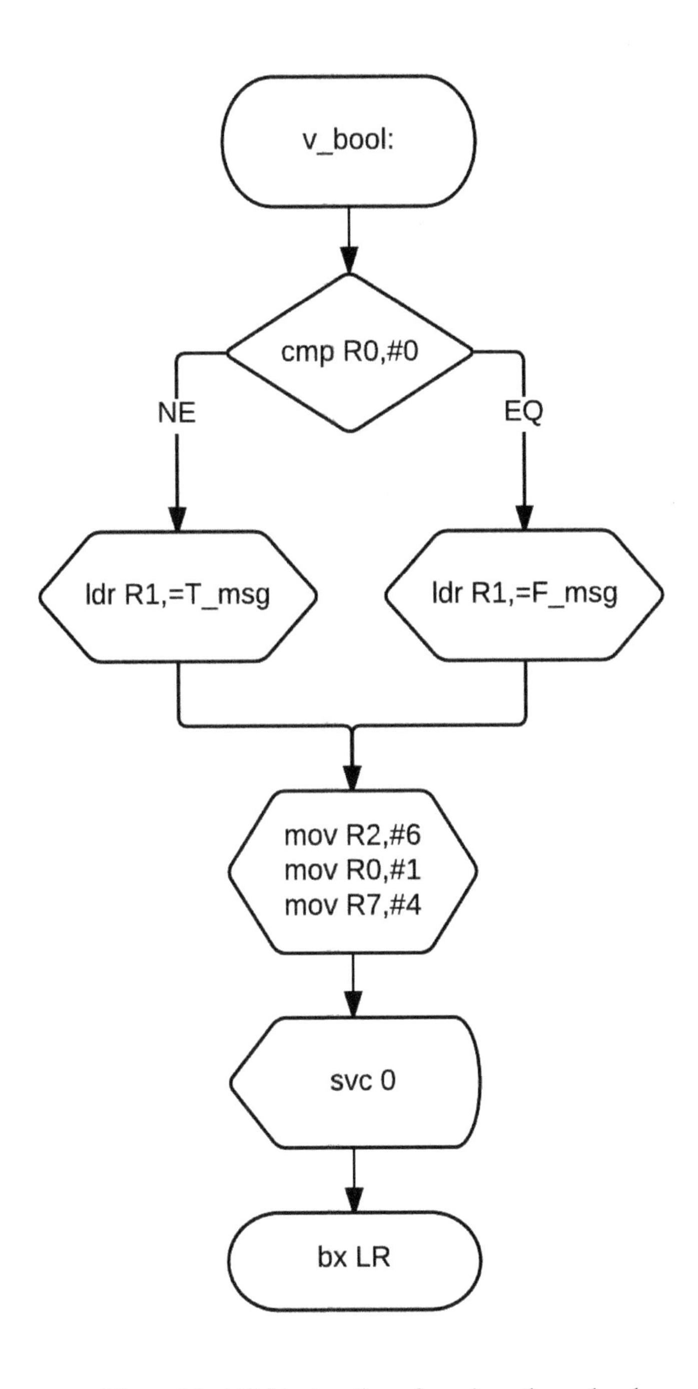

Figure 4.1: ARM instructions for subroutine v_bool

The flowchart in Figure 4.1 illustrates what ARM instructions are to be used, but where are the branch instructions, and how do we perform that reconnect after the two LDR instructions? Flowcharts have an implicit advantage over a real computer program: each symbol has a built-in branch directive with an arrow indicating what is to be done next. In an actual assembly language program, the "next instruction" is the one that immediately follows it in memory (except for a branch instruction because it points to a different location).

Program Counter (PC) Register

The PC register contains the memory address of the next instruction to be executed. On approximately one half of all CPUs, it is referred to as the IP (Instruction Pointer) instead of the PC, but they perform the same function. For every instruction except branch instructions, the address of the next instruction will be the one immediately following it in memory. During the execution of an instruction, the CPU automatically increments the PC to the address of the instruction immediately following it in memory. This is an increment by four in the ARM because all instructions (except for Thumb code, see Appendix N) are four bytes long.

In the ARM, the PC can also be changed in other ways because the PC is also register R15, which is a general purpose register. So you could skip over a few instructions with "ADD R15,#8" or even "ADD PC,#8." You could return from a subroutine with a "MOV R15,R14," instead of a "BX LR" instruction, but I generally don't recommend any of these. If you do something cute, and it saves a worthwhile amount of execution time, please document it in your code. Remember: The object of writing good code does not include trying to confuse and aggravate the person maintaining the code a few years later.

In most CPUs, we need two branch instructions to implement the above flowchart: one to branch to the desired code (LDR instruction in this case) and a second to branch back to a common point. If we modify the flowchart, we can achieve the desired effect with one branch instruction. This gives us four programming approaches to load the correct text message address into register R1:

1. If the value is false, branch to the code that sets R1 pointing to "False," and then branch over where R1 is set to "True."
2. If the value is true, branch to the code that sets R1 pointing to "True," and then branch over where R1 is set to "False."
3. Assume false, set R1 pointing to "False," and if the assumption is correct, then branch over instruction resetting R1 to "True."
4. Assume true, set R1 pointing to "True," and if the assumption is correct, then branch over instruction resetting R1 to "False".

The ARM also has a fairly unique feature that enables a fifth very efficient approach to be taken. All five approaches begins by setting the Z(ero) status bit depending on the contents of register R0.

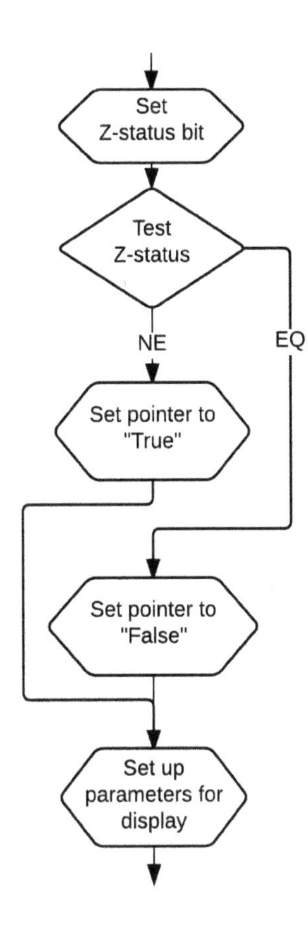

Figure 4.2: Flowchart symbols stacked to illustrate "branch around."

The first two approaches require two branch instructions. Figure 4.2 illustrates the code for the first approach which appears in Listing 4.1. I've stacked the flowchart symbols vertically like the instructions would appear in memory to better illustrate the branching around of the two LDR instructions that are not wanted.

```
        cmp     R0,#0
        beq     setf
        ldr     R1,=T_msg
        b       comtf
setf:   ldr     R1,=F_msg
comtf:  mov     R2,#6
```

Listing 4.1: Approach 1: Branch if false

Listing 4.2 is the second approach and is basically the same except a BNE is taken instead of a BEQ.

```
        cmp     R0,#0
        bne     sett
        ldr     R1,=F_msg
        b       comtf
sett:   ldr     R1,=T_msg
comtf:  mov     R2,#6
```

Listing 4.2: Approach 2: Branch if true

Approaches three and four eliminate one of the two branch instructions. In figure 4.3, the third technique is illustrated where the code assumes the value is false, thereby setting R1 pointing to "False," and if the assumption is correct then branches over the instruction resetting R1 to "True."

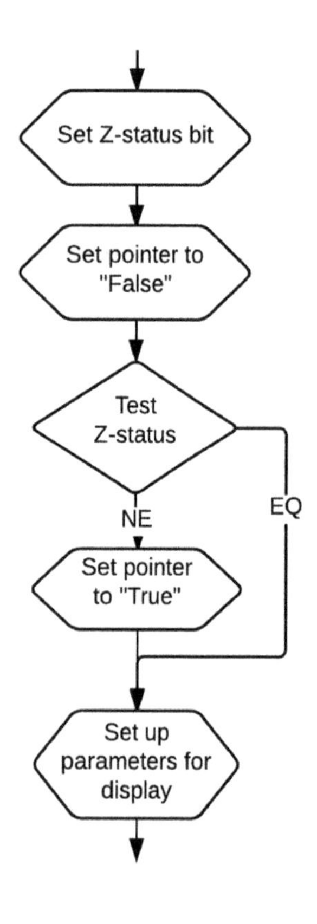

Figure 4.3: Use only one "branch around."

```
        cmp     R0,#0
        ldr     R1,=F_msg
        beq     comtf
        ldr     R1,=T_msg
comtf:  mov     R2,#6
```

Listing 4.3: Approach 3: Branch if false

Approach four is the same as three except the value is assumed to be "True."

```
        cmp     R0,#0
        ldr     R1,=T_msg
        bne     comft
        ldr     R1,=F_msg
comtf:  mov     R2,#6
```

Listing 4.4: Approach 4: Branch if true

The third and fourth approaches execute a little faster than the first two. The ARM instruction format enables a fifth approach which is not only faster, but it has "cleaner" looking code.

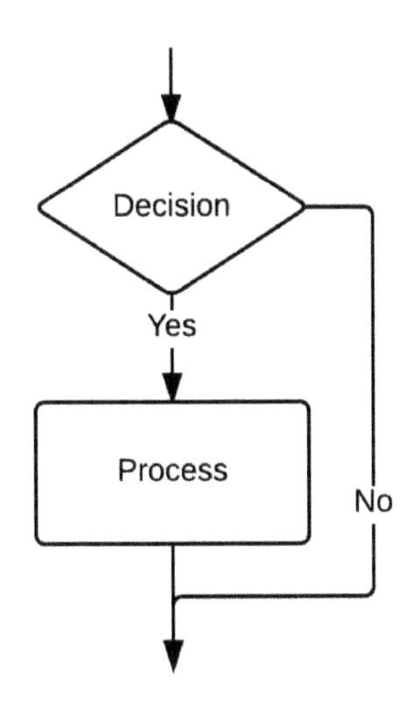

Figure 4.4: ARM instructions

ARM instructions are conditionally executed. Most computer architectures have a few instructions, such as JE (Jump if Equal) and JC (Jump if Carry), whose execution depends on the results from a previous instruction, but almost every ARM instruction can be set up to be "ignored" if a particular status condition is present. In other words, an ADD can be executed or ignored depending on the value of the status bits (carry, equal/zero, negative, overflow) that were set from the execution of a previous instruction.

```
cmp       R0,#0
ldrne     R1,=T_msg
ldreq     R1,=F_msg
mov       R2,#6
```

Listing 4.5: No branch instructions used

The best approach using the ARM instruction set:

1. As in each of the previous approaches, begin by setting a status flag bit in the ARM CPU if register R0 contains a value of zero.
2. If the Z-flag (zero status) is not set, then load the address of the "True" message into register R1.
3. If the Z-flag (zero status) is set, then load the address of the "False" message into register R1.
4. Continue execution by setting up the other registers for the Linux service call to display the message pointed to by register R1.

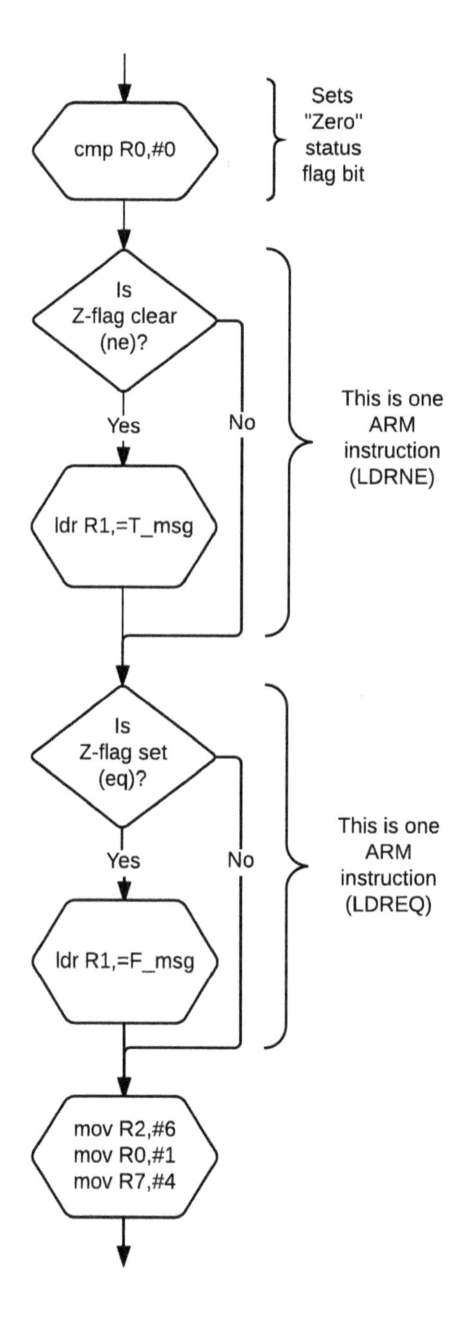

Figure 4.5: No branch instructions used

What's so bad about using branch instructions? Obviously branch instructions are extremely important, but there is such a thing as too much of a good thing. A program having fewer branch instructions provides the following benefits:

1. **Easier to maintain**: Programs that flow linearly from top to bottom are easy to read and understand by us humans.
2. **Spaghetti code**: Programs that have a lot of branches connecting several small

segments of code get as tangled up as spaghetti on a dinner plate. The code might work fine, but it is very difficult to modify with new features.

3. **Efficiency**: Not only does each branch instruction take up time and space, it also clears the instruction pipeline. I don't want to get into CPU electronics design, but pipelining is basically a method where the CPU can do more than one thing at the same time. Therefore, it provides more processing power for the same clock speed.

Actually, this loss of efficiency is not usually a disaster. Except, why are you using assembly language instead of C or Java or Python? Perhaps, you're working on a real-time application where every nanosecond counts. Spaghetti code, however, is never good, no matter what language you're coding. It literally leads to a software maintenance nightmare.

Current Program Status Register (CPSR)

The Z-flag is just one of the arithmetic status bits available in the ARM CPU. Nearly every CPU design has a Processor State Register which provides information regarding previous instructions that were executed (as well as other status info):

1. Was the previous result positive or negative?
2. Was the previous result zero?
3. Did the instruction end in error (like the sum of two positive numbers resulting in a negative number)?

In the ARM, the CPSR (Current Program Status Register) has N, Z, C, and V status flags:

N	Negative: Previous operation result was negative (i.e., bit 31 = 1)
Z	Zero: Previous operation result was zero (i.e., bits 31..0 = 0)
C	Carry: Previous operation resulted in a value that exceeded 32 bit register.
V	Overflow: Previous operation resulted in an error such as wrong positive/negative value.

Table 4.1: Status bits in the CPSR

However, the ARM architecture is very unique in which instructions modify and check the condition flags of the CPSR:

1. Each data processing instruction has the option of whether to change the values of the NZCF flags or leave them as they were. On almost every other CPU

design, arithmetic and logic functions (add, sub, and, or, …) always change the NZCV flags, and load/store instructions (mov, ldr, …) never change the NZCV flags.
2. Almost every ARM instruction can be conditionally executed depending on the value in the NZCF flags. Almost all other CPUs only have branch (also known as jump) instructions that examine the flags.

Let's use the subtraction instruction as an example for showing the multiple formats available in the ARM:

```
sub     R1,R2   @ Subtract the value in R2 from R1, but do not change
                NZCV flags
subs    R1,R2   @ Subtract the value in R2 from R1, and change flags
                depending on the result
subeq   R1,R2   @ Like the first example, but only execute it if the Z-flag
                is set (i.e., a previous result was zero).
                @ Combination of previous two examples: only do the
subeqs  R1,R2   subtraction if the Z-flag is set, and also change NZCV
                based on subtraction results.
```

Listing 4.7: Subtraction instruction examples using status bits

The general format of an ARM assembly instruction is **OpCode{cond}{s}**

1. The "OpCode" is simply the mnemonic for the operation to be performed by the current instruction (i.e., add, sub, mul. mov, …).
2. The "cond" is optional and indicates which combination of condition status bits has to be set (or clear) for the current instruction to be executed or ignored. Note: In the following list, "Z" means the Z-flag must be set (value=1), "!Z" means the Z-flag must be clear (value=0), "C" means the C-flag must be set, "!C" means the C-flag must be clear, etc.
3. The "s" is optional and indicates whether the NZCV condition status bits are to be modified by the execution of the current instruction.

```
EQ          Z           Equal (equals zero)
NE          !Z          Not equal
CS or HS C              Carry set / unsigned higher or same
CC or LO !C             Carry clear / unsigned lower
MI          N and !C    Minus / negative N set
PL          !N          Plus / positive or zero
VS          V           Overflow
VC          !V          No overflow
HI          !C and !Z   Unsigned higher
LS          !C or Z     Unsigned lower or same
```

GE	N = V	Signed greater than or equal
LT	N != V	Signed less than
GT	!Z and (N = V)	Signed greater than
LE	Z or (N != V)	Signed less than or equal
AL	Always (default)	Typically use blank (i.e., use b rather than bal)

Table 4.2: List of ARM assembly language condition codes

Examine the following sequence of code that shows variations in the OpCodes which demonstrate the condition status flag use. It begins by loading initial values of 2 and 4 into registers R4 and R5, respectively.

```
mov     R5,#4    @ Always load: [R5] = 4 and do not change NZCV
mov     R4,#2    @ Always load: [R4] = 2 and do not change NZCV
subs    R5,R4    @ Always subtract: [R5] = [R5] – [R4] = 2 and change
                 NZCV = 0000
moveq   R4,#10   @ Do nothing because Z=0 {ne}
subgts  R5,R4    @ Because Z=0 and N = V {gt}, subtract: [R5] = [R5] –
                 [R4] = 0 and change NZCV = 0100
subeqs  R5,R4    @ Because Z=1 {eq}, subtract: [R5] = [R5] – [R4] = -2
                 and change NZCV = 1000
addmi   R5,R4    @ Because N=1 and C=0 {mi}, add: [R5] = [R5] + [R4] =
                 0 and leave NZCV unchanged
addne   R5,R4    @ Because Z=0 {ne}, do [R5] = [R5] + [R4] = 2 and
                 leave NZCV unchanged
```

Listing 4.8: Sample code sequence that sets and tests status bits NZCV

Stack Pointer (SP) Register

The SP register points to the "top" of the stack, an area of memory where temporary data may be stored.

At the beginning of the subroutines we've seen thus far, the internal documentation has included the following:

1. What the subroutine does
2. What and where are its input parameters
3. Where are its results returned if any
4. Which registers did it use and therefore destroy the previous contents

It's time to get rid of the last one because it's unnecessary and easily remedied. We will save the contents in memory on the "stack" before we use the registers and restore the original values just before we return. Although the concept of a memory stack has been used from the early days, instructions that explicitly use one didn't become popular until appearing on many computers like the DEC PDP 11 in the mid-1970s. Characteristics of a stack:

- A common metaphor of stack operation is the placing and getting of cafeteria trays and plates. You place new trays on top and also remove trays from the top. Who would try to take the tray on the very bottom or from the middle of a stack?
- Data is stored onto and retrieved from the stack in a LIFO (Last In, First Out) manner.
- Stack usage is very easy: You "push" new data onto the stack and "pop" the most recent data from the top of the stack. The pushing and popping user does not have to know the details of where in memory the stack is actually located and exactly how it works.
- The area of memory allocated to the stack and the SP pointer contents are set up by Linux when it starts each program.
- In the ARM, the SP is also register R13, which is a general purpose accumulator.
- The stack is a great way to save or allocate memory for temporary variables in a subroutine.
- A stack enables the construction of recursive subroutines that call themselves.
- It is possible to "blow" the stack by pushing more data onto it than it was reserved to handle.

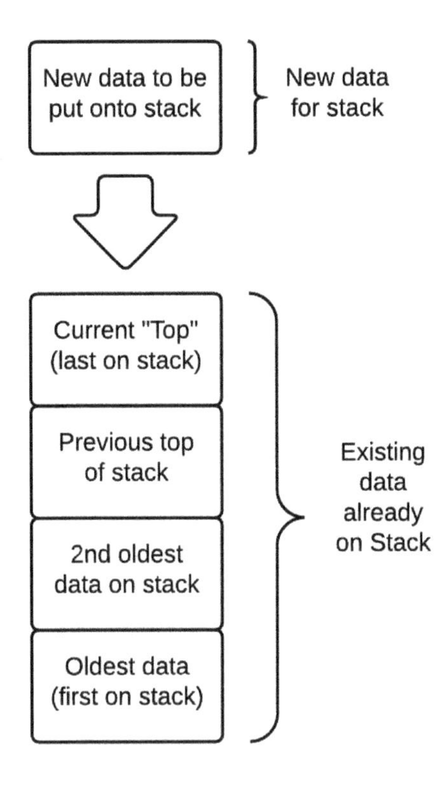

Figure 4.6: Stack data structure

Figure 4.6 illustrates pushing a value onto a stack that already contains four values. The size of each value can vary among applications and CPU architectures. In the Coding and Debugging section, we will enhance the v_bool subroutine from the Prototype section by saving register contents onto the stack.

Coding and Debugging

We'll be using the same commands (compile, link, debug) as we have in previous labs as shown in Listing 4.9. The purpose of the first half of this lab is to study subroutine calling and return addresses. The following procedure is taken:

1. Compile, link, debug, and list the first few instructions of the prototype program in Listing 4.0.
2. Set breakpoints to suspend program execution just before c_bool is called and just after it returns.
3. Start the program (gdb "run" command), and it will stop at the first breakpoint.
4. Examine the register contents before the call, then continue ("c" command) the program until it is suspended by the second breakpoint.
5. Compare the contents in the LR and compare that to the PC (Program counter) address of the current instruction.

Start the debugger like you did in the previous labs (~$ gdb model), and then enter the l(ist) command to display the first part of the program, and then set breakpoints.

```
pi@raspberrypi ~ $ as -g -o model.o model.s
pi@raspberrypi ~ $ ld -o model model.o
pi@raspberrypi ~ $ gdb model
```

Listing 4.9: Compile, link, debug

```
(gdb) list
1              .global          @ Provide program starting address to linker
               _start
2    _start:   mov     R0,#1    @ A value of 1 indicates "True"
3              bl      v_bool   @ Call subroutine to display "True" or "False".
4              mov     R0,#0    @ Exit Status code of 0 for "normal completion"
5              mov     R7,#1    @ Service command 1 terminates this program
6              svc     0        @ Issue Linux command to terminate program
7
(gdb) b 3
Breakpoint 1 at 0x8078: file model.s, line 3.
(gdb) b 4
Breakpoint 2 at 0x807c: file model.s, line 4.
(gdb)
```

Listing 4.10: Setting breakpoints before and after subroutine call

Start the main program using the gdb "run" command. It will suspend execution when it reaches the first breakpoint just before v_bool is called. Examine the register dump taken using the "i r" command and note the following:

- [R0] = 1 @ Value of True
- R1 through R7 are all zero
- Link register (LR) is also zero
- PC (address of current instruction) = 0x8078

```
(gdb) run
Starting program: /home/pi/model

Breakpoint 1, _start () at model.s:3
3          bl          v_bool          @ Call subroutine to display "True" or "False".
(gdb) i r
r0         0x1         1
r1         0x0         0
r2         0x0         0
r3         0x0         0
r4         0x0         0
r5         0x0         0
r6         0x0         0
r7         0x0         0
r8         0x0         0
r9         0x0         0
r10        0x0         0
r11        0x0         0
r12        0x0         0
sp         0x7efff860  0x7efff860
lr         0x0         0
pc         0x8078      0x8078 <_start+4>
cpsr       0x10        16
(gdb)
```

Listing 4.11: Breakpoints (and register dump) before calling v_bool

Continue the program execution with the gdb "c" command, and execution will now be suspended by the second breakpoint just after v_bool has returned. Note that on the third line of Listing 4.12, the word "True" was output by v_bool before the breakpoint was reached. Examine the register dump taken using the "i r" command and note the following:

- R0, R1, R2, R7 have different values returned than when v_bool was called
- [LR] = 0x807C (ARM instructions are 4 bytes. Hexadecimal 8078 + 4 = 807C, address of return is address of call plus 4)
- [PC] = 0x807C

```
(gdb) c
Continuing.
True
Breakpoint 2, _start () at model.s:4
4         mov              R0,#0          @ Exit Status code of 0 for "normal completion"
(gdb) i r
r0        0x6              6
r1        0x100b0          65712
r2        0x6              6
r3        0x0              0
r4        0x0              0
r5        0x0              0
r6        0x0              0
r7        0x4              4
r8        0x0              0
r9        0x0              0
r10       0x0              0
r11       0x0              0
r12       0x0              0
sp        0x7efff860       0x7efff860
lr        0x807c           32892
pc        0x807c           0x807c <_start+8>
cpsr      0x20000010       536870928
(gdb)
```

Listing 4.12: Breakpoint after return form v_bool

While we're here, we might as well take a look at what's in register R1 which is the pointer to either "True" or "False" left over from subroutine v_bool. Note that we can use the gdb "x" command with either the decimal (65712) or hexadecimal (0x100b0) as shown on the top four lines of Listing 4.13. The output, 0x65757254, represents the ASCII characters "e" (hex 65), "u" (hex 75),"r" (hex 72), and "T" (hex 54). More appropriate commands for dumping a character string use /4 followed by xb, db, or cb. Why are the first /x displays backwards compared to the others? It is related to byte and word addressing of memory, and I'll delay that explanation until the discussion in Lab 8 on little endian format.

```
(gdb) x 65712
0x100b0
<t_msg>:          0x65757254
(gdb) x 0x100b0
0x100b0
<t_msg>:          0x65757254
(gdb) x/4xb 0x100b0
0x100b0
<t_msg>:          0x54 0x72 0x75 0x65
```

```
(gdb) x/4db 0x100b0
0x100b0
<t_msg>:        84 114 117 101
(gdb) x/4cb 0x100b0
0x100b0
<t_msg>:        84 'T' 114 'r' 117 'u' 101 'e'
(gdb)
```

Listing 4.13: The gdb debugger's x command for displaying memory contents

In Listing 4.14, we'll basically perform the same exercise, but this time with a "False" output. The gdb "set $r0=0" command is used to initialize R0 to zero before the call to b_bool.

```
(gdb) run
Starting program: /home/pi/model

Breakpoint 1, _start () at model.s:3
3       bl          v_bool        @ Call subroutine to display "True" or "False".
(gdb) set $r0 = 0
(gdb) i r
r0      0x0         0
r1      0x0         0
r2      0x0         0
r3      0x0         0
r4      0x0         0
r5      0x0         0
r6      0x0         0
r7      0x0         0
r8      0x0         0
r9      0x0         0
r10     0x0         0
r11     0x0         0
r12     0x0         0
sp      0x7efff860  0x7efff860
lr      0x0         0
pc      0x8078      0x8078 <_start+4>
cpsr    0x10        16
(gdb)
```

Listing 4.14: Breakpoint before calling v_bool: set [R0]=0 (False)

Now let's single step from the breakpoint to verify that the Z-flag is set. Listing 4.15 shows that the CPSR has a hexadecimal value of 60000010, where the 6 (binary 0110) represents the NZCV status bits. Therefore, the both the Z and C status bits are set, while N and V are clear.

```
(gdb) s
v_bool () at model.s:13
13        v_bool: cmp    R0,#0            @ Set condition flags for True or False.
(gdb) s
14        ldrne          R1,=T_msg        @ Load pointer to "True" if z-flag clear.
(gdb) i r
r0        0x0            0
r1        0x0            0
r2        0x0            0
r3        0x0            0
r4        0x0            0
r5        0x0            0
r6        0x0            0
r7        0x0            0
r8        0x0            0
r9        0x0            0
r10       0x0            0
r11       0x0            0
r12       0x0            0
sp        0x7efff860     0x7efff860
lr        0x807c         32892
pc        0x808c         0x808c <v_bool+4>
cpsr      0x60000010     1610612752
(gdb)
```

Listing 4.15: Shows Z-flag set before LDRNE instruction executes

```
(gdb) c
Continuing.
False
Breakpoint 2, _start () at model.s:4
4         mov       R0,#0       @ Exit Status code of 0 for "normal completion"
(gdb)
```

Listing 4.16: Allowing program to continue to display "False" and finish

Push, Pop, and the Stack

In the second half of this lab we'll enhance the v_bool subroutine and examine how the stack works:

1. Modify the source code from listing 4.0 to include the push and pop instructions as illustrated below on lines 13 and 22 of listing 4.17
2. Compile, link, debug, and list the first few instructions.
3. Set breakpoints to suspend program execution before and after calling subroutine v_bool.
4. When the first breakpoint is reached, note the register contents. Also look at the SP (stack pointer) register.
5. When the second breakpoint is reached, compare all the registers, even the SP.

```
1.              .global   _start        @ Provide program starting address to linker
2. _start:      mov       R0,#1         @ A value of 1 indicates "True"
3.              bl        v_bool        @ Call subroutine to display "True" or "False".
4.              mov       R0,#0         @ Exit Status code of 0 for "normal completion"
5.              mov       R7,#1         @ Service command code 1 terminates program.
6.              svc       0             @ Issue Linux command to terminate program
7.
8. @            Subroutine v_bool will display "True" or "False" on the monitor
9. @                      R0: contains 0 implies false; non-zero implies true
10. @                     LR: Contains the return address
11. @                     All register contents will be preserved.
12.
13. v_bool:     push      {R0-R7}       @ Save contents of registers R0 through R7.
14.             cmp       R0,#0         @ Set condition flags for True or False.
15.             ldrne     R1,=T_msg     @ Load pointer to "True" if z-flag clear.
16.             ldreq     R1,=F_msg     @ Load pointer to "False" if z-flag set.
17.             mov       R2,#6         @ Number of characters to display.
18.             mov       R0,#1         @ Code for stdout (standard output, monitor)
19.             mov       R7,#4         @ Linux service command code to write.
20.             svc       0             @ Call Linux command.
21.
22.             pop       {R0-R7}       @ Restore saved register contents.
23.             bx        LR            @ Return to the calling program
24.
25.             .data
26. T_msg:      .ascii    "True "       @ ASCII string to display if true
27. F_msg:      .ascii    "False "      @ ASCII string to display if false
28.             .end
```

Listing 4.17: Subroutine v_bool enhanced to preserve register contents

```
(gdb) l
1                   .global _start     @ Provide program starting address to linker
2       _start:     mov     R0,#1       @ A value of 1 indicates "True"
3                   bl      v_bool      @ Call subroutine to display "True" or "False".
4                   mov     R0,#0       @ Exit Status code of 0 for "normal completion"
5                   mov     R7,#1       @ Service command code 1 terminates program.
6                   svc     0           @ Issue Linux command to terminate program
7
8       @           Subroutine v_bool will display T for True or F for False
(gdb) b 3
Breakpoint 1 at 0x8078: file model.s, line 3.
(gdb) b 4
Breakpoint 2 at 0x807c: file model.s, line 4.
(gdb)
```

Listing 4.18: Setting breakpoints before and after v_bool call

```
(gdb) run
Starting program: /home/pi/model

Breakpoint 1, _start () at model.s:3
3       bl          v_bool          @ Call subroutine to display "True" or "False".
(gdb) i r
r0      0x1         1
r1      0x0         0
r2      0x0         0
r3      0x0         0
r4      0x0         0
r5      0x0         0
r6      0x0         0
r7      0x0         0
r8      0x0         0
r9      0x0         0
r10     0x0         0
r11     0x0         0
r12     0x0         0
sp      0x7efff860  0x7efff860
lr      0x0         0
pc      0x8078      0x8078 <_start+4>
cpsr    0x10        16
(gdb)
```

Listing 4.19: Breakpoint before calling new v_bool (note the register contents)

The preceding listing (4.19) shows the register contents before the call, and the following listing (4.20) shows the contents after the call. Notice that all registers are the same except for the CPSR and the LR (which was set by the BL instruction to make the call and used by the BX instruction as the return address). The stack was used by v_bool (PUSH), but it was also restored (POP).

```
(gdb) c
Continuing.
True
Breakpoint 2, _start () at model.s:4
4        mov              R0,#0          @ Exit Status code of 0 for "normal completion"
(gdb) i r
r0       0x1              1
r1       0x0              0
r2       0x0              0
r3       0x0              0
r4       0x0              0
r5       0x0              0
r6       0x0              0
r7       0x0              0
r8       0x0              0
r9       0x0              0
r10      0x0              0
r11      0x0              0
r12      0x0              0
sp       0x7efff860       0x7efff860
lr       0x807c           32892
pc       0x807c           0x807c <_start+8>
cpsr     0x20000010       536870928
(gdb)
```

Listing 4.20: Breakpoint after the call. Are the registers the same?

CPSR Contents

Another important point of this lab is the CPSR and conditional execution of instructions. We took a peek at it earlier, but I'll leave a more thorough investigation to the reader/student, but provide the following guidelines and suggestions:

1. Compile, link, debug, and list the first few instructions of the program in Listing 4.17.
2. Set a breakpoint to suspend program execution at the compare instruction (line 14 in Listing 4.17)

3. Start the program (gdb "run" command), and it will stop at the breakpoint.
4. Use the gdb "i r" info registers and "s" single step commands to watch how the CPSR and register R1 change for the next few instructions.
5. The NZCV status bits are at the left side of the CPSR (bits 31 down to 28). You'll have to convert the leftmost hex digit to binary to see the changes. The Z-flag is bit position 30.

Maintenance

Review Questions

1. Use the Internet to discover other "standards" for representing True and False in computer systems.
2. * The CMP instruction sets the NZCV status bits. Why do you think its mnemonic isn't cmps and would cmps work also?
3. Subroutine v_bool uses the CMP instruction to set the status bits. Name at least two other instructions that would have done the same thing as CMP did here.
4. Although it's not recommended as good programming style, subroutine v_bool could be "called" using the B instruction. What additional instruction would be needed to complete the call?
5. How can the addition of two positive numbers result in a negative number? Which status bit would be set indicating this error occurred?

Programming Exercises

1. Modify v_bool to display "True" only when R0 is 1, "False" only when R0 is 0, and display "Unknown" for all other values.
2. Run this application in debug mode and examine the NRCV status bits before and after the CMP instruction.
3. Modify the v_bool subroutine and the main program that calls it to verify the answers to review questions 3 and 4 from above.

Lab 5
Display Numbers in Binary

Lab 5 will cover two very important aspects of assembly language programming: program loops and ARM CPU instruction format. Computers are great for performing the same task over and over and over again: thousands of times, millions of times, or more, and program loops enable this task. The ARM's "machine code" instruction format will be described to help explain the programming techniques. We will also get a handy subroutine for displaying binary numbers.

Prototype

In Listing 5.0, a first version of a new subroutine (v_bin) is presented that will display the contents of register R0 as a series of 32 binary digits (bits). This output is very useful for displaying the results from Boolean operations. The specifications for using v_bin are the following:

1. The subroutine name is v_bin and it's purpose is to display a 32-bit register in binary digits
2. It's input (i.e., resources it needs) is a number contained in 32-bit register R0.
3. The return from subroutine v_bin is to the address in the LR (Link Register). Although, v_bin doesn't have to be called with a BL (branch and link) instruction, it is the best and most common technique to load the return address into the LR.
4. The contents of registers R0 though R7 may be altered by subroutine v_bin, so the calling program must save any valuable information in them before v_bin is called. We will "fix" that in the Coding and Debugging section using the same PUSH and POP instructions used in Lab 4.

```
1.            .global   _start          @ Provide program starting address to linker
2. _start:    mov       R0,#0b101101    @ An arbitrary sample binary value for testing
3.            bl        v_bin           @ Call subroutine to view binary value of R0 in
                                        ASCII.
4.            mov       R0,#0           @ Exit Status code of 0 for "normal
                                        completion"
5.            mov       R7,#1           @ Service command code 1 terminates this
                                        program.
6.            svc       0               @ Issue Linux command to terminate program
7.
8. @          Subroutine v_bin will display a 32-bit register in binary digits
9. @                    R0: contains the 32 bit value to be displayed in binary
10. @                   LR: Contains the return address
11. @                   Registers R0 through R7 will be used by v_bin and not saved
12.
13. v_bin:    mov       R3,R0           @ R3 will hold a copy of input word to be
                                        displayed.
14.           mov       R6,#31          @ Number of bits to display (-1)
15.
16. @         Loop through single bits and output each to the standard output (stdout)
              display.
17.
18. nxtbit:   mov       R4,#1           @ used to mask off 1 bit at a time for display
19.           and       R1,R4,R3,lsr R6 @ Select next 0 or 1 to be displayed.
20.           ldr       R5,=dig         @ Pointer to the "01" string of ASCII
                                        characters.
21.           add       R1,R5           @ Set R1 pointing to "0" or "1" in memory
22.           mov       R2,#1           @ Number of characters to be displayed at a
                                        time.
23.           mov       R0,#1           @ Code for stdout (standard output, i.e.,
                                        monitor)
24.           mov       R7,#4           @ Linux service command code to write string.
25.           svc       0               @ Linux service command code to write string.
26.           subs      R6,#1           @ Decrement number of bits remaining to
                                        display
27.           bge       nxtbit          @ Go display next bit until all 32 displayed
28.
29.           bx        LR              @ Return to the calling program
30.
31.           .data
32. dig:      .ascii    "01"            @ ASCII string of binary digits 0 and 1
33.           .end
```

Listing 5.0: Subroutine to output binary number in ASCII

Go ahead and run the program in Listing 5.0. Try some different values on line 2 such as #0b11100, #67, and #0x17. Use the edit, compile, link, and execute steps like in previous labs. You could also use the gdb debugger introduced in Lab 2 if you like.

Introductions

There will be no new instructions, assembler directives, Linux services, or batch commands introduced in Lab 5, but modifications to three instructions already seen (AND, SUB, B, MOV) are present.

	List of ARM instructions introduced in Lab 5		
5.0.19.	and	R1,R4,R3,lsr R6	@ Logical AND with 4 registers and a shift
5.0.26.	subs	R6,#1	@ Subtract and set condition codes
5.0.27.	bge	nxtbit	@ Branch if greater or equal
5.8.18.	movhi	R6,#0	@ If bad range, default to displaying only 1 bit.

Principles

Loops have been part of computer programming from the beginning. The objective of Charles Babbage's computer from the 1830s was to calculate large tables of numbers. The objective of Herman Hollerith's tabulating machine from the 1890s was to count large numbers of people as part of the census.

Program Flow

Computers are great for doing repetitive operations. A loop is a "process" that can be performed multiple times until a "decision" is made to move onto something else. Examples of processes and decisions:

- Process: Eating one mouthful of food at lunch
- Decision: Is there any more food on my plate?

- Process: Grading one student's exam
- Decision: Are there any more exams to grade?

- Process: Display one bit ("0" or "1") on the monitor
- Decision: Are there any more bits remaining to display?

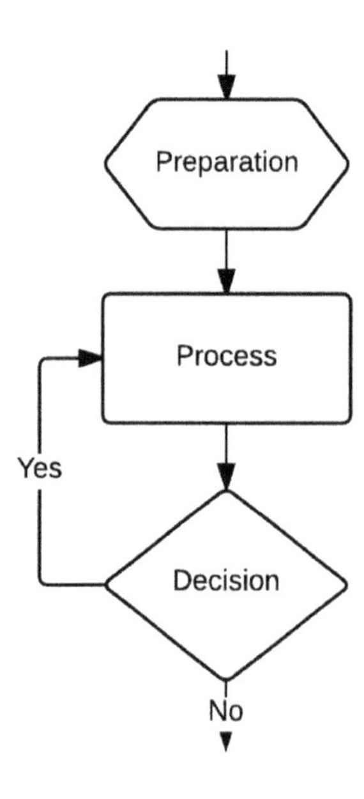

Figure 5.0: Program loop

A loop consists of three parts:

1. Preparation: Set initial values for a) variables to be modified during each pass of the loop and b) variables, like counters, that will determine when to exit the loop.
2. A process to be repeated multiple times: Examples include adding numbers to a running total, searching a table for a particular value, and calling the same set of subroutines multiple times.
3. Decision when to exit the loop: Some loops such as those used in a medical device performing real-time life-support are not intended to stop. But most loops do have an exit objective such as a) all of the desired sets of numbers have been added, b) the entire table has been searched, c) the desired value has been located, etc.

Why Binary?

What's wrong with decimal? Babbage's Analytical Engine computer design was decimal. Have we digressed in the past 200 years? Actually, there have been many decimal-based computers, but why are almost all of today's computers based on binary? The simple answer is that the logical building blocks (i.e., electronics in today's systems) are simpler and more efficient in binary than they are in decimal. If you are new to binary numbers, please see Appendix B or search the Internet to get some background.

A decimal number is really a short notation for a polynomial of powers of 10. For example: 137 is $1\times10^2 + 3\times10^1 + 7\times10^0$. Likewise, a binary number is really a short notation for a polynomial of powers of 2. For example: 110101 is $1\times2^5 + 1\times2^4 + 0\times2^3 + 1\times2^2 + 0\times2^1 + 1\times2^0$. By the way, this polynomial structure is the main reason we count bits from right to left starting with zero.

A Loop Through 32 Binary Digits (Bits)

How does this first version of subroutine v_bin work? We build a loop which counts down from 31 (the position of the leftmost bit) to 0 (the position of the rightmost bit). As show in Listing 5.0, the loop is between lines 18 and 27. Register R6 not only counts down from 31 to 0, but also indicates which bit is examined on each pass through the loop.

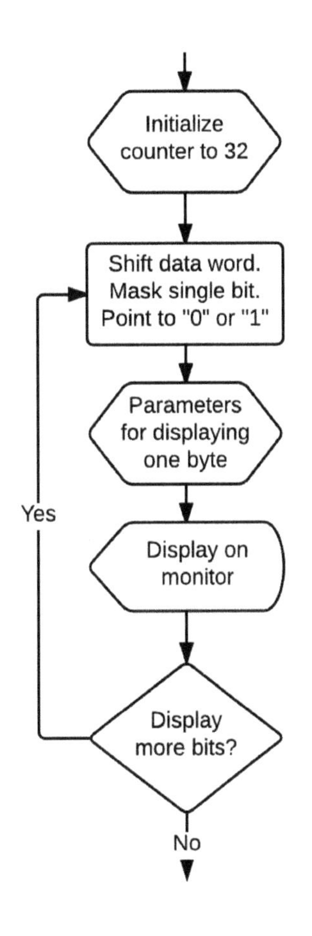

Figure 5.1: Program loop to display 32 bits

Instructions in the loop select a particular bit, convert it to ASCII, and call a Linux service routine to output the one character to the monitor screen.

1. Preparation: Set initial value of "count down" register R6 to 31. R6 also points to the first bit in the data (loaded into register R3).
2. A process to be repeated multiple times: Select the next bit, indicated by value in R6, and display it as either an ASCII "0" or "1" on the monitor.
3. Decision when to exit the loop: Register R6 is decremented by one on each pass through the loop which allows it to point to all bit positions 31 through 0. When R6 is decremented from 0 to -1, then an exit from the loop is taken because all 32 bit positions have been displayed.

```
nxtbit:   and    R1,R4,R3,lsr R6  @ Select next 0 or 1 to be displayed.
          subs   R6,#1            @ Decrement number of bits remaining to display
          bge    nxtbit           @ Go display next bit until all 32 displayed
```

Listing 5.1: New instructions

Variations of three ARM instructions are introduced in this loop (SUBS, BGE, and a rather complicated looking AND) that we haven't used in previous programs. The easiest one to explain is the SUBS subtraction instruction. It is identical to the SUB instruction, except it also sets status flags as a result of the subtraction. If R6 is decremented from 0 to -1, then the "N" flag will be set indicating a negative number. The BGE instruction (branch if greater or equal) then uses this "N" flag to know whether to branch back to continue the loop ("N" is clear implying greater or equal to zero) or not branch and fall out of the loop.

AND R1,R4,R3,LSR R6

Now we'll examine that new format of the AND instruction. You're probably saying, "Wait a minute. That AND instruction looks pretty complicated." I've heard comments like, "I thought the ARM processor has a RISC (Reduced Instruction Set Computer) architecture containing a relatively small number of instructions, and those instructions are relatively simple compared to those of a CISC (Complex Instruction Set Computer) architecture. That AND instruction doesn't appear to be so simple."

If you thought the ARM was unique in the way it handles the setting and testing of the condition status bits, just look at how it works on most of its arithmetic and logic operations. First, didn't we already use the AND in Lab2, and didn't it use only two registers like all the other operations that did logic or arithmetic? Didn't it work by doing a bit-by-bit logical "and" of two registers and leave the result in one of those two registers? That format which we used in Lab2 is the format of the AND used on almost every CPU manufactured in the past 50 years. Why and how does the ARM's AND use four registers? Also, what's that shift instruction doing inside the AND instruction? Take note: It's not just the AND that works this way, it's almost all the arithmetic and logical instructions.

The ARM Machine Code

It's time to look behind the curtain. I've been hiding some of the complexity of the ARM instruction format until you got more comfortable working with assembly language. But now, it would be difficult to explain the AND instruction used in this lab without describing the bit-by-bit format of the 32-bit machine code instruction that the ARM executes.

- Most ARM instructions are conditionally executed (depends on the NZCV flags)
- Most ARM instructions have the option of modifying the NZCV flags
- Most ARM instructions can use four registers
- Most ARM instructions include a shift operation

Although there is no single format that covers all ARM instructions, the following description covers the move (MOV), arithmetic (ADD, SUB, ...), and logical (AND, ORR, EOR) instructions. Note that all the shift instructions (LSL, …) are actually built into the move, arithmetic, and logical instructions and are not really individual instructions of their own.

Figure 5.2 shows what the assembler must do to translate the format we enter as assembly language text to the binary format machine code required by the ARM processor. Let's use subtraction as an example because its operands are not commutative and the algebra is traditional. Let's start with something simple, like subtracting 4 from the contents of register R1. The assembler constructs the machine code instruction 0x02511004 by "filling in" the fields of the instruction word with "sub" = 0010, "eq" = 0000, "s" = 1, "#" = 1, and R1 is both the source as well as the destination register.

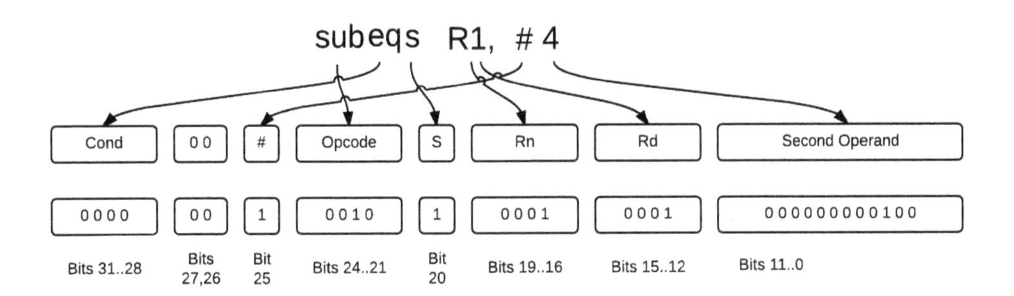

Figure 5.2: Convert SUBEQS R1,#4 to ARM machine code 0x02511004

bits	name	Contents
31..28	Cond	Only execute this instruction on condition of the value in the NZCF flags
27..26	00	These two bits are always zero for this instruction type
25	#	Immediate operand flag
24..21	Opcode	Which operation (add, sub, and, orr, eor, …)
20	S	Indicates that this instruction will modify the condition codes
19..16	Rn	ID number of register containing the minuend (first operand)
15..12	Rd	ID number of register to receive the result
11..0	op2	Three formats possible for second operand

Table 5.0: General bit layout for ARM instructions (mov, add, sub, and, orr, eor, …)

In general, the subtraction instruction looks like the following algebra:

D = N - M×2^S^ where 2^S is not really an exponent, but a shift count.

- **D**: Destination register to hold the 32-bit result
- **N**: First operand register: In subtraction, it is the minuend (quantity from which another quantity is to be subtracted)
- **M×2^S^**: Second operand: In subtraction, it is the subtrahend (quantity to subtract). There are three possible formats for M and S:
    1. **M** and **S** are both in registers: The contents of register M are shifted (logical, algebraic, or circular) by the value in register S.
    2. **M** is a register and **S** is a constant: The contents of register M are shifted (logical, algebraic, or circular) by a constant (range of 0 through 31).
    3. **M** and **S** are both constants: This is somewhat like scientific notation where M is a constant (0 through 255) and S is a shift count (0 through 30, even integers).

The above subtraction example uses the third format for the second operand while the following AND instruction from v_bin uses the first format for the second operand. Note in the following layout of the AND instruction that the op-code for AND = 0000, condition code "always" = 1110, "s" = 0, "#" = 0, and the LSR shift code is 01.

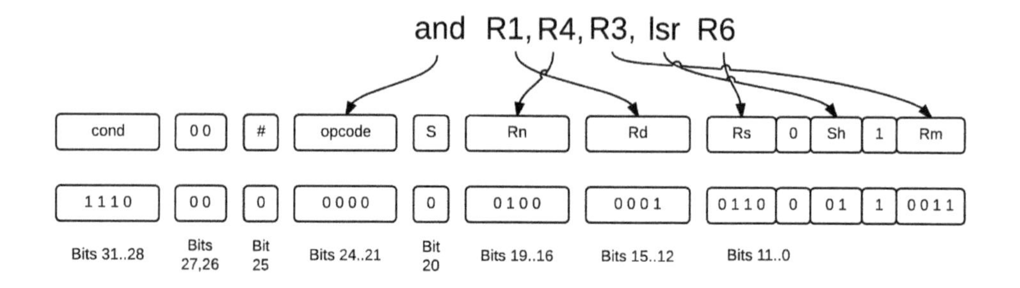

Figure 5.3: "Assemble" AND with shift instruction to ARM machine code 0xE0041633

Here's what that one AND instruction does:

1. Takes the value from register R3 and shifts it right the number of bit positions specified in register R6
2. Then it combines that shifted value with the contents of register R4 using a Boolean "and" function
3. And finally, it stores the result into register R1. The contents of registers R3, R4, and R6 are unchanged.

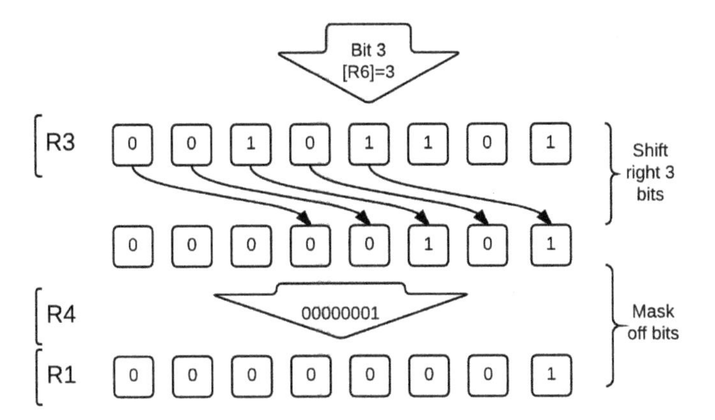

Figure 5.4: AND R1,R4,R3,LSR R6 with [R6] = 3

So what's this one instruction really doing (i.e., why is it doing it)? It's selecting a particular bit from R3 and putting it into the rightmost bit position in R1 all by itself.

1. Not only is R6 counting down the number of passes through the loop, but it also points to the next bit to be displayed. For the first pass through the loop [R6]=31. This 31 indicates there are 31 more passes through the loop It also indicates that bit 31 (the leftmost bit) will be displayed on this first pass through the loop. For the last pass through the loop, [R6]=0 meaning there are no more passes through the loop and bit 0 (the rightmost bit) will be displayed. The R3 LSR R6 portion of the instruction simply shifts the desired bit into the rightmost position (bit 0).
2. Because R4 contains only a binary 1, the AND portion of the instruction masks off all bits except for the rightmost bit.
3. The result of the shift with AND is then stored in a separate register, R1, so that the working registers R3, R4, and R6 are not changed and will be ready for the next pass through the loop.

Table 5.1 shows the AND instruction selecting each bit from the 0b101101 sample data value as R6 is decremented from 31 to 0 through the loop.

R6	R3 lsr R6	R1
31	0	0
30	0	0
29	0	0
...	...	...
6	0	0
5	1	1
4	10	0
3	101	1
2	1011	1
1	10110	0
0	101101	1

Table 5.1: AND instruction selects bit to be displayed

Other Instructions

This book is focused on learning ARM assembly language and is not a detailed bit-by-bit description of the ARM's architecture. Labs 8 and 9 describe the instruction format somewhat deeper than here in Lab 5. There are also several fine reference books and Internet sites that will go deeper into that area.

The multiply (MUL) instruction is very similar to the other arithmetic instructions, except it only works with registers (i.e., you cannot multiply a register by a constant) and it may have a 64-bit result (using two 32-bit registers).

Let's continue this discussion with the branch instruction. As you might expect, the BGE, BGT, BEQ, and B instructions are really just the same branch instruction, but it relies on those four high-order NZCV condition bits (the same as all other 32-bit ARM instructions) to indicate whether the branch is taken.

Three other instruction sets are present in the Raspberry Pi: 16-bit Thumb mode (Lab 17 and Appendix N), VFPv3 floating point (Labs 13, 14, 15, and 18), and NEON which provides parallel processing of floating point, logical, and integer values (Labs 15, 18, and 19).

Is there anything else I'm hiding? Well, it's not the end of the book. Is it? As far as the standard ARM instructions go, there won't be many more surprises, but we do have upcoming complexity in floating point operations. Actually, all this stuff is pretty easy once you know it, but so is swimming or driving an automobile.

Coding and Debugging

Thirty two passes are made through the nxtbit loop with register R6 counting down from 31 to 0. We will now run v_bin as it appeared in the Prototype section. As you may have suspected, there are a few enhancements we can make to improve performance and utility, and we will incorporate them shortly.

Register	Contains	Value
R0	Output to stdout display	Always = 1
R1	Current bit to display	Value is either 0 or 1
R2	Number of characters to output	Always = 1
R3	Binary number being displayed	Copy of original R0
R4	Masks off low-order bit (bit 0)	Always = 1
R5	Points to string: "01"	Always the same
R6	Selects which bit of R3 to display	Counts down from 31 to 0
R7	Linux service command to write file/device	Always = 4

Table 5.2: Registers used in nxtbit loop:

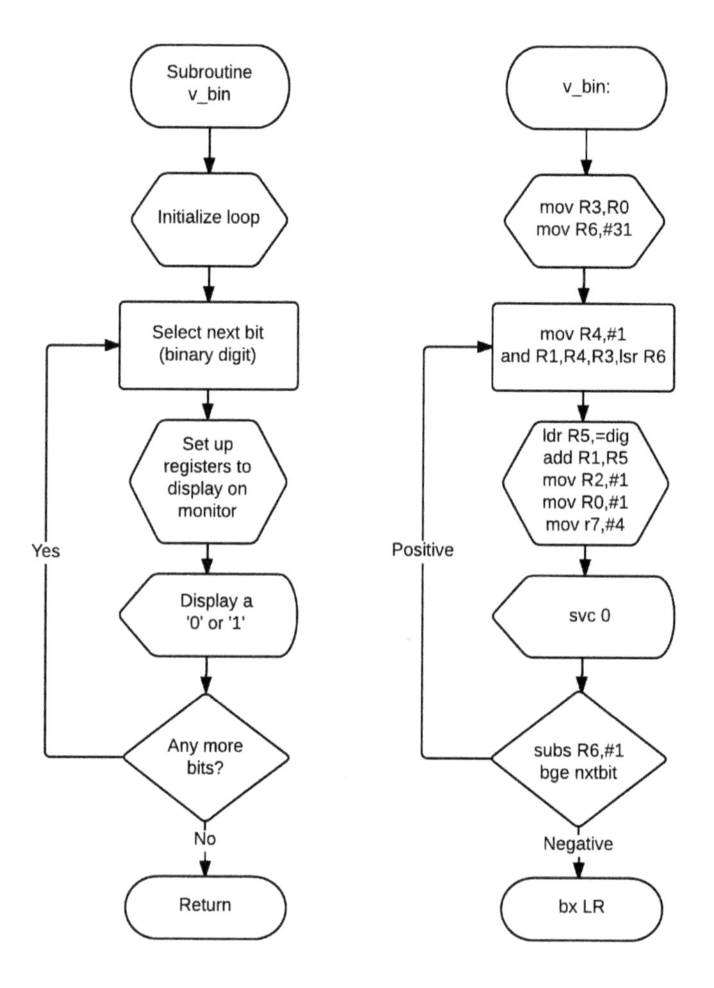

Figure 5.5: Subroutine v_bin displays 32 bits

Let's take a pass through the loop and take a look at register contents at a few interesting locations. Start the debugger and set breakpoints at lines 20, 27, and 29.

```
(gdb) b 20
Breakpoint 1 at 0x8098: file model.s, line 20.
(gdb) b 27
Breakpoint 2 at 0x80b4: file model.s, line 27.
(gdb) b 29
Breakpoint 3 at 0x80b8: file model.s, line 29.
(gdb)
```

Listing 5.2: Set breakpoints after starting the gdb debugger.

In Listing 5.3, we start the program with the "run" command, and the debugger suspends execution upon reaching the first breakpoint which is located immediately after the AND instruction. Notice the following contents in the register dump produced by the "i r" command.

- [R0] = 0x2d (binary 101101) which is the "arbitrary" test value set on line 2 of the main program
- [R1] = 0 which is the value of bit 31 of R3 calculated by the AND instruction
- [R3] = 0x2d which is a copy of the value in R0. Why didn't we just use it from R0? We need R0 in the loop for the service call to Linux to display a value.
- [R4] = 1 is the mask that selects just one bit use the logical AND instruction
- [R6] = 0x1f (decimal 31) is where we are in the loop
- [LR] = 0x807c is the return address which is not needed until all 32 passes are made through the loop.

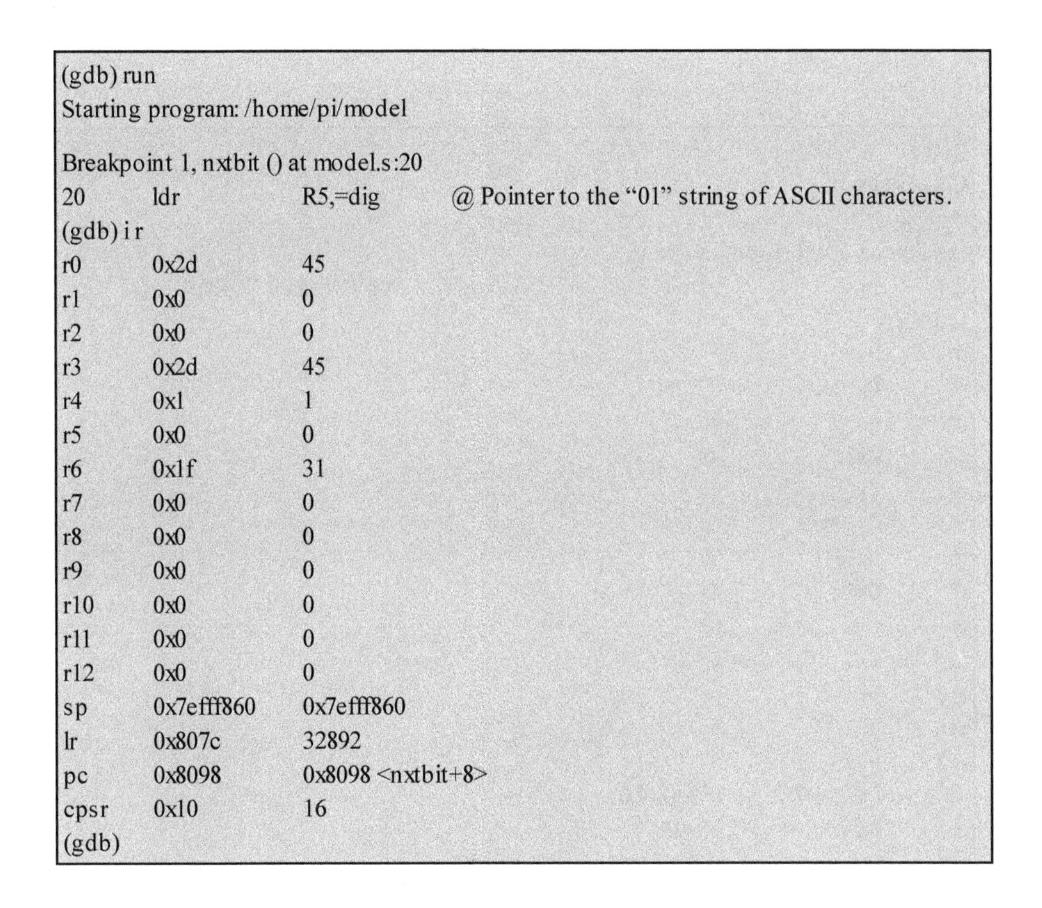

```
(gdb) run
Starting program: /home/pi/model

Breakpoint 1, nxtbit () at model.s:20
20      ldr         R5,=dig         @ Pointer to the "01" string of ASCII characters.
(gdb) i r
r0      0x2d        45
r1      0x0         0
r2      0x0         0
r3      0x2d        45
r4      0x1         1
r5      0x0         0
r6      0x1f        31
r7      0x0         0
r8      0x0         0
r9      0x0         0
r10     0x0         0
r11     0x0         0
r12     0x0         0
sp      0x7efff860  0x7efff860
lr      0x807c      32892
pc      0x8098      0x8098 <nxtbit+8>
cpsr    0x10        16
(gdb)
```

Listing 5.3: Register contents immediately after the AND instruction

In Listing 5.4, we continue the program execution with the "c" command, and the debugger suspends execution upon reaching the second breakpoint which is after the zero has been output (it is displayed immediately below the word "continuing.") and after the loop counter R6 has been decremented, and just before the branch is taken to continue the loop. Notice the following contents in the register dump produced by the "i r" command.

- [R0] = 1 tells Linux to display on device stdout
- [R1] = 0x100c0 which is the address of a "0" character in memory that was just displayed
- [R2] = 1 indicates only one character is to be displayed
- [R3] = 0x2d — same as before.
- [R4] = 1 — same as before
- [R5] = 0x100c0 which is the address of a two-character string "01"
- [R6] = 0x1e (decimal 30) is where we will be for the next pass through the loop
- [R7] = 4 which tells Linux that an output service is requested
- [CPSR] = 0x00000010 has bits 31 down to 28 all zero indicating N=0, Z=0, C=0, V=0.

```
(gdb) c
Continuing.
0
Breakpoint 2, nxtbit () at model.s:27
27      bge             nxtbit          @ Go display next bit until all 32 displayed
(gdb) i r
r0      0x1             1
r1      0x100c0         65728
r2      0x1             1
r3      0x2d            45
r4      0x1             1
r5      0x100c0         65728
r6      0x1e            30
r7      0x4             4
r8      0x0             0
r9      0x0             0
r10     0x0             0
r11     0x0             0
r12     0x0             0
sp      0x7efff860      0x7efff860
lr      0x807c          32892
pc      0x80b4          0x80b4 %lt;nxtbit+36>
cpsr    0x20000010      536870928
(gdb)
```

Listing 5.4: Bottom of loop in subroutine v_bin

In Listing 5.5, we will first remove the first two breakpoints and then continue the program execution with the “c” command, and the debugger suspends execution upon reaching the third breakpoint which is just before the BX instruction that will return control to the main program. Notice the binary display from the loop appearing immediately below the word “continuing.” Notice also the following contents in the register dump produced by the “i r” command.

- [R6] = 0xffffffff (decimal -1) indicates the entire loop was completed
- [LR] = 0x807c is the return address which the BX instruction will soon use.
- [PC] = 0x80b8 is the address of next instruction to execute.
- [CPSR] = 0x0x80000010 has bit 31 set indicating N=1 which terminated the loop.

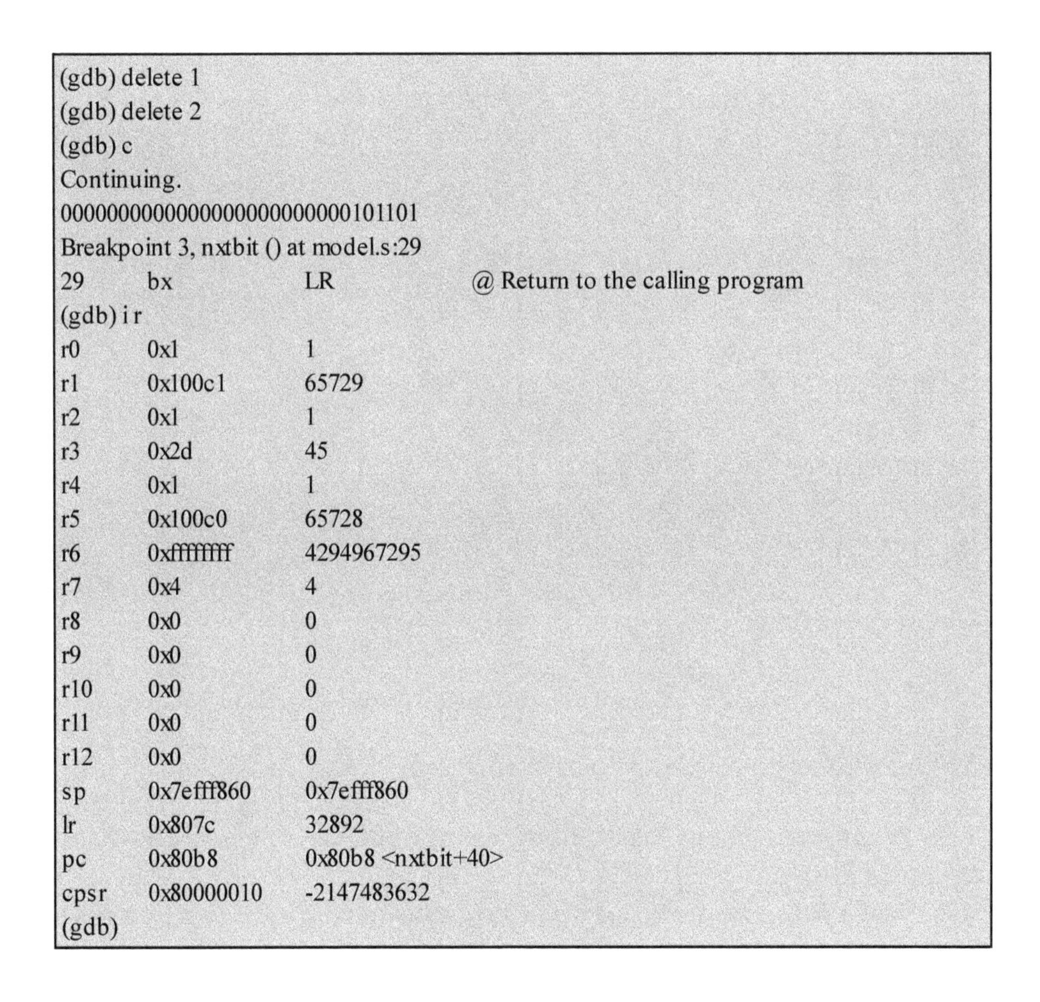

```
(gdb) delete 1
(gdb) delete 2
(gdb) c
Continuing.
00000000000000000000000000101101
Breakpoint 3, nxtbit () at model.s:29
29      bx              LR              @ Return to the calling program
(gdb) i r
r0      0x1             1
r1      0x100c1         65729
r2      0x1             1
r3      0x2d            45
r4      0x1             1
r5      0x100c0         65728
r6      0xffffffff      4294967295
r7      0x4             4
r8      0x0             0
r9      0x0             0
r10     0x0             0
r11     0x0             0
r12     0x0             0
sp      0x7efff860      0x7efff860
lr      0x807c          32892
pc      0x80b8          0x80b8 <nxtbit+40>
cpsr    0x80000010      -2147483632
(gdb)
```

Listing 5.5: Subroutine v_bin is ready to return.

In Listing 5.6, we will "single step" the BX instruction using the "s" command, and the debugger suspends execution after the one instruction. Notice the binary display produced by the "i r" command.

- [LR] = 0x807c is the return address which the BX instruction used for the return.
- [PC] = 0x807c is the address of next instruction to execute.

```
(gdb) s
_start () at model.s:4
4         mov        R0,#0        @ Exit Status code of 0 for "normal completion"
(gdb) i r
r0        0x1          1
r1        0x100c1      65729
r2        0x1          1
r3        0x2d         45
r4        0x1          1
r5        0x100c0      65728
r6        0xffffffff   4294967295
r7        0x4          4
r8        0x0          0
r9        0x0          0
r10       0x0          0
r11       0x0          0
r12       0x0          0
sp        0x7efff860   0x7efff860
lr        0x807c       32892
pc        0x807c       0x807c <_start+8>
cpsr      0x80000010   -2147483632
(gdb)
```

Listing 5.6: Control has returned to the main program.

Let's provide three enhancements to subroutine v_bin:

1. Performance can be improved by moving fixed values in registers outside the loop.
2. Add a parameter informing v_bin how many bits are to be displayed. When all we want to see is the low-order (on the right side) six bits, who wants all those leading zeroes on the left?
3. Preserve register contents as was done for the final version of v_bool.

```
@           Subroutine v_bin will display a 32-bit register in binary digits
@                     R0: contains the 32 bit value to be displayed in binary
@                     LR: Contains the return address
@                     Registers R0 through R7 will be used by v_bin and not saved

v_bin:      mov       R3,R0               @ R3 will hold a copy of input word to be displayed.
            mov       R6,#31              @ Number of bits to display (-1)

@           Loop through single bits and output each to the standard output (stdout)
            display.

nxtbit:     mov       R4,#1               @ used to mask off 1 bit at a time for display
            and       R1,R4,R3,lsr R6     @ Select next 0 or 1 to be displayed.
            ldr       R5,=dig             @ Pointer to the "01" string of ASCII characters.
            add       R1,R5               @ Set R1 pointing to "0" or "1" in memory
            mov       R2,#1               @ Number of characters to be displayed at a time.
            mov       R0,#1               @ Code for stdout (standard output, i.e., monitor)
            mov       R7,#4               @ Linux service command code to write string.
            svc       0                   @ Linux service command code to write string.
            subs      R6,#1               @ Decrement number of bits remaining to display
            bge       nxtbit              @ Go display next bit until all 32 displayed

            bx        LR                  @ Return to the calling program
```

Listing 5.7: Original v_bin from Prototype section

Thirty-two passes are made through the nxtbit loop with register R6 counting down from 31 to 0. Notice that most of the registers (R0, R2, R4, R5, R7) are loaded with the same data on each pass through the loop. These really only need to be loaded once as illustrated by the flowchart in Figure 5.6. In this particular example, the savings in time will be negligible because the Linux service call will consume the vast majority of the time through each pass of the loop. However, there are many examples where moving constants outside the loop will provide significant real-time savings. Many times there's a tradeoff between performance and maintenance. Code that has been enhanced for performance should be clearly documented so that those performing future maintenance will be less likely to introduce new bugs.

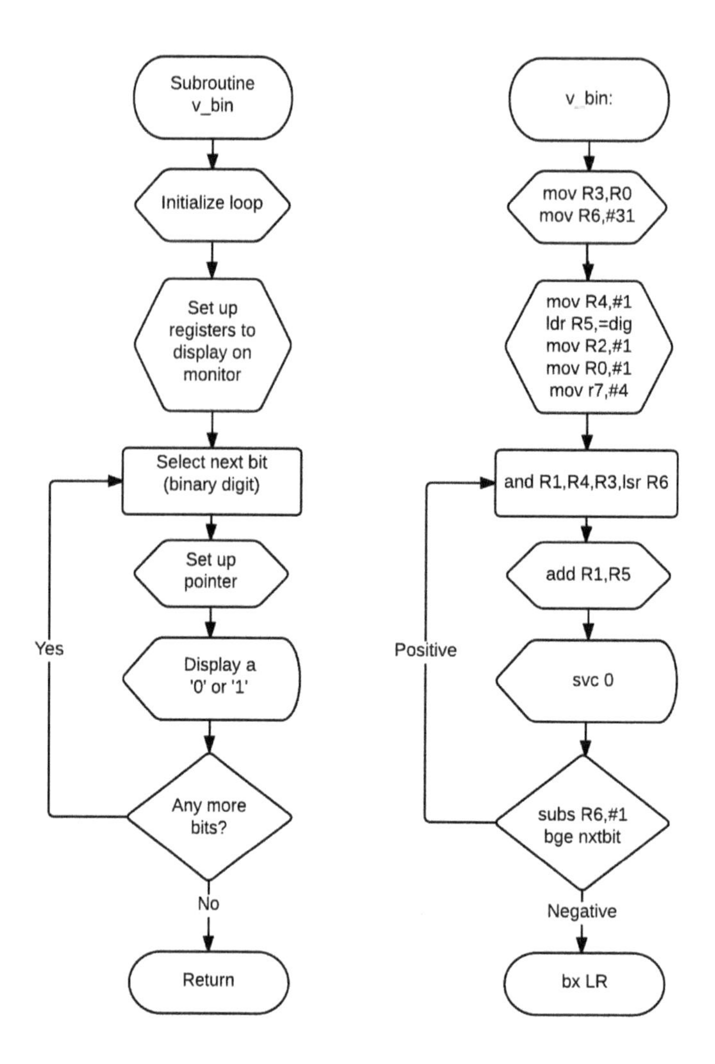

Figure 5.6: Constants moved outside loop

Thirty-two bits is a pretty wide number to examine. Sometimes we need to see all 32 bits, but many times, 8 or 16 or even 1 bit is all we want. So, subroutine v_bin is modified in Listing 5.8 so that the calling program can specify how many bits (from the right) will be displayed. We're also relieving the calling program from having "to know" which register contents will be destroyed by v_bin (i.e., we will save what we use and restore them in v_bin before returning).

1.	.global	_start	@ Provide program starting address to linker
2. _start:	mov	R0,#0b101101	@ An arbitrary sample binary value for testing
3.	mov	R2,#6	@ Display 6 bits (bits 5 down to 0).
4.	bl	v_bin	@ Call subroutine to view binary value of R0 in ASCII.
5.	mov	R0,#0	@ Exit Status code of 0 for "normal completion"

```
6.              mov     R7,#1           @ Service command code 1 terminates this
                                        program.
7.              svc     0               @ Issue Linux command to terminate program
8.
9. @            Subroutine v_bin will display a 32-bit register in binary digits
10. @                   R0: contains a number to be displayed in binary
11. @                   R2: Number of bits to be displayed (from right side of R0)
12. @                   LR: Contains the return address
13. @                   All register contents will be preserved
14.
15. v_bin:      push    {R0-R7}         @ Save contents of registers R0 through R8,
                                        LR
16.             sub     R6,R2,#1        @ Number of bits to display (-1)
17.             cmp     R6,#31          @ Test error value entered (bit 31 is leftmost
                                        bit).
18.             movhi   R6,#0           @ If bad range, default to displaying only 1
                                        bit.
19.             mov     R3,R0           @ R3 will hold a copy of input word to be
                                        displayed.
20.             mov     R4,#1           @ Used to mask off 1 bit at a time for display
21.             ldr     R5,=dig         @ Pointer to the "01" string of ASCII
                                        characters.
22.             mov     R2,#1           @ Number of characters to be displayed at a
                                        time.
23.             mov     R0,#1           @ Code for stdout (standard output, i.e.,
                                        monitor)
24.             mov     R7,#4           @ Linux service command code to write string.
25.
26. @           Loop through single bits and output each to the standard output (stdout)
                display.
27.
28. nxtbit:     and     R1,R4,R3,LSR R6 @ Select next 0 or 1 to be displayed.
29.             add     R1,R5           @ Set R1 pointing to "0" or "1" in memory
30.             svc     0               @ Linux service command code to write string.
31.             subs    R6,#1           @ Decrement number of bits remaining to
                                        display
32.             bge     nxtbit          @ Go display next bit until all 32 displayed
33.
34.             pop     {R0-R7}         @ Restore saved register contents
35.             bx      LR              @ Return to the calling program
36.
37.             .data
38. dig:        .ascii  "01"            @ ASCII string of binary digits 0 and 1
39.             .end
```

Listing 5.8: Subroutine v_bin with saved registers and variable display width

Maintenance

Review Questions

1. Why are loops important in a computer program?
2. What's an infinite loop? How does one happen? Why would we intentionally create an infinite loop?
3. What's the largest immediate value that can be loaded into a register (without doing any shifting; i.e. multiplying by a power of 2)? In other words, for an instruction like mov R1,#13, what is the largest value (instead of #13)?
4. * The instruction mov R1,#5120 should not be possible (due to the limitations noted in the previous question). What machine code instruction does the assembler generate to make it work? Clue: 5120 = 4096+1024.
5. Why does the ARM include an RSB (reverse subtraction) instruction? That is, why can't you just change the order of the operands (instead of [R4]-[R5], use [R5]-[R4])?
6. Name two differences between the architecture of the ARM and almost every other CPU.

Programming Exercises

1. Use v_bin to display the results in Lab 2 for the logical instruction examples. This will be much more convenient than the decimal display.
2. Modify v_bin to omit leading (leftmost) zeroes (i.e., instead of displaying 001011, display 1011 instead. Note: there is an ARM instruction that counts leading zeroes.
3. Modify v_bin to precede the binary number with "0b" (i.e., 0b00101 instead of just 00101).
4. If the number of bits to display is invalid (i.e., it is greater than 32), I have v_bin defaulting to display only one bit. Perhaps you would like it to default to the full 32 bits or even do something silly like display 50 bits (18 leading zeroes). Modify one line of code in v_bin to put in your change.

Lab 6
Display Numbers in Hexadecimal

In the previous lab, we developed subroutine v_bin to display a register's contents in binary (base 2), and in the next lab we will develop subroutine v_dec which displays integers in our everyday decimal (base 10). The usefulness of both is obvious: Binary displays the native form used in the computer system while decimal displays the form commonly used by us humans, but why is hexadecimal (base 16) popular? In Lab 6, we will see the importantance of hexadecimal as well as construct subroutine v_hex using two loops to display a number in hexadecimal. It will build upon the looping and branching techniques introduced in Lab 5.

Prototype

Hexadecimal is a compact form of binary representation where we have sixteen symbols {0,1,2,3,4,5,6,7,8,9,A,B,C,D,E,F} to represent numbers. A hexadecimal "number" is really a short notation for a polynomial of powers of 16. For example: $5A732C_{16}$ is $5\times16^5 + 10\times16^4 + 7\times16^3 + 3\times16^2 + 2\times16^1 + 12\times16^0$ where A and C are digits representing values of 10 and 12, respectively. Please see Appendix C if you need more background on hexadecimal.

The programming of subroutine v_hex will be very similar to that of subroutine v_bin with the following exceptions:

1. Each character displayed will represent 4 bits rather than only 1 (i.e., hexadecimal rather than binary). This will affect the shift in the AND instruction as well as the count down in register R6. Sixteen different digits "0123456789ABCDEF" will be displayed rather than only two "01."
2. An option has been included where leading zeroes may be omitted. There is a loop to search for the first non-zero hex digit.

```
1.          .global   _start      @ Provide program starting address to linker
2. _start:  mov       R0,#0xA7    @ An arbitrary sample hex value for testing
3.          mov       R2,#0       @ Don't display any leading zeroes.
4.          bl        v_hex       @ Call subroutine to view hexadecimal value of
                                  R0 in ASCII.
```

6.	mov	R7,#1	@ Service command code 1 terminates this program.
7.	svc	0	@ Issue Linux command to terminate program
8.			
9. @	Subroutine v_hex will display a 32-bit register in binary digits		
10. @		R0: contains a number to be displayed in hexadecimal	
11. @		R2: Number of nibbles to be displayed (from right side of R0)	
12. @		Note: If R2=0 or R2>8, leading zeroes (on left) will not be displayed.	
13. @		LR: Contains the return address	
14. @		All register contents will be preserved	
15.			

Listing 6.0a: "Main program" with v_hex calling parameters

16. v_hex:	push	{R0-R7}	@ Save contents of registers R0 through R8, LR
17.	mov	R3,R0	@ R3 will hold a copy of input word to be displayed.
18.	mov	R4,#0b1111	@ Used to mask off 4 bits at a time for display
19.	mov	R6,R2,lsl#2	@ Load number of bits to display (4 bits for each nibble)
20.			
21. @	Set up registers for calling Linux to display 1 character on the display monitor.		
22.			
23.	ldr	R5,=dig	@ Pointer to the "012...EF" string of ASCII characters.
24.	mov	R2,#1	@ Number of characters to be displayed at a time.
25.	mov	R0,#1	@ Code for stdout (standard output, i.e., monitor)
26.	mov	R7,#4	@ Linux service command code to write string.
27.			
28. @	Determine number of bits to be output (R6 has that value if its between 4 and 32).		
29.			
30.	cmp	R6,#32	@ Test error value entered (there's only 32 bits in register).
31.	movhi	R6,#0	@ Default to omitting leading zeroes if value > 32.
32.	subs	R6,#4	@ Set R6 point to "right" side of first nibble to output.
33.	bge	nxthex	@ If proper range, then skip over finding first non-zero.

Listing 6.0b: Initialize registers for loops and display

```
34.
35.@         Skip over leading zeroes (on left)
36.
37.         mov     R6,#28            @ Number of bits in register - number of bits per hex digit
38.nxtzer:  ands    R1,R4,R3,lsr R6   @ Select next hex digit (0 ... F) to be displayed.
39.         bne     nxthex            @ Go write first non-zero hex digit.
40.         subs    R6,#4             @ Decrement number of nibbles remaining to display
41.         bgt     nxtzer            @ Go check next nibble, but not 1's column (i.e., value = 0)
42.
```

Listing 6.0c: First loop skips over leading zeroes.

```
43.@         Loop through groups of 4-bit nibbles and output each to stdout (monitor).
44.
45.nxthex:  and     R1,R4,R3,lsr R6   @ Select next hex digit (0 ... F) to be displayed.
46.         add     R1,R5             @ Set R1 pointing to "0", "1", ... or "F" in memory
47.         svc     0                 @ Linux service command code to write string.
48.         subs    R6,#4             @ Decrement number of bits remaining to display
49.         bge     nxthex            @ Go display next nibble until max bit-count reached
50.
51.         pop     {R0-R7}           @ Restore saved register contents
52.         bx      LR                @ Return to the calling program
53.
54.         .data
55.dig:     .ascii  "0123456789"      @ ASCII string of digits 0 through 9
56.         .ascii  "ABCDEF"          @ ASCII string of digits A through F
57.         .end
```

Listing 6.0d: Second loop to convert each 4-bit nibble to a hex digit.

Listings 6.0c and 6.0d contain program loops. Both loops use an AND instruction similar to that used in Lab 5, except four bits are shifted and masked instead of one.

Figure 6.0 illustrates the case where [R6] = 4 which selects the second nibble from the right in register R3, masks off the lower 4 bits using register R4 and places the result in register R1.

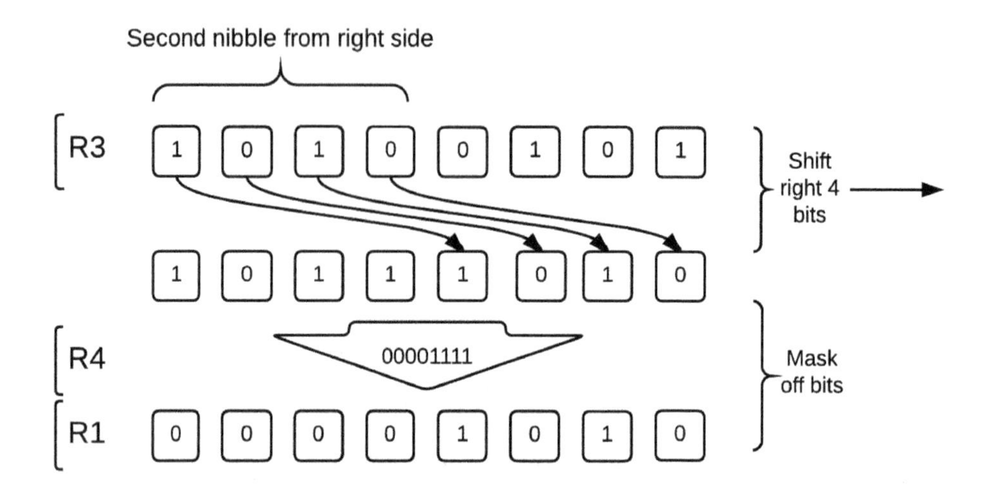

Figure 6.0: AND instruction selects and right justifies a 4-bit nibble

Go ahead and run this hexadecimal-display sample program contained in listings 6.0 through 6.3. Use the same command lines from the binary prototype in Lab 5 to assemble, link, and execute. Try it first with R0 initialized to hexadecimal A7 (line 2), but also try other values, even some in decimal or binary. Using the gdb debugger "set" command (first introduced in lab 2) will not only make entering new values for R0 quicker, but also enable a full 32 bits to be input if desired.

Introductions

There will be no new ARM instructions, assembler directives, Linux services, or command lines introduced in Lab 6.

Principles

We have three issues to deal with in Lab 6:

1. Why are we interested in displaying in hexadecimal?
2. What algorithm will we use to convert the binary number in a register to a series of hexadecimal digits?
3. How do we get some more assembly language programing experience using loops?

Why Hexadecimal?

Do you really like a number being expressed like 3AF4D or 0A1BD? It almost seems as awkward as Roman numerals like MCCIX or LVII. Why don't we use decimal all the time? Whatever happened to octal which uses the digits 0 through 7? The popularity and usefulness of hexadecimal come from the number of bits in a byte.

How Big is a Byte?

When I ask my students, "How big is a byte?," I almost always get the same answer: "Eight bits." If I had asked that question in the 1960s or 1970s, I would have received answers ranging from 5 bits to 9 bits. What's changed in computer hardware or software that has locked down that variable answer to an almost unanimous 8 bits? The answer lies in the question, "What do we typically store in a byte?"

In the early days of computing, computers were specialized for either science and engineering or business. By the way, that's an exclusive "or" in most cases, not an inclusive "or." Science and engineering concentrated on arithmetic problems: calculations involving numbers. Business applications focused on lists of names and addresses composed of characters. Numbers used in calculations are typically stored in words while characters used in names and addresses are typically stored in bytes.

Minimum set of characters useful for mailing lists (a total of at least 39 characters):

- Letters (26)
- Digits (10)
- Punctuation (at least 3: blank, period, comma)

So how many bits will it take to hold at least 39 unique character symbols? The maximum number of characters in various "byte" sizes follows:

1. $2^1 = 2$
2. $2^2 = 4$
3. $2^3 = 8$
4. $2^4 = 16$
5. $2^5 = 32$
6. $2^6 = 64$
7. $2^7 = 128$
8. $2^8 = 256$

In the early 1960s, "bytes" were 6 bits which could accommodate 64 possible symbols, more than enough to hold our essential 39 characters. Octal was very popular at that time because 6 bits could be represented by two 3-bit octal digits (000, 001, 010, 011, 100, 101, 110, 111). However, a more reasonable set of characters for mailing lists and other business applications in the English language follows:

- Upper case letters (26)
- Lower case letters (26)
- Digits (10)
- Punctuation (at least 10)

The above set of 72 characters will not fit in six bits, so seven bits is needed. The ASCII character set (see Appendix D) is a 7-bit code that has 128 characters defined. Eight-bit bytes became almost universal partly due to the rounding-up of 7 bits to an even 8 bits and partly by the influence of the IBM 360 mainframe with its 8-bit EBCDIC character set. Eight bit values are displayed as two 4-bit hex digits. Yes, some people initially represented 8-bit values as three octal digits, but that required 50% more digits than hexadecimal, and what are you going to do with that extra 9th bit?

How Big is a Word?

Although it's not needed in this lab, let's comment on the size of a "word." Now we have a question that will currently have multiple answers: 16 bits, 32 bits, and 64 bits. In the days of mainframe computing in the 1960s, we would also get answers like 36 bits (IBM 7094, Univac 1108, and DEC PDP 10), 48 bits (Burroughs B5500), and 60 bits (Control Data CDC 6600). Why the variation? Well, what's typically stored in a word? A single precision integer or floating point number is stored in a word. Different CPU architectures offered different levels of precision and range of values in their calculations.

Word Addressable or Byte Addressable?

Although a computer system could be designed with separate memory for bytes and words, it would be extremely wasteful, and that hasn't been done in any popular modern design. So how do we get an overlap of bytes and words?

- Word addressable: The CPU instructions address whole words. Then how do we get to a particular byte within any particular word?
- Byte addressable: The CPU instructions address individual bytes. Then where does a word start? Is it any group of four bytes (assuming 32-bit words and 8-bit bytes), and in what order do we fill a word from bytes: left to right or right to left?

Several mainframe computers in the 1960s were word-addressable. The CDC 6600 addressed 60-bit words and had ten 6-bit bytes within each word. The Univac 1108 addressed 36-bit words and supported both six 6-bit bytes and four 9-bit bytes within each word. The PDP-10 addressed 36-bit words with five 7-bit bytes plus an extra bit left over in each word. The IBM 360, as well as the mini-computers of the 1970s and microcomputers of the 1980s were byte addressable. Like almost all microprocessors, the ARM is byte addressable, and that's how we'll access characters in memory in this lab. In Lab 7, we will be working with words, but have to address them on 4-byte boundaries.

Decimal	Binary (in groups of 4 bits)	Hexadecimal
7094	1 1011 1011 0110	1BB6
1620	110 0101 0100	654
1108	100 0101 0100	454
6600	1 1001 1100 1000	19C8
3033	1011 1101 1001	BD9
7800	1 1110 0111 1000	1E78

Table 6.0: Decimal converted to binary and then converted to hexadecimal

Nibbles or Hex Digits

So what's a nibble? The first time a friend mentioned it to me, it was a joke because the term was not in general use. He used "nibble" to represent a "little byte," which is what it currently means today: four bits or half a byte.

In the examples in this lab, I will generally use the term "nibble" to represent the container size (4 bits) and "hex digit" to represent the contents (a symbol in the range of "0" through "F"). Likewise, "byte" would be the container and "ASCII character" would be the contents. Following that, we would have “word” as the container size and “integer” or “floating point” number as the contents. Of course, there are more types of contents than just characters, integers, and floating point numbers, but those are the basics.

Conversion to Any Base

A popular way to convert a number to a particular base is successive division. The remainders from each division will provide the digits (i.e., symbols) beginning with the rightmost digit. For example, converting the number 3274 to decimal follows:

1. 3274 / 10 = 327 Remainder 4
2. 327 / 10 = 32 Remainder 7
3. 32 / 10 = 3 Remainder 2
4. 3 / 10 = 0 Remainder 3

So the “number” 3274 is represented in decimal as the sequence of remainders “3” “2” “7” and “4” (order reversed). By the way, this technique of successively dividing a number by the desired base works regardless of how the “computer” internally stores numbers. It could be binary, decimal, or any conceivable internal structure that would permit division.

Converting the same number 3274 to binary follows:

1. 3274 / 2 = 1637 Remainder 0
2. 1637 / 2 = 818 Remainder 1
3. 818 / 2 = 409 Remainder 0
4. 409 / 2 = 204 Remainder 1
5. 204 / 2 = 102 Remainder 0
6. 102 / 2 = 51 Remainder 0
7. 51 / 2 = 25 Remainder 1
8. 25 / 2 = 12 Remainder 1
9. 12 / 2 = 6 Remainder 0
10. 6 / 2 = 3 Remainder 0
11. 3 / 2 = 1 Remainder 1
12. 1 / 2 = 0 Remainder 1

So the “number” 3274 is represented in binary as the sequence of remainders “1” “1” “0” “0” “1” “1” “0” “0” “1” “0” “1” and “0” (order reversed). As an exercise, try converting 3274 to base five by successively dividing by five until the quotient is zero (3274/5 = 654 remainder 4, ...). The answer will be 101044_5.

Multiplying and Dividing by Shifting

If we want to multiply by ten "in our heads" in our everyday decimal system, we just append a zero. For example to multiply 709 by 10, we append "0" to "709" and get "7090." Likewise, when we multiply by 100 (i.e., 10^2), we append two zeroes, and for 1000, we append 3 zeroes, etc. For dividing by powers of ten, we do the reverse: we remove zeroes on the right. What if there are not enough zeros present on the right? Then we move the decimal point. For example to divide 36040 by 100, we move the decimal point to the left two places giving us 360.40.

When we shift to the left in base two, we are multiplying by a power of two, and when we shift to the right, we are dividing by a power of two. This means that conversion into and from binary format is done very efficiently using shifting as we did in Lab 5 rather than division. Converting the same number 3274 (110011001010_2) to binary by shifting is below. Note: The notation ">> 1" means shift 1 bit position to the right, and the "Carry out" refers to the rightmost bit that is lost when the value is shifted.

1. 110011001010 >> 1 = 11001100101 with Carry out 0
2. 11001100101 >> 1 = 1100110010 Carry out 1
3. 1100110010 >> 1 = 110011001 Carry out 0
4. 110011001 >> 1 = 11001100 Carry out 1
5. 11001100 >> 1 = 1100110 Carry out 0
6. 1100110 >> 1 = 110011 Carry out 0
7. 110011 >> 1 = 11001 Carry out 1
8. 11001 >> 1 = 1100 Carry out 1
9. 1100 >> 1 = 110 Carry out 0
10. 110 >> 1 = 11 Carry out 0
11. 11 >> 1 = 1 Carry out 1
12. 1 >> 1 = 0 Carry out 1

Conversion to Hexadecimal

We can convert a number to base 16 by successive divisions by 16. The remainders from each division will provide the digits (i.e., symbols) beginning with the rightmost digit. For example, converting the number 3274 to hexadecimal follows:

1. 3274 / 16 = 204 Remainder 10 (A in hexadecimal)
2. 204 / 16 = 12 Remainder 12 (C in hexadecimal)
3. 12/ 16 = 0 Remainder 12 (C in hexadecimal)

So the "number" 3274 is represented in hexadecimal as the sequence of remainders "C" "C" and "A."

Dividing by 16 by Shifting

When we shift to the right 1 bit position, we are dividing by 2^1. Likewise, when we shift to the right 2 bit positions, we are dividing by $2^2=4$. A 3-bit shift to the right is dividing by $2^3=8$, and of course a 4-bit shift will be a division by $2^4=16$. Converting the same number 3274 (110011001010_2) to hexadecimal by shifting is below. Note: The notation ">> 4" means shift 4 bit positions to the right, and the "Carry out" refers to the rightmost four bits that are lost when the value is shifted.

1. 110011001010 >> 4 = 11001100 with Carry out 1010 (A in hexadecimal)
2. 11001100 >> 4 = 1100 Carry out 1100 (C in hexadecimal)
3. 1100 >> 4 = 0 Carry out 1100 (C in hexadecimal)

The ARM processor resident on the Raspberry Pi does not have a divide instruction available, so we have no choice but to use the shifting technique. However, even if we were using a CPU that did have an integer divide instruction, we would probably still use the shift anyway because it is much faster than the divide. Note: The Raspberry Pi does have a divide in its floating point processor, but there are multiple reasons why we would not take that approach: more complicated, speed, etc.

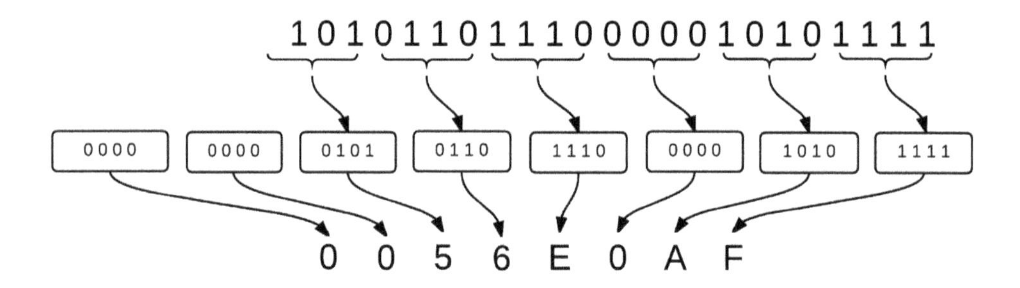

Figure 6.1: Hexadecimal representation of binary by grouping 4 bits at a time

Arrays

An array is an ordered list of adjacent storage locations in memory. It could be a list of bytes, words, or even a more complicated combination of bytes and words. A register is typically used to "index" into the array as well as sequentially stepping through each value from the beginning to the end. In the previous lab, we used a very short byte array of "0" followed by "1" to convert the single binary bit to its corresponding ASCII character value to be displayed. In base 16, we will use the same technique, but the array will have to contain all sixteen digits "0123456789ABCDEF" needed for hexadecimal.

```
        .data
dig:    .ascii     "0123456789"     @ ASCII string of digits 0 through 9
        .ascii     "ABCDEF"         @ ASCII string of digits A through F
```

Listing 6.1: Array containing hexadecimal digits in ASCII

Coding and Debugging

The first seven lines of Listing 6.0 (repeated below) form the "main" program that calls subroutine v_hex with a test value. Note: We will not be adding any enhancements to the v_hex subroutine that appeared in the Prototype section. The line numbers from Listing 6.0 can be used for setting your breakpoints in the gdb debugger. I will not walk you through breakpoints and register dumps in this lab. You may want to consider setting a breakpoint at line 4, then use the gdb "set" command to try various inputs to subroutine v_hex.

```
        .global  _start      @ Provide program starting address to linker
_start: mov      R0,#0xA7    @ An arbitrary sample hex value for testing
        mov      R2,#0       @ Don't display any leading zeroes.
        bl       v_hex       @ Call subroutine to view hexadecimal value of R0 in
                             ASCII.
        mov      R0,#0       @ Exit Status code of 0 for "normal completion"
        mov      R7,#1       @ Service command code 1 terminates this program.
        svc      0           @ Issue Linux command to terminate program

@       Subroutine v_hex will display a 32-bit register in binary digits
@                    R0: contains a number to be displayed in hexadecimal
@                    R2: Number of nibbles to be displayed (from right side of R0)
@                    Note: If R2=0 or R2>8, leading zeroes (on left) will not be displayed.
@                    LR: Contains the return address
@                    All register contents will be preserved
```

Listing 6.2: Main test program and subroutine v_hex calling specifications

Subroutine v_hex is divided into three parts:

1. Set up and prepare registers for the display and loops.
2. Loop to search for the first non-zero hexadecimal digit.
3. Loop to display each hexadecimal digit in the data word.

Figure 6.2 illustrates the first part which ends with two off-page connectors. According to the input argument specifications: If Register R2 contains a value between one and eight, inclusive, then that's how many digits will be displayed. If R2 has any other value, such as zero, then subroutine v_hex will skip over all leading zeroes.

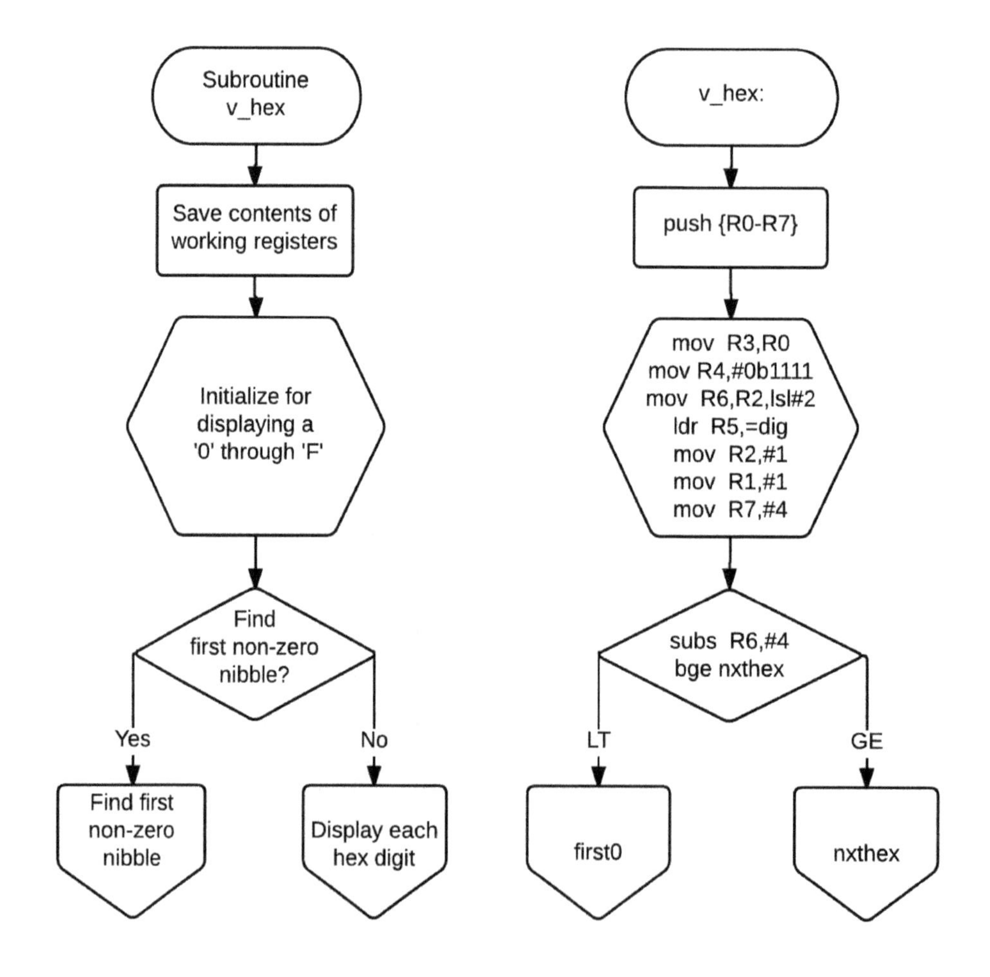

Figure 6.2: Initialization in subroutine v_hex

```
v_hex:  push    {R0-R7}         @ Save contents of registers R0 through R7.
        mov     R3,R0           @ R3 will hold a copy of input word to be displayed.
        mov     R4,#0b1111      @ Used to mask off 4 bits at a time for display
        mov     R6,R2,lsl#2     @ Load number of bits to display (4 bits for each nibble)
@       Set up registers for calling Linux to display 1 character on the display monitor.
        ldr     R5,=dig         @ Pointer to the "012...EF" string of ASCII characters.
        mov     R2,#1           @ Number of characters to be displayed at a time.
        mov     R0,#1           @ Code for stdout (standard output, i.e., monitor)
        mov     R7,#4           @ Linux service command code to write string.
@       Determine number of bits to be output (R6 has that value if its between 4 and 32).
        cmp     R6,#32          @ Test error value entered (there's only 32 bits in register).
        movhi   R6,#0           @ Default to omitting leading zeroes if value > 32.
        subs    R6,#4           @ Set R6 point to "right" side of first nibble to output.
        bge     nxthex          @ If proper range, then skip over finding first non-zero.
```

Listing 6.3: Assembly code corresponding to flowchart in Figure 6.2 (initialization)

The following flowchart in Figure 6.3 illustrates the second part of the code that searches for the first non-zero nibble. This code may or may not be executed depending on the v_hex input argument value in register R2. Since at least one zero must be displayed, this loop has two exits: a non-zero nibble reached or only one digit left to examine.

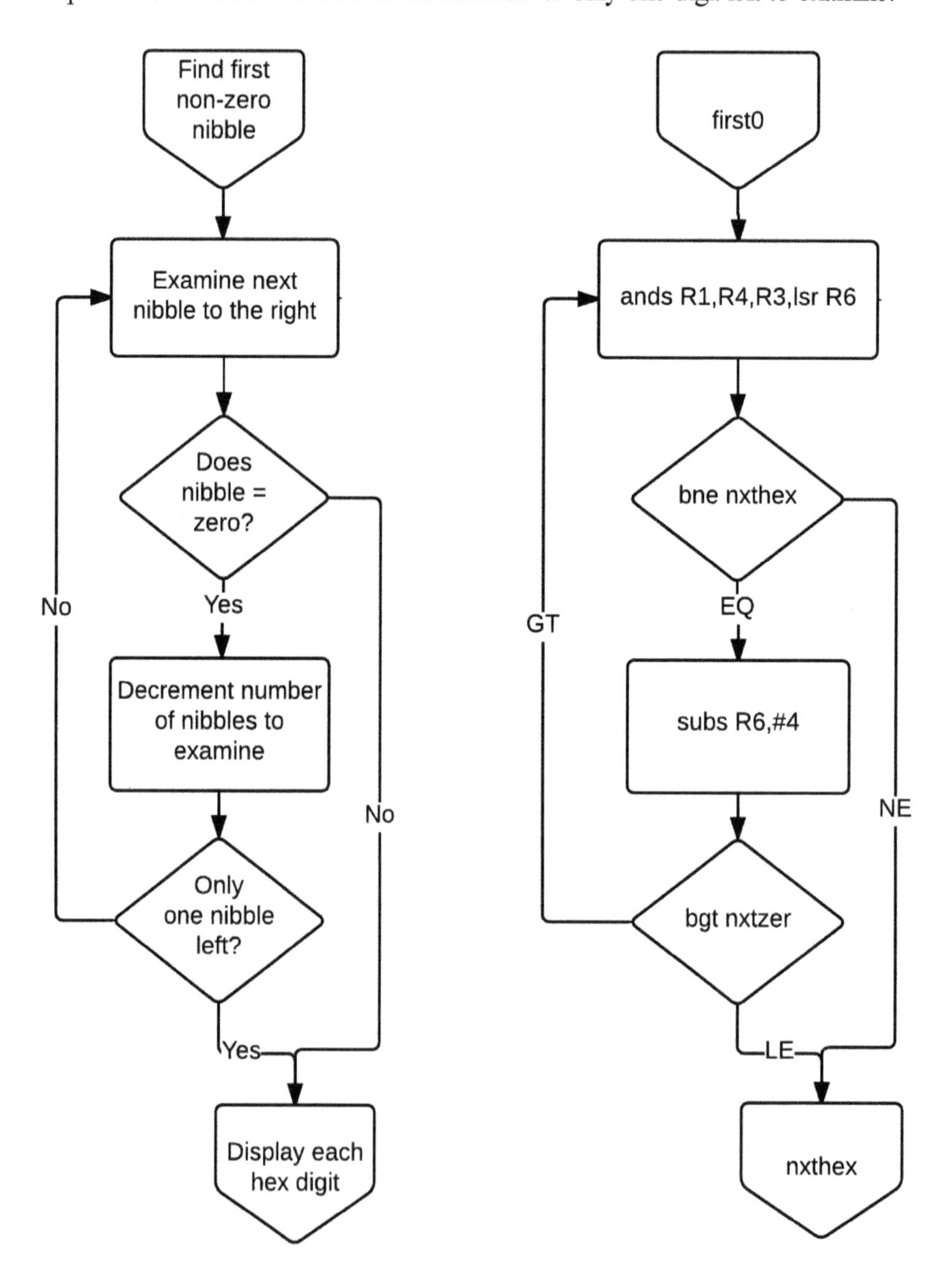

Figure 6.3: Skip over leading zeroes

```
@        Skip over leading zeroes (on left)
         mov     R6,#28             @ Number of bits in register - number of bits per hex
                                    digit
nxtzer:  ands    R1,R4,R3,lsr R6   @ Select next hex digit (0 ... F) to be displayed.
         bne     nxthex             @ Go write first non-zero hex digit.
         subs    R6,#4              @ Decrement number of nibbles remaining to
                                    display
         bgt     nxtzer             @ Go check next nibble, but not 1's column (i.e.,
                                    value = 0)
```

Listing 6.4: Assembly code corresponding to flowchart in Figure 6.3

A maximum of seven passes is made through the nxtzer loop with register R6 counting down from 28 to 4. Notice that most of the registers (R0, R2, R4, R5, R7) are constant and were already loaded outside the loop (Figure 6.2 and Listing 6.4) like what was done in Lab 5 for binary conversion.

R1	Hex digit currently being examined	Value is in range of 0 through 15
R3	32-bit number being displayed	Copy of original R0
R4	Masks off low-order nibble (bits 3-0)	Always = 0xF
R6	Selects which nibble of R3 to display	Counts down from 28 to 4

Table 6.1: Registers used by ANDS instruction in the nxtzer loop

Display Hex Digits

Now that we are positioned at the first nibble to display, and the constant registers are loaded, it's time to perform the job that v_hex was called to do by looping through each 4-bit nibble to be displayed as a hex digit. A maximum of eight passes wiil be made through the nxthex loop with register R6 counting down from 28 to 0. As in subroutine v_bin, the "AND R1,R4,R3,LSR R6" is the important instruction that pulls out the unique data value needed on each pass through the loop.

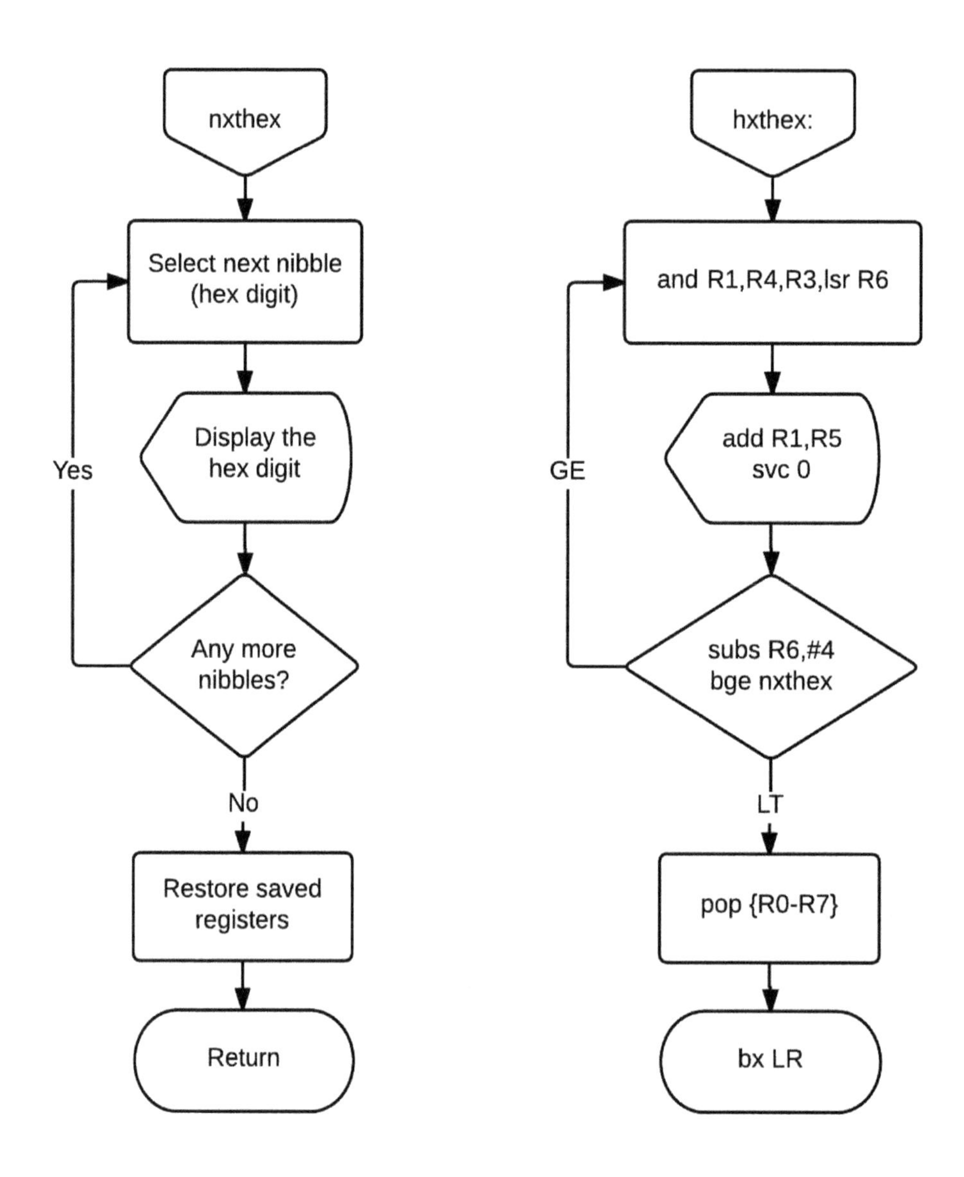

Figure 6.4: Loop to display hex digits

```
@         Loop through groups of 4-bit nibbles and output each to stdout (monitor).

nxthex:   and    R1,R4,R3,lsr R6  @ Select next hex digit (0 ... F) to be displayed.
          add    R1,R5            @ Set R1 pointing to "0", "1", ... or "F" in memory
          svc    0                @ Linux service command code to write string.
          subs   R6,#4            @ Decrement number of bits remaining to display
          bge    nxthex           @ Go display next nibble until max bit-count
                                  reached

          pop    {R0-R7}          @ Restore saved register contents
          bx     LR               @ Return to the calling program
```

Listing 6.6: Assembly code corresponding to flowchart in figure 6.4 (display hex digits)

Notice that most of the registers (R0, R2, R3, R4, R5, R7) are constant and were already loaded outside the loop (Figure 6.2 and Listing 6.6) like what was done in Lab 5 for binary conversion. In both examples used in this book, the savings in time is negligible because the Linux service call will consume the vast majority of the execution time through each pass of the loop. However, there are many examples where moving constants outside the loop will provide significant real-time savings.

R0	Output to stdout display	Always = 1
R1	Selected hex digit	Value is 0, 1, 2, ..., or F
R2	Number of characters to output	Always = 1
R3	32-bit number being displayed	Copy of original R0
R4	Masks off low-order nibble (bits 3-0)	Always = 0xF
R5	Points to string: "0123456789ABCDEF"	Always the same
R6	Selects which nibble of R3 to display	Counts down from 28 to 0
R7	Linux service command code to write file or device	Always = 4

Table 6.2: Registers used in nxthex loop

Maintenance

Review Questions

1. As an exercise, convert 3274 to base seven by successively dividing by seven until the quotient is zero (3274/7 = 467 remainder 5, ...).
2. * Compare the merits of a byte-addressable computer architecture to one that is word-addressable.
3. Subroutine v_hex can be easily modified to display a number in octal or even base 4 or base 32. Why can't it be easily modified to display a number in decimal?

Programming Exercises

1. Modify v_hex to precede the binary number with "0x" (i.e., 0x23B4 instead of just 23B4).
2. Modify v_hex to create a subroutine v_oct that will display the number in octal (base 8).
3. Modify v_hex to create a subroutine v_32 that will display the number in base 32. Hint: Table "dig" will have to be doubled in length, 16 more letters from G through V.
4. In v_hex, the loop to find the first non-zero hex digit to display is very similar to the loop to display the hex digits. Modify hex_bin to combine them in a single loop. Note: Be careful not to output "nothing at all" if the value is zero.
5. Generally, a signed number is negative if its high order bit is a one (bit 31 = 1). Modify v_hex to output a negative number in hexadecimal. Hint: Take a look at how it will be done in the next lab on decimal (Listing 7.1, line 23).

Lab 7
Display Numbers in Decimal

Lab 7 builds upon the looping and branching techniques introduced in the previous two labs. Here we will construct a subroutine using three loops to display a number in decimal. One of the loops will be nested within another. We will also be searching data tables in memory.

Prototype

The format of the "main" program that calls subroutine v_dec is almost identical to what you have seen in the previous labs. The only argument is the data itself contained in register R0. In the Maintenance section, I will suggest some programming exercises having alterations such as inserting commas, a decimal point, currency symbol, etc.

```
1.            .global   _start     @ Provide program starting address to linker
2. _start:    mov       R0,#125    @ An arbitrary sample decimal value for testing
3.            bl        v_dec      @ Call subroutine to view decimal value of R0 in
                                   ASCII.
4.            mov       R0,#0      @ Exit Status code of 0 for "normal completion"
5.            mov       R7,#1      @ Service command code 1 terminates this
                                   program.
6.            svc       0          @ Issue Linux command to terminate program
7.
8. @          Subroutine v_dec will display a 32-bit register in decimal digits
9. @                    R0: contains a number to be displayed in decimal
10. @                   If negative (bit 31 = 1), then a minus sign will be output
11. @                   LR: Contains the return address
12. @                   All register contents will be preserved
13.
```

Listing 7.0a: "Main program" and v_dec calling arguments

```
14. v_dec:   push   {R0-R7}        @ Save contents of registers R0 through R7.
15.
16.          mov    R3,R0          @ R3 will hold a copy of input word to be displayed.
17.          mov    R2,#1          @ Number of characters to be displayed at a time.
18.          mov    R0,#1          @ Code for stdout (standard output, i.e., monitor)
19.          mov    R7,#4          @ Linux service command code to write string.
20.
21. @        If bit-31 is set, then register contains a negative number and "-" should be output.
22.
23.          cmp    R3,#0          @ Determine if minus sign is needed.
24.          bge    absval         @ If positive number, then just display it.
25.          ldr    R1,=msign      @ Address of minus sign in memory
26.          svc    0              @ Service call to write string to stdout device
27.          rsb    R3,R3,#0       @ Get absolute value (negative of negative) for display.
```

Listing 7.0b: Initialize registers and display minus sign if needed

```
28. absval:  cmp    R3,#10         @ Test whether only one's column is needed
29.          blt    onecol         @ Go output "final" column of display
30.
31. @        Get highest power of ten this number will use (i.e., is it greater than 10?, 100?, ...)
32.
33.          ldr    R6,=pow10+8    @ Point to hundred's column of power of ten table.
34. high10:  ldr    R5,[R6],#4     @ Load next higher power of ten
35.          cmp    R3,R5          @ Test if we've reached the highest power of ten needed
36.          bge    high10         @ Continue search for power of ten that is greater.
37.          sub    R6,#8          @ We stepped two integers too far.
38.
```

Listing 7.0c: Determine number of decimal digits to display

```
39. @        Loop through powers of 10 and output each to the standard output (stdout) monitor display.
40.
41. nxtdec:  ldr    R1,=dig-1      @ Point to 1 byte before "0123456789" string
42.          ldr    R5,[R6],#-4    @ Load next lower power of 10 (move right 1 dec column)
43.
44. @        Loop through the next base ten digit column to be displayed.
```

```
45.
46. mod10:   add     R1,#1         @ Set R1 pointing to the next higher digit '0'
                                    through '9'.
47.          subs    R3,R5         @ Do a count down to find the correct digit.
48.          bge     mod10         @ Keep subtracting current decimal column value
49.          addlt   R3,R5         @ We counted one too many (went negative)
50.          svc     0             @ Write the next digit to display
51.          cmp     R5,#10        @ Test if we've gone all the way to the one's
                                    column.
52.          bgt     nxtdec        @ If 1's column, go output rightmost digit and
                                    return.
53.
```

Listing 7.0d: Loop to display each decimal digit except one's column

```
54. @        Finish decimal display by calculating the one's digit.
55.
56. onecol:  ldr     R1,=dig          @ Pointer to "0123456789"
57.          add     R1,R3            @ Generate offset into "0123456789" for one's
                                       digit.
58.          svc     0                @ Write out the final digit.
59.
60.          pop     {R0-R7}          @ Restore saved register contents
61.          bx      LR               @ Return to the calling program
62.
63.          .data
64. pow10:   .word   1                @ 10^0
65.          .word   10               @ 10^1
66.          .word   100              @ 10^2
67.          .word   1000             @ 10^3 (thousand)
68.          .word   10000            @ 10^4
69.          .word   100000           @ 10^5
70.          .word   1000000          @ 10^6 (million)
71.          .word   10000000         @ 10^7
72.          .word   100000000        @ 10^8
73.          .word   1000000000       @ 10^9 (billion)
74.          .word   0x7FFFFFFF       @ Largest integer in 31 bits (2,147,483,647)
75. dig:     .ascii  "0123456789"     @ ASCII string of digits 0 through 9.
76. msign:   .ascii  "-"              @ needed for negative decimal numbers.
77.          .end
```

Listing 7.0e: Display one's column and the data tables

Go ahead and run this decimal-display sample program contained in Listing 7.0. Use the same command lines from the binary and hexadecimal prototypes in the previous two labs to assemble, link, and start the debugger. Try it first with R0 initialized to decimal 125, but also try other values, even some that are negative or expressed in binary or hexadecimal. In Listing 7.1, I set a breakpoint at line 3, so that I can modify R0 using the gdb debugger "set" command before calling subroutine v_dec.

```
(gdb) b 3
Breakpoint 1, _start () at model.s:3
3           bl          v_dec
(gdb) set $r0 = 6502
(gdb) c
Continuing.
6502[Inferior 1 (process 2664) exited normally]
(gdb) run
Starting program: /home/pi/model

Breakpoint 1, _start () at model.s:3
3           bl          v_dec
(gdb) set $r0 = -6501
(gdb) c
Continuing.
-6501[Inferior 1 (process 2666) exited normally]
```

Listing 7.1: Setting breakpoint and testing with decimal values of positive 6502 and negative 6501

In Listing 7.2, I set R0 to binary 101010101010 and hexadecimal ABCDEF allowing subroutine v_dec to convert them to the decimal values of 2730 and 11259375, respectively.

```
(gdb) run
Starting program: /home/pi/model

Breakpoint 1, _start () at model.s:3
3           bl          v_dec
(gdb) set $r0 = 0b101010101010
(gdb) c
Continuing.
2730[Inferior 1 (process 2762) exited normally]
(gdb) run
Starting program: /home/pi/model

Breakpoint 1, _start () at model.s:3
3           bl          v_dec
(gdb) set $r0 = 0xabcdef
(gdb) c
```

```
Continuing.
11259375[Inferior 1 (process 2765) exited normally]
(gdb)
```

Listing 7.2: Input values are binary 101010101010 and hexadecimal ABCDEF with display in decimal

Introductions

The big change in Lab 7 is the use of data tables.

	List of ARM instructions introduced in Lab 7		
7.0.27.	rsb	R3,R3,#0	@ Reverse Subtract [R3] = 0 - {R3].
7.0.34.	ldr	R5,[R6],#4	@ Load next 32-bit word from table

	List of assembler directives introduced in Lab 7
.word	Generate a 32-bit binary integer

Principles

In the previous two labs, we displayed a number in binary or hexadecimal using an algorithm of repeated division by two or sixteen, respectively. Because it's faster and especially since we do not have a divide instruction, we used shifting to perform the divisions. Of course, the same algorithm of repeated divisions would work for decimal, but we don't have a divide instruction, and we can't even shift because 10 is not an integer power of 2. For example, the way we would like to display 3274 in base 10 is the following:

1. 3274 / 10 = 327 Remainder 4
2. 327 / 10 = 32 Remainder 7
3. 32 / 10 = 3 Remainder 2
4. 3 / 10 = 0 Remainder 3

In the above technique, we get the digits from the remainders starting with the one's column and moving left. An alternate technique using repeated divisions would start from the leftmost column and move right. A decimal number is really a short notation for a polynomial of powers of 10. For example: 3274 is $3\times10^3 + 2\times10^2 + 7\times10^1 + 4\times10^0$. We can also get the decimal digits from the quotients and moving right as in the following:

1. 3274 / 1000 = 3 Remainder 274
2. 274 / 100 = 2 Remainder 74
3. 74 / 10 = 7 Remainder 4
4. 4 / 1 = 4 Remainder 0

We don't have a division instruction, but just like multiplication is really repeated additions, division is really repeated subtractions. Our algorithm in this lab will be a series of subtractions for each power of ten, and we will start from the high-order side (leftmost digit). For example, if we were going to display the number 3274, we'd begin by subtracting 1000 from 3274 until we were less than 1000. That would give us the following 3 subtractions:

- 3274 - 1000 = 2274
- 2274 - 1000 = 1274
- 1274 - 1000 = 274

We would then continue with the next column by repeatedly subtracting 100 (2 subtractions), and continue from there by repeatedly subtracting 10 (7 subtractions) as follows:

- 274 - 100 = 174
- 174 - 100 = 74
- 74 - 10 = 64
- 64 - 10 = 54
- 54 - 10 = 44
- 44 - 10 = 34
- 34 - 10 = 24
- 24 - 10 = 14
- 14 - 10 = 4

Once we're less than ten, we then have the one's column digit and certainly don't need a loop to divide by 10^0. We'll have to do the following to implement this decimal display algorithm:

1. First determine the high order column (10^3 or 1000 for example of 3274)
2. Build an outer loop to step down through powers of ten (10^3, 10^2, 10^1 for this example)
3. Build an inner loop to perform the "division" by repeated subtractions within each pass through the outer loop
4. A table of powers of ten: $10^1 = 10$, $10^2 = 100$, $10^3 = 1000$,...

Nested Loops

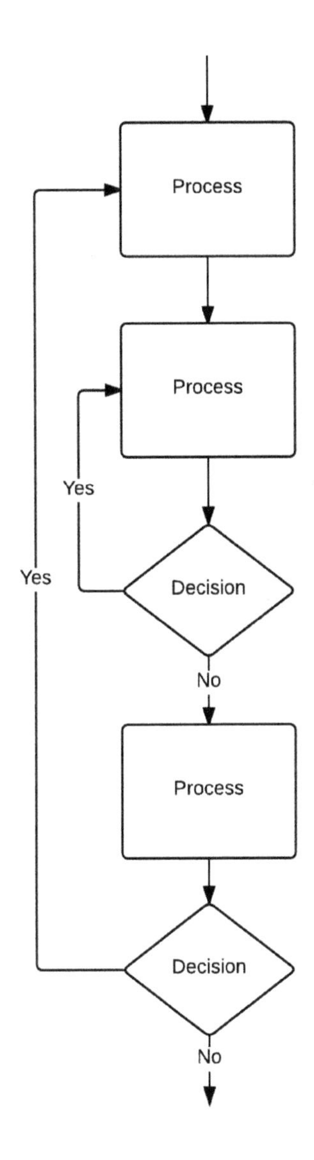

Figure 7.0: Nested loops

A very common programming technique is one loop nested within another. Each loop will have its own exit condition. In this lab, the nested loops will calculate the digits of the decimal number.

- The outer loop will step down by powers of ten. It will exit when it reaches the one's column.
- The inner loop will perform the division by a power of ten using repeated subtractions. It will exit when the result is less than the power of ten.

Although nested loops are a powerful technique, it's very easy to write large nested loops with confusing code where one loop's data and counters interfere with that of the other.

Arrays

A key component used by the v_dec algorithm is the array containing the list of powers of ten (one, ten, hundred, thousand, ..., billion). Each value in this table will be a 32-bit integer taking up 4 bytes of memory. The assembler .WORD directive will generate the binary value for each value starting from 1 and increasing to 10^9 which is the largest power of ten that will fit in 32 bits. The final value in the table is 2,147,483,647 which is the largest positive number fitting in a signed 32 bit word, and it will mark the table's end.

```
        .data
pow10:  .word   1               @ 10^0
        .word   10              @ 10^1
        .word   100             @ 10^2
        .word   1000            @ 10^3 (thousand)
        .word   10000           @ 10^4
        .word   100000          @ 10^5
        .word   1000000         @ 10^6 (million)
        .word   10000000        @ 10^7
        .word   100000000       @ 10^8
        .word   1000000000      @ 10^9 (billion)
        .word   0x7FFFFFFF      @ Largest integer in 31 bits (2,147,483,647)
```

Listing 7.3: Powers of ten table

Coding and Debugging

In the Prototype section, we set a breakpoint at line 3 enabling different test values to be sent to subroutine v_dec. At this point, you should be fairly comfortable using the debugger, so in the remainder of this lab, I will suggest locations for breakpoints, but not provide all the details. Typically once the breakpoint is reached and the program execution is suspended, you will need to use the "i r" info registers, the "s" single step, and the "c" continue commands of the debugger. Because some of these breakpoints are located in loops, you will probably want to use the gdb "d" delete command to remove a breakpoint after a couple of passes through the loop. Note: I will not be adding any enhancements to the v_dec subroutine that appeared in the Prototype section so the line numbers from Listing 7.0 are the same as those in the remainder of this lab for setting your breakpoints in the gdb debugger.

```
1.          .global   _start     @ Provide program starting address to linker
2. _start:  mov       R0,#125    @ An arbitrary sample decimal value for testing
3.          bl        v_dec      @ Call subroutine to view decimal value of R0 in
                                 ASCII.
4.          mov       R0,#0      @ Exit Status code of 0 for "normal completion"
5.          mov       R7,#1      @ Service command code 1 terminates this
                                 program.
6.          svc       0          @ Issue Linux command to terminate program
7.
8. @        Subroutine v_dec will display a 32-bit register in decimal digits
9. @                  R0: contains a number to be displayed in decimal
10. @                 If negative (bit 31 = 1), then a minus sign will be output
11. @                 LR: Contains the return address
12. @                 All register contents will be preserved
13.
```

Listing 7.4: Main Test program for v_dec with v_dec input parameters

Subroutine v_dec is divided into five parts:

1. Prepare registers for the display and loops.
2. Display a minus sign if the number is negative.
3. Find the highest power of ten needed; equivalent to determining the number of decimal places needed
4. Use a nested loop to display the ten's, hundred's, thousand's, ... columns
5. Display the one's column

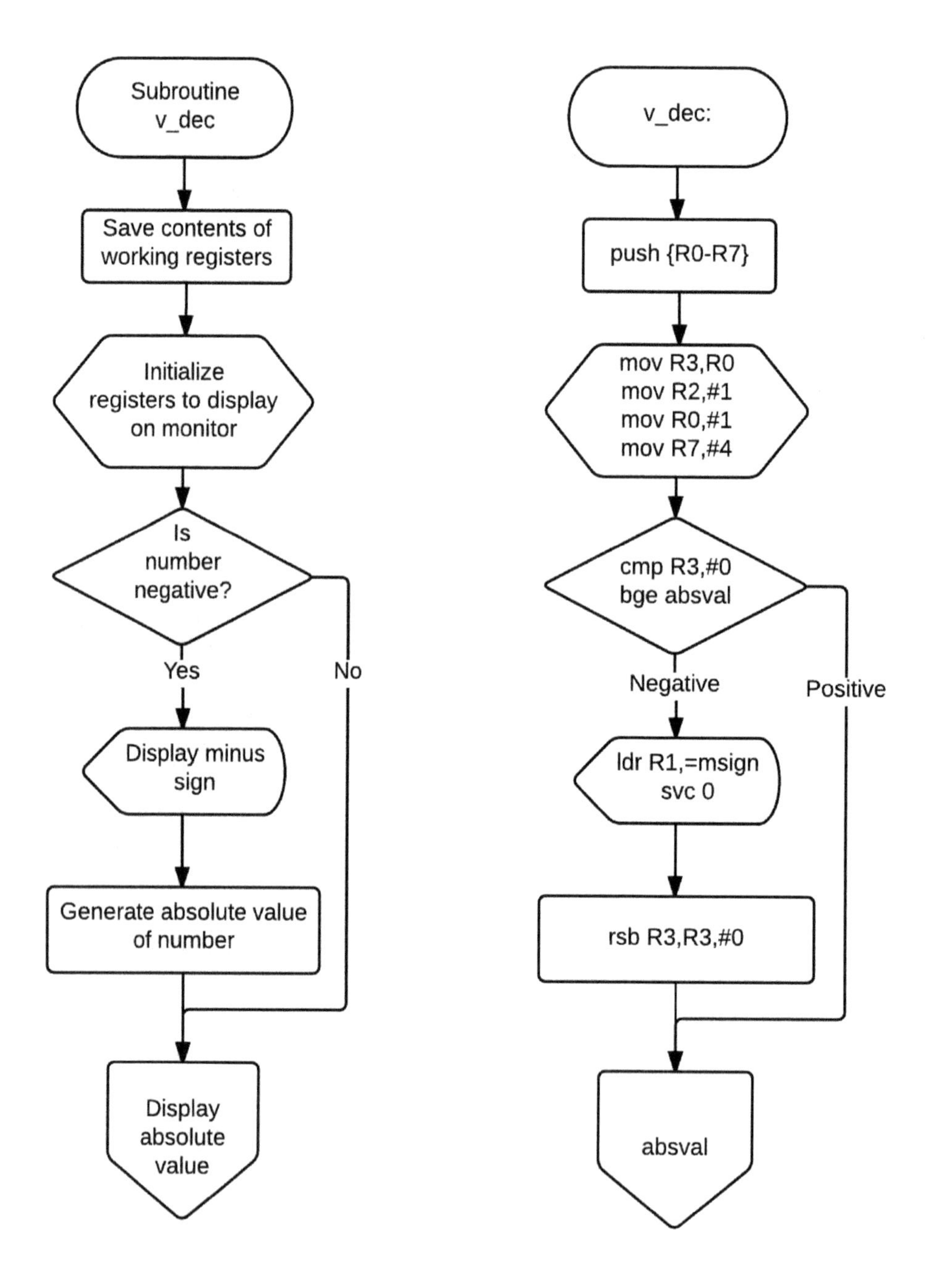

Figure 7.1: Parts 1 and 2: Initialize registers and display the minus sign

```
14. v_dec:   push   {R0-R7}    @ Save contents of registers R0 through R7.
15.
16.          mov    R3,R0      @ R3 will hold a copy of input word to be
                               displayed.
17.          mov    R2,#1      @ Number of characters to be displayed at a time.
18.          mov    R0,#1      @ Code for stdout (standard output, i.e., monitor)
```

19.	mov	R7,#4	@ Linux service command code to write string.
20.			
21. @	If bit-31 is set, then register contains a negative number and "-" should be output.		
22.			
23.	cmp	R3,#0	@ Determine if minus sign is needed.
24.	bge	absval	@ If positive number, then just display it.
25.	ldr	R1,=msign	@ Address of minus sign in memory
26.	svc	0	@ Service call to write string to stdout device
27.	rsb	R3,R3,#0	@ Get absolute value (negative of negative) for display.

Listing 7.5: Initialize registers and display minus sign if negative

Parts 1 and 2 are illustrated in Figure 7.1 and Listing 7.5. A breakpoint suggestion is "b 27": Examine the absolute value of a negative number using the reverse subtraction instruction.

Powers of Ten Table

Part 3 of v_dec involves finding the highest power of ten which is equivalent to determining the number of decimal places needed. For values less than 10, we will just index into the decimal digits table because no "divisions" will be needed.

The assembler generates one 32-bit binary integer for each .WORD directive in the pow10 table. We will search the table until a power of ten is reached that is greater than the value of the number to be displayed, and then we will drop back one level. The search will begin at 100 (R6 pointing to "pow10+8").

63.	.data		
64. pow10:	.word	1	@ 10^0
65.	.word	10	@ 10^1
66.	.word	100	@ 10^2
67.	.word	1000	@ 10^3 (thousand)
68.	.word	10000	@ 10^4
69.	.word	100000	@ 10^5
70.	.word	1000000	@ 10^6 (million)
71.	.word	10000000	@ 10^7
72.	.word	100000000	@ 10^8
73.	.word	1000000000	@ 10^9 (billion)
74.	.word	0x7FFFFFFF	@ Largest integer in 31 bits (2,147,483,647)

Listing 7.6: Powers of Ten Table

LDR R5,[R6],#4

On each pass through the high10 loop, R6 will be pointing to the next higher power of ten. How is that done? There is no explicit "ADD R6,#4" instruction in the loop. The ARM processor, like almost every microcomputer, minicomputer, and mainframe CPU manufactured in the past 50 years, supports an auto-increment feature in its memory load and store instructions for stepping through arrays. The format of the LDR instruction on line 34 is in what's called "post indexed addressing" mode in the ARM, where the register's contents is incremented or decremented after the data word has been loaded. I'll have a lot more to say about memory addressing in the Lab 8.

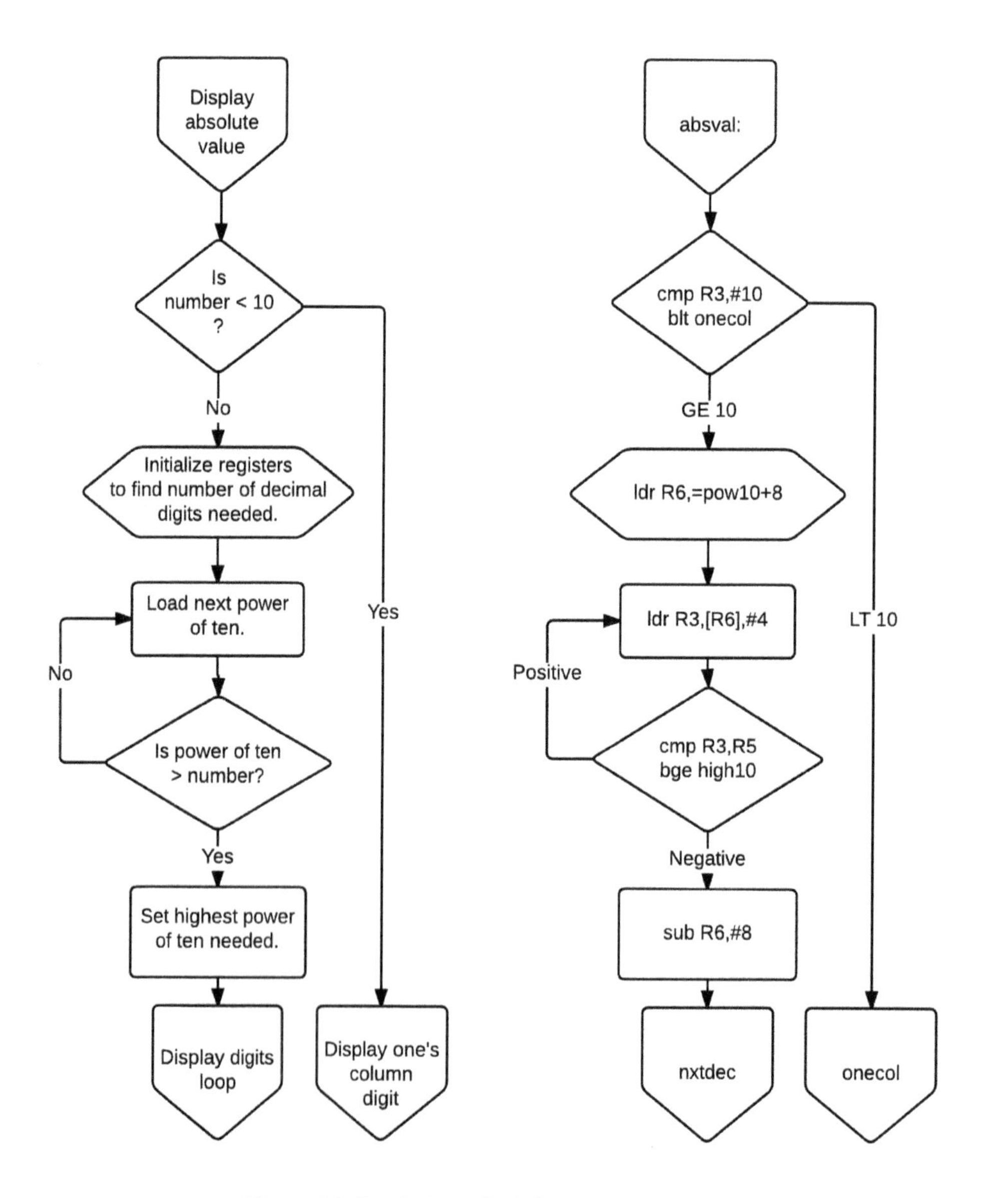

Figure 7.2: Part3: Hunt for leftmost decimal digit

28. absval:	cmp	R3,#10	@ Test whether only one's column is needed
29.	blt	onecol	@ Go output "final" column of display
30.			
31. @	Get highest power of ten this number will use (i.e., is it greater than 10?, 100?, ...)		
32.			
33.	ldr	R6,=pow10+8	@ Point to hundred's column in power table.
34. high10:	ldr	R5,[R6],#4	@ Load next higher power of ten
35.	cmp	R3,R5	@ Test if we've reached the highest power of ten needed
36.	bge	high10	@ Continue search for power that is greater.
37.	sub	R6,#8	@ We stepped two integers too far.
38.			

Listing 7.7: Set R6 pointing to highest power of ten for number being displayed

Breakpoint suggestion (b 35): Examine power of ten loaded into R5 on each pass of the loop. Also note how register R6 is automatically incremented by 4 during the execution of the "LDR R5,[R6],#4" instruction.

R3	Integer being displayed (absolute value)	Absolute value of original R0
R5	Current power of ten to be compared to contents of R3	A value from 100 to 1 billion
R6	Points into power of 10 table	Points to next value in 1 to 1 billion pow10 table

Table 7.1: Registers used in high10 loop in Listing 7.10

Nested Loops Code Example

Part 4 of v_dec uses a nested loop to display all decimal digits except the one's column.

- The outer loop: In the previous loop in part 3, we marched up through the powers of ten table searching for the high order column using the "LDR R5, [R6],#4" instruction. In this outer loop, we will reverse that direction and march back down to the ten's column using the "LDR R5,[R6],–#4" instruction.
- The inner loop: We will perform the division by a power of ten using repeated subtractions. It will exit when the result is less than the power of ten, and the number of times through this inner loop will be the quotient needed to provide the decimal digit to be displayed.

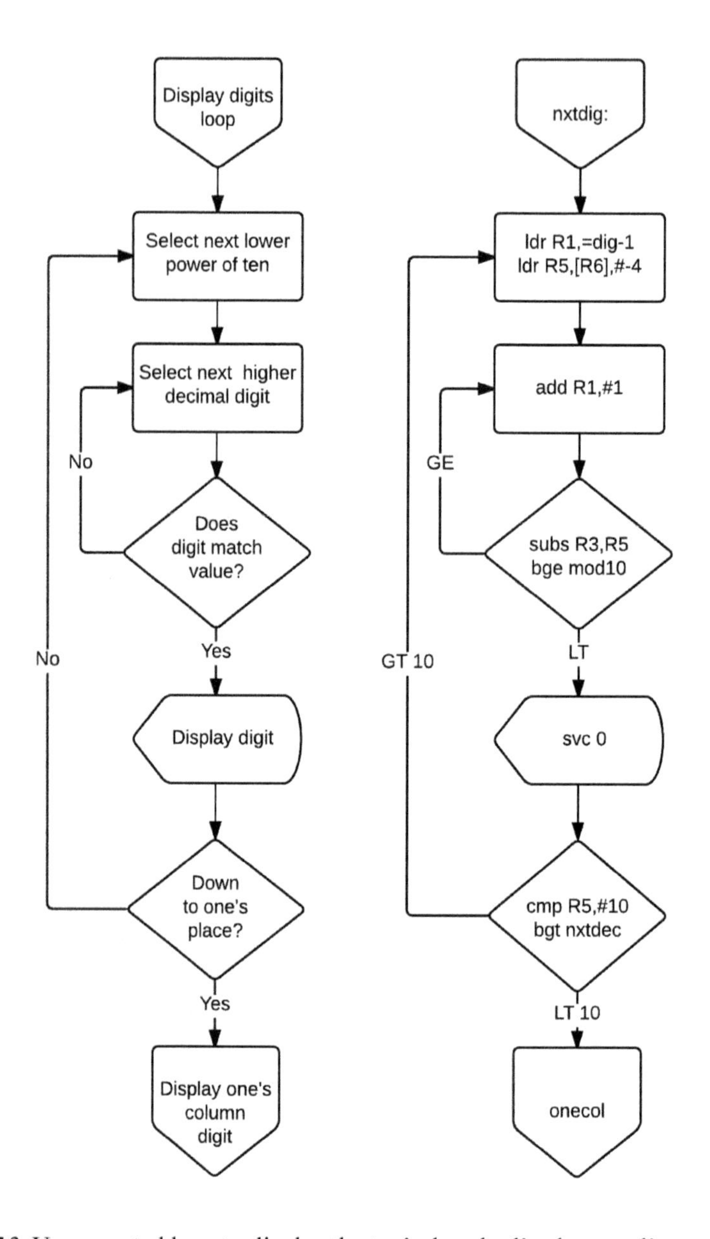

Figure 7.3: Use a nested loop to display the ten's, hundred's, thousand's, ... columns

```
39. @        Loop through powers of 10 and output each to the standard output
             (stdout) monitor display.
40.
41. nxtdec:  ldr   R1,=dig-1      @ Point to 1 byte before "0123456789" string
42.          ldr   R5,[R6],#-4    @ Load next lower power of 10 (move right 1 dec
                                  column)
43.
44. @        Loop through the next base ten digit to be displayed (i.e., thousands,
             hundreds, ...)
45.
```

```
46. mod10:  add    R1,#1     @ Set R1 pointing to the next higher digit '0'
                             through '9'.
47.         subs   R3,R5     @ Do a count down to find the correct digit.
48.         bge    mod10     @ Keep subtracting current decimal column value
49.         addlt  R3,R5     @ We counted one too many (went negative)
50.         svc    0         @ Write the next digit to display
51.         cmp    R5,#10    @ Test if we've gone all the way to the one's
                             column.
52.         bgt    nxtdec    @ If 1's column, go output rightmost digit and
                             return.
53.
```

Listing 7.8: Nested loop to display all decimal digits except one's column

Breakpoint suggestion (b 50): Examine each decimal digit being displayed.

R0	Output to stdout display	Always = 1
R1	Points to ASCII character to output	Either points to "0","1","2",..,"9"
R2	Number of characters to output	Always = 1
R3	Integer being displayed	Copy of original R0
R5	Points to string: "0123456789"	Always the same
R6	Points to power of 10 in table	Counts down to 10
R7	Linux service command to write file/device	Always = 4

Table 7.2: Registers used in nxtdec loop

One's Column

The fifth and final part of the v_dec program execution is the display of the one's column. No divisions are needed. We simply index into the table of ASCII digits just like in the binary and hexadecimal subroutines in the previous two labs.

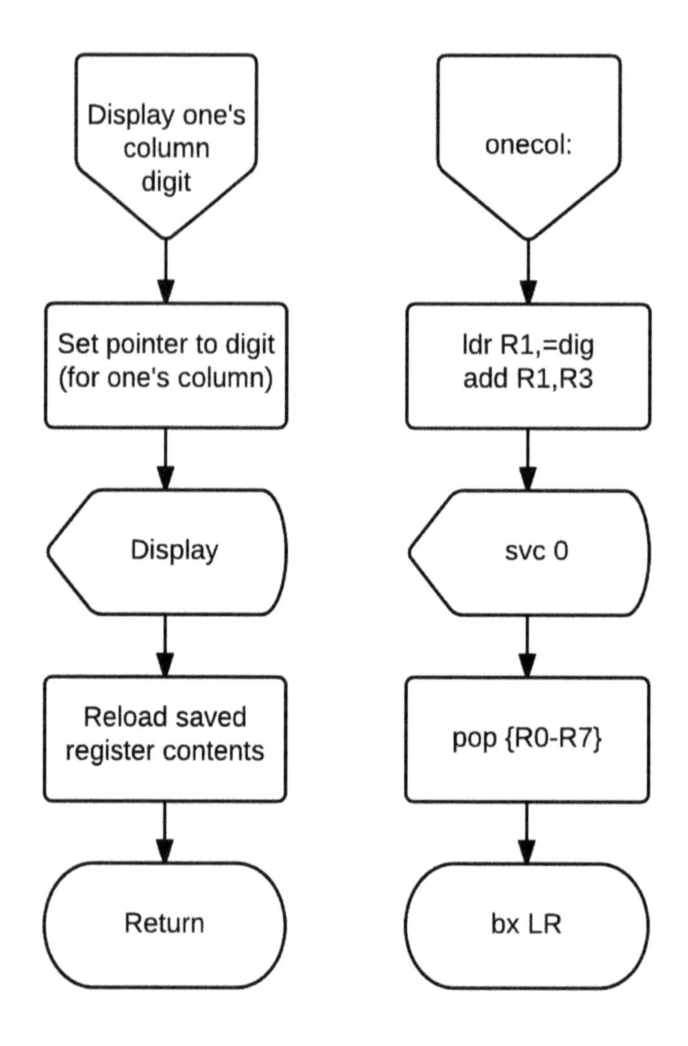

Figure 7.4: Display one's digit and return

```
54. @           Finish decimal display by calculating the one's digit.
55.
56. onecol:     ldr     R1,=dig      @ Pointer to "0123456789"
57.             add     R1,R3        @ Generate offset into "0123456789" for one's digit.
58.             svc     0            @ Write out the final digit.
59.
60.             pop     {R0-R7}      @ Restore saved register contents
61.             bx      LR           @ Return to the calling program
62.
```

Listing 7.9: Just index into "0123456789" for one's digit

Maintenance

Review Questions

1. What's a possible reason why the ARM doesn't contain a divide instruction?
2. The Raspberry Pi does have a divide, but it's not in the ARM CPU. Where is it?
3. Why is binary to decimal conversion not as easy as binary to hexadecimal?
4. * In Listing 7.5, the CMP R3,#0 instruction on line 23 can be eliminated if line 16 is modified. What modification is this and why is it probably a bad idea even though it would function perfectly?

Programming Exercises

1. Modify v_dec to output negative numbers enclosed in parenthesis. Instead of -123, output (123) which is the way many accountants prefer it.
2. Modify v_dec to make a subroutine v_hep which outputs in base 7. Hint: The pow10 table will have to be rebuilt using powers of 7.
3. Modify v_dec to make a subroutine v_base which will output in any base between 2 (binary) and 16 (hexadecimal). The particular base will be input to subroutine v_base through register R1. Hint: The pow10 table will have to be built "on the fly" by multiplying the contents of register R1 by itself. You will need an instruction similar to "STR R5,[R4],#4" to write the table in memory. The "dig" array will also have to be expanded to '0' though 'F.'
4. Modify v_dec to output leading blanks so that the number being displayed will be "right justified." For example, if we always want 12 "columns" of characters output, then the number 328 would be output as 9 blanks followed by '3', '2', and '8'.; -1003 would be output as 7 blanks follower by '-', '1', '0', '0', and '3'. The total number of columns could be input to v_dec in register R2.
5. Assume there is an implied decimal point. For example, if the number is in meters and the output should be displayed in kilometers, insert a decimal point in the display (i.e., display 3.270 rather than 3270). If the number is currency, display $3.20 instead of 320. The number of decimal places could be input to v_dec in register R2.
6. Modify v_dec to display a comma every three decimal places (i.e., display 10,327,101 instead of 10327101).

Lab 8
Data Types and Storage

The focus of the preceding labs has been more on program structure than data storage structure. In Lab 8, we will lean more toward the details of data formats and the instructions that move data between registers and memory. The warm-up exercises in the Prototype section will introduce the "string" data type as well as linking multiple independent subroutines. Block transfer instructions, indexed addressing modes, sign bit extension, big and little endian formats, and the beginnings of a disassembler are also covered in this lab.

Prototype

The warmup exercises in this Prototype section build a couple of programs having multiple subroutines that must be linked together, including a new one named v_ascz, which displays "string" format data.

In computer programming languages, there are various "types" of variables: integers, real numbers, characters, and strings. Integers are stored in words, double words, half words, and bytes. Real numbers are stored in words and double words. Characters are typically stored in bytes, but can also be stored in words or half words. Strings are stored in byte arrays, and their lengths vary. It has become a very common practice to terminate a character string with a null byte (binary zero). Subroutine v_ascz will be similar to v_asc from Lab 3 except it will not need argument R2 containing the length of the string. Subroutine v_ascz will display all ASCII characters from the first one pointed to by register R1 until a null byte is found.

```
1.          .global    v_ascz         @ External entry location for subroutine v_ascz
2.
3. @        Subroutine v_ascz will display a string of characters
4. @                   R1: Points to beginning of ASCII string
5. @                   End of string will be marked by a null byte
6. @                   LR: Contains the return address
7. @                   All register contents will be preserved
8.
9. v_ascz:  push       {R0-R8,LR}     @ Save contents of registers R0 through R8, LR
10.         sub        R2,R1,#1       @ R2 will be index while searching string for null.
11. hunt4z: ldrb       R0,[R2,#1]!    @ Load next character from string (and increment
                                      R2 by 1)
12.         cmp        R0,#0          @ Set Z status bit if null found
13.         bne        hunt4z         @ If not null, go examine next character.
14.         subs       R2,R1          @ Get number of bytes in message (not counting
                                      null)
15.         mov        R0,#1          @ Code for stdout (standard output, i.e.,
                                      monitor)
16.         mov        R7,#4          @ Linux service command code to write string.
17.         svcne      0              @ Issue command to display string on stdout
18.
19.         pop        {R0-R8,LR}     @ Restore saved register contents
20.         bx         LR             @ Return to the calling program
21.         .end
```

Listing 8.0: Subroutine v_ascz will output string terminated with a null and is in source file v_ascz.s.

Since subroutine v_ascz is not provided the length of the string, it must now calculate it. The length of the string is found by searching for a null byte using the loop beginning on line 11 of Listing 8.0. There are many ways subroutine v_ascz could have been programmed. Instead of finding the length and then using one service call to Linux, we could have output each character one at a time in the hunt4z loop. It would not have been as efficient due to the multiple Linux service calls, but the v_ascz subroutine as it is presented is not as efficient as it could have been either. Actually, the calling program does not need to know how v_ascz is programmed, just what its specifications say it will do and its calling arguments.

There are a couple of other changes in subroutine v_ascz compared to the original v_asc:

1. The .global assembler directive is needed to make the v_ascz label external so that the linker will be able to connect the subroutine call in the main program to the subroutine's starting location.
2. I'm saving a couple more registers onto the stack than I'm using at this time in the subroutine. Basically, I'm setting it up for some more flexibility that can be added to all of the display subroutines at a later time.

```
1.          .global    _start          @ Provide program starting address to linker
2._start:   ldr        R1,=msgtxt      @ Set R1 pointing to message to be displayed
3.          bl         v_ascz          @ Call subroutine to view text string in ASCII.
4.          mov        R0,#0           @ Exit Status code 0 for "normal completion
5.          mov        R7,#1           @ Service command 1 terminates the program
6.          svc        0               @ Issue Linux command to terminate program
7.
8.          .data
9. msgtxt:  .asciz     "Hey there\n"   @ Message terminated by a null (all 8 bits = 0).
10.         .end
```

Listing 8.1: Source code to test subroutine v_ascz is in file model.s.

The main program in Listing 8.1 is similar to that of Listing 3.0 which introduced the ASCII display subroutine v_asc, except for the following:

1. Register R2 does not need to be initialized with the length of the string for subroutine v_ascz.
2. The text string to be displayed is built using the .asciz assembler directive that puts a null byte after the string of text characters.
3. The main program that calls subroutine v_ascz remains in the model.s source file while the actual subroutine code has been moved to its own separate file v_ascz.s.

Since we have broken the source code into two files instead of one, we now have a new work flow provided in Listing 8.2. Both source files must be edited and compiled, and now the linker is really doing its job of "linking." The "ld" command line now contains the name of the executable file followed by each of the object files to link, all separated by one or more blanks. The linker step will be needed for each program change, but only the source code file that is changed will need to be recompiled (i.e., reassembled). We will still be naming the executable file "model" even though it's built from both the model.o and v_ascz.o object files.

```
~$ nano model.s
~$ as -o model.o model.s
~$ nano v_ascz.s
~$ as -o v_ascz.o v_ascz.s
~$ ld -o model model.o v_ascz.o
~$ ./model
```

Listing 8.2: Sequence of edit, compile, link, and execute

Go ahead and run the program and try different strings of different lengths on line 9 of the main program. If you decide to use the gdb debugger, you will need the -g option on the assembler command lines. An interesting breakpoint location is line 11 of v_ascz.

A second sample program that links multiple subroutines is shown in Listing 8.3. Here, the main program calls three subroutines to display the contents of register R0 in three different formats: decimal, hexadecimal, and binary.

1.	.global	_start	@ Provide program starting address to linker
2. _start:	ldr	R1,=msgtxt	@ Document what is being displayed on monitor
3.	bl	v_ascz	@ Display text line on monitor
4.	ldr	R1,=blank	@ Set pointer to "blank string" used as number separator.
5.	mov	R0,#123	@ Use decimal 123 for test example (could be anything)
6.	bl	v_dec	@ Call subroutine to view decimal value of R0 in ASCII.
7.	bl	v_ascz	@ Put a blank after the decimal number output to monitor.
8.	mov	R2,#0	@ Tell subroutine v_hex to not display leading zeroes.
9.	bl	v_hex	@ Call subroutine to view hexadecimal value of R0 in ASCII.
10.	bl	v_ascz	@ Put a blank after the hexadecimal number output to monitor.
11.	mov	R2,#8	@ Number of bits desired for binary display
12.	bl	v_bin	@ Call subroutine to view binary value of R0 in ASCII.
13.	ldr	R1,=newln	@ Set pointer to "new line" character string.
14.	bl	v_ascz	@ Put a carriage return / line feed after binary display.
15.			
16.	mov	R0,#0	@ Exit Status code set to 0 indicates "normal completion
17.	mov	R7,#1	@ Service command code 1 terminates this program.
18.	svc	0	@ Issue Linux command to terminate program
19.			
20.	.data		
21. msgtxt:	.asciz	"Decimal, hexadecimal, and binary: " @ Label for output line to monitor	
22. blank:	.asciz	" "	@ Separator for display (blank)
23. newln:	.asciz	"\n"	@ End of line (It will be 0x0A)
24.	.end		

Listing 8.3: Display register contents in multiple formats

Each subroutine to be called should be separately compiled as its own source file resulting in its own object file, as was done for subroutine v_ascz. Each should also contain a .global statement exposing its entry point for the linker.

Listing 8.4 shows the beginning of file v_bin.s containing subroutine v_bin. Similar modifications are required for subroutines v_hex and v_dec.

- Listing 5.8: v_bin subroutine moved to text file v_bin.s
- Listing 6.0: v_hex subroutine moved to text file v_hex.s
- Listing 7.0: v_dec subroutine moved to text file v_dec.s

```
1.          .global     v_bin       @ Provide subroutine entry point to linker
2.
3. @        Subroutine v_bin will display a 32-bit register in binary digits
4. @                    R0: contains the 32 bit value to be displayed in binary
5. @                    LR: Contains the return address
6. @                    Registers R0 through R7 will be used by v_bin and not saved
7.
8. v_bin:   mov         R3,R0       @ R3 will hold a copy of input word.
```

Listing 8.4: First few lines of subroutine v_bin with .global statement

As was pointed out in Lab 1 and Appendix A, all of the source code for this book is available in directory RPi_Asm_Bridge, which can be downloaded from the GitHub Internet site. Listing 8.5 contains the work flow to generate the new test program by copying the listing files, editing each where necessary to add the .global directive, compiling each separately, and finally linking everything together. If any changes are later performed, only the source file that is changed needs to be reassembled. The linker will use previously compiled object files that did not have to be altered.

```
~$ cp RPi_Asm_Bridge/Listing_8_3.txt model.s
~$ as -o model.o model.s
~$ cp RPi_Asm_Bridge/Listing_8_0.txt v_ascz.s
~$ as -o v_ascz.o v_ascz.s
~$ cp RPi_Asm_Bridge/Listing_5_8.txt v_bin.s
~$ nano v_bin.s
~$ as -o v_bin.o v_bin.s
~$ cp RPi_Asm_Bridge/Listing_6_0.txt v_hex.s
~$ nano v_hex.s
~$ as -o v_hex.o v_hex.s
~$ cp RPi_Asm_Bridge/Listing_7_0.txt v_dec.s
~$ nano v_dec.s
~$ as -o v_dec.o v_dec.s
~$ ld -o model model.o v_ascz.o v_bin.o v_hex.o v_dec.o
~$ ./model
```

Listing 8.5: Linking main program with multiple subroutines

Go ahead and run the program and try different values on line 5 of the main program. If you decide to use the gdb debugger, you will need the -g option on the assembler command lines. Note: The linker command line will be simplified in Lab 9 when we build a "library" containing all of the display subroutines.

Introductions

The Prototype section introduced a slight variation to the LDR instruction for moving one byte from memory into a register. It also used an .asciz directive for generating an ASCII string terminated with a null byte. In the Principles section, we will examine in detail the four groups of instructions that move data between registers and memory: LDR and STR for moving a single byte or word and LDM and STM for moving up to sixteen registers at a time.

	List of ARM instructions introduced in Lab 8		
8.0.11.	ldrb	R0,[R2,#1]!	@ Load next character from string (and increment R2 by 1)
8.10.1.	stmeqdb	R8,{R0-R7}	@ Store R0-R7 only if Z-flag is set
8.10.2.	ldmgtdb	R8,{R0-R7}	@ Load R0-R7 only if !Z and (N = V)
8.12.17.	tst	R3,#1<<20	@ Test if this is a load or store instruction.
8.15.45.	bleq	v_ascz	@ Output "]" to close post-indexed format.

	List of assembler directives introduced in Lab 8
.align	Set memory address to half-word, word, double-word, ... border
.asciz	Put string of ASCII characters into memory with null at end

Principles

What do we know about memory from the preceding labs?

- Arrays can store lists and tables of data as either words or bytes.
- Load and store instructions can efficiently step through arrays using an auto-increment or auto-decrement option.
- The "stack" is a place for temporary storage.
- The program instructions area and data area can be located together, or they can be separated (.TEXT and .DATA directives).

Harvard and Von Neumann

The early computer designs had the instructions located in a physical medium that was independent of the data storage. Babbage's analytical engine had the instructions on punched cards modeled after the Jacquard loom of the 1800s. Some computers in the 1940s had their programs entered and stored in plug boards. This design, having the instructions and data in separate media or at least on separate I/O buses or data channels, is referred to as a Harvard architecture.

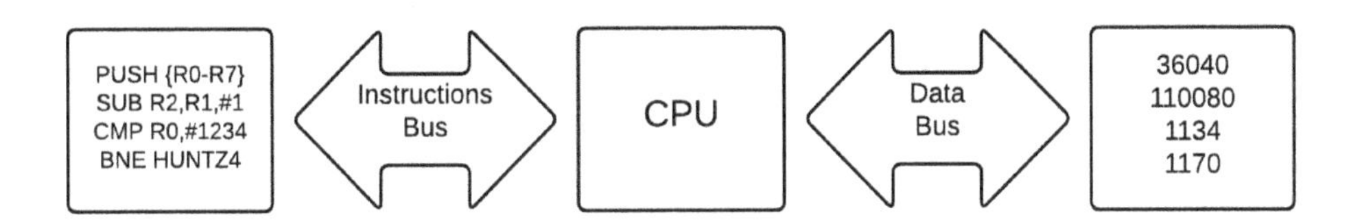

Figure 8.0: Harvard architecture has independent instructions bus and data bus.

The Harvard architecture is still very popular today, but mostly with micro-controllers used in embedded systems, rather than microprocessors used for general computing. A micro-controller is a small System On a Chip (SOC) containing a CPU, some memory, and specialized I/O data lines. The term microprocessor is usually reserved for integrated circuit chips that contain only the CPU, and it usually has only digital I/O lines. Some advantages of the Harvard architecture follow:

- Improved performance because instructions and data can be fetched from the two separate buses at exactly the same time.
- The instruction format does not depend on the byte and word sizes.
- Programs cannot be accidentally written over as if they were data.

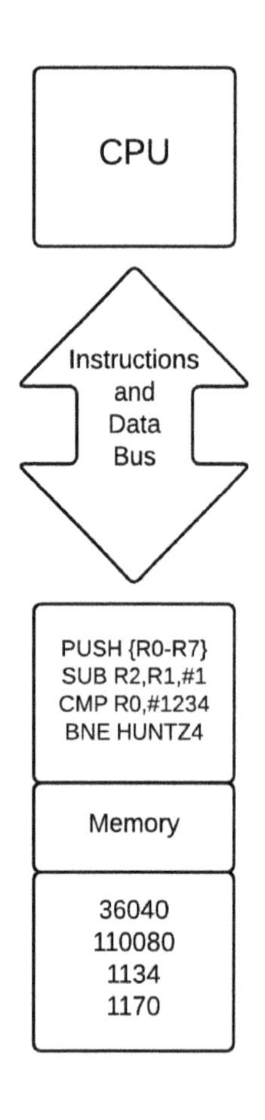

Figure 8.1: Von Neumann architecture

A different approach is attributed to John Von Neumann in which the instructions and data are in the same memory or at least on the same memory bus. Most general purpose computers, including the mainframes from the 1960s, minicomputers from the 1970s, and microcomputers from the 1980s have the Von Neumann architecture. Some advantages of the Von Neumann design follow:

- Because a program is simply data stored on the computer's disk, this architecture enables a large variety of programs to be selected to be run.
- Multiple independent applications can be run on the same computer at the same time, and an operating system can easily share the computing resources among these applications.
- Only one memory bus is needed, thereby reducing costs.
- Programs can be developed on the same machine in which they will eventually run.

Like almost all microprocessors, the ARM 7 which is in the Raspberry Pi, has a Von Neumann architecture. The ARM 9 is unique among the family of ARM processors in that it has a separate address bus and data bus, and it therefore falls within the general characteristics of a Harvard architecture.

Addressing Memory

The addressing of computer memory has been a nightmare for decades. Unpredictable cost has been the main culprit, but ironically, it has also been a good problem. Over the past 50 years, the price per byte of memory has dropped by a factor of over one billion at the same time memory speed has increased by a factor of over one thousand. To say that this price decrease and performance increase is astonishing would be an understatement. Flabbergasted is probably a better word to describe it. The physical size has also diminished: The memory that once would have filled a cabinet the size of a small kitchen can now get lost in a shirt pocket.

Some computer applications are simple and don't require much memory, but then there are other applications that require much more memory than has been available. These large applications have employed complicated memory paging, segmenting, and overlay schemes in order to run. Only parts of the program could be in memory at a time, and the application or a sophisticated hardware paging scheme would continuously move parts of the program between disk files and memory.

How do these great changes in memory price lead to computer programmers' nightmares? Large address spaces generally require a lot of bits in the instruction word, and CPU designers were reluctant to address much more memory than what seemed practical. In every CPU design, at least some of the instructions have to access memory. When the amount of available memory was small because it was extremely expensive, a reasonable maximum memory size was 65,536 (65K or 64K if you say K = 1024). This memory size requires a 16-bit address. Many mainframes, minicomputers, and microcomputers had a 65K addressing limit built into their instruction formats. As memory prices dropped much faster than CPU designs evolved, computer hardware systems designers and systems programers deployed a variety of memory management techniques to shoehorn in much larger memories into a restricted addressing space design. Fortunately those days seem to be behind us with the ARM 7 in the Raspberry Pi, and we now have a 32-bit address space which is 65K times greater than those earlier 65K limits.

LDR R1,[PC, #56]

A RISC architecture has only a few instructions that access memory while most of its instructions work on data that is already in registers. Since the ARM has a 32-bit address space, how does it get a single 32-bit instruction to load a 32-bit address into a register? Let's use the debugger to tell us what code the assembler actually generates for line 2 in Listing 8.1 to load the 32-bit address of the ASCII message. Set a breakpoint at line 3, then run the program and dump the instruction word at address 0x8074. This address comes from the PC register contents from the "i r" command (minus 4 bytes to get the memory address of the preceding instruction on line 2). Your value may be different from 0x8074 due to your particular configuration and version of Linux, so be sure to check the contents of the PC.

```
(gdb) b 3
Breakpoint 1, _start () at model.s:3
(gdb) run
Breakpoint 1, _start () at model.s:3
3          bl          v_ascz          @ Call subroutine to view text string in
                                       ASCII.
```

```
(gdb) i r
r1              0x100b8         65720
sp              0x7efff860      0x7efff860
pc              0x8078          0x8078 <_start+4>
cpsr            0x10            16
(gdb) x/wi 0x8074
0x8074 <_start>: ldr            r1, [pc, #56] ;    0x80b4  <hunt4z+36>
(gdb) x/w 0x80b4
0x80b4 <hunt4z+36>:     0x000100b8
gdb) x/s 0x100b8
0x100b8 <msgtxt>:       "Hey there\n"
```

Listing 8.6: The assembler generates "LDR R1,=msgtxt" using a modified indirect addressing

The x/wi memory dump in Listing 8.6 tells us that the assembler really loaded R1 with the data stored at [PC] + 56. On the same line, the debugger also tells us this is address 0x80b4. Dumping address 0x80b4 gives us address 0x100b8, and dumping address 0x100b8 as a string shows the message to be displayed. You can see why this is referred to as "indirect addressing." The assembler generated an extra "hidden" word for us in order to load the 32-bit address. Where is this word? If you do a further memory dump (x/wi), you will find it is placed immediately after the last instruction in the program (line 29, BX LR).

If you do a little subtraction, 0x80b4 - 0x8074 = 0x40 which is decimal 64, which is not 56 as provided on the same line. Why the discrepancy? Indexing off of the PC register can be a little tricky, and I recommend you don't do it unless you're comfortable with the CPU on which your program will be running. The difference of 8 in this example is due to pipelining where the ARM CPU is already fetching future instructions from memory before the current one has been completed.

By the way: If we had not placed the ASCII data string in another section of memory using the .DATA directive, the assembler would not have needed that extra hidden word. The offset from the PC contents would have been sufficient. In many embedded systems that contain the program in read-only memory (ROM) and the data in a separate read/write memory, we would have included the ASCII data in the program .TEXT space.

Arrays, Tables, Vectors, Matrices

An array is an ordered list of adjacent storage locations in memory. It could be a list of bytes, words, or even a more complicated combination of bytes and words. Tables, vectors, and matrices are other names commonly associated with arrays, and many times only differ by the number of dimensions (number rows, number of columns, etc.). A register is typically used to "index" into an array as well as sequentially step through each value from the beginning to the end.

```
dig:      .ascii   "0123456789"    @ ASCII string of digits 0 through 9
          .ascii   "ABCDEF"        @ ASCII string of digits A through F
pow10:    .word    1               @ 10^0
          .word    10              @ 10^1
          .word    100, 1000       @ 10^2 and 10^3
```

Listing 8.7: Examples of character and word arrays

Base Registers

On most computer systems, instructions that load or store memory generate the absolute memory address from the addition of the contents of a base register plus some offset contained within the instruction. In the ARM architecture, any of the registers R0 through R15 can be used as a base register in an instruction. Remember that R13, R14, and R15 are also used as the stack pointer SP, link register LR, and program counter PC, respectively. Why not just have the entire absolute address simply inside the instruction?

- The instructions would be too big: All ARM instructions are 32 bits in size. To "build in" the 32-bit address would require making all instructions 64 bits or have instructions of various sizes which is typical of some CISC architectures, but not RISC architectures. Yes, I know that most versions of the ARM CPU architecture have a Thumb code mode allowing a 16-bit abridged version of some of the 32-bit instructions (see Lab 17 and Appendix N).
- Index into arrays: A large portion of both business and engineering computer applications works with arrays, tables, vectors, and matrices, and these are best accessed using registers anyway.
- Multiprogramming: The ability of an operating system to have multiple application programs running simultaneously and sharing resources (CPU time, memory, I/O devices) is referred to as multiprogramming. An application program developer does not know where in memory the program will be loaded and run because the operating system makes those relocation choices when it fits as many applications into memory as needed. When a program is loaded into memory for execution, the appropriate base registers, stack pointer, and PC are initialized depending on where the program and data are loaded.

Indexed Addressing

There are three addressing modes in the ARM for moving data between a single register and a memory location:

- Indirect: The base register contains the complete memory address. Example: LDR R0,[R4].
- Pre-Indexed: The memory address is calculated by adding the contents of a base register with an immediate constant. Example: LDR R0,[R4,#56].
- Post-Indexed: The base register contains the complete memory address, but then the base register contents will be updated after the address is used in the current instruction. Example: LDR R0,[R4],#56.

We've already been using all three modes in the lab examples. It's easy to remember which assembler syntax distinguishes "pre" from "post" because it is similar to common algebraic expressions. The brackets enclose what is to be done first. In A×[B+56], we add B and 56 before (pre) using it to multiply by A. While in [A×B]+56, we first multiply A and B, and next we add 56 (post).

LDRB R0,[R2,#1]!

On line 11 in Listing 8.0, pre-indexed mode is used to load the next byte from the string into the lower 8 bits of register R0. The upper 24 bits of R0 will be set to zero. The [R2,#1] expression indicates pre-indexed format where the address of the byte to be loaded is determined by the sum of the contents of base register R2 and immediate value 1. Let's examine Figure 8.2 which shows the machine code instruction that is generated by the assembler. How large of an immediate offset can be added to or subtracted from a base register in pre-indexed format?

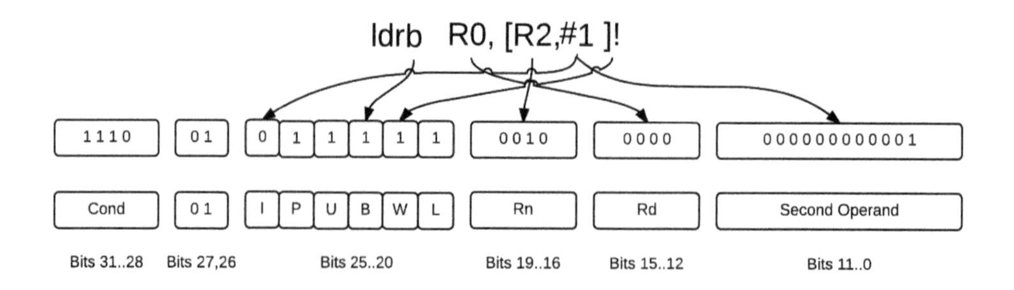

Figure 8.2: Machine code generated for pre-indexed mode load byte instruction

As seen in Figure 8.2, twelve bits are available in the second operand which provide a maximum immediate offset of 4095 bytes. The field definitions of the LDR/STR machine code instructions depicted in figures 8.2 and 8.3 are provided below:

- Cond (bits 31..28): Status indicating whether instruction should be executed (1110 indicates always execute).
- 0 1 (bits 27,26): Opcode indicating that this is an LDR/STR instruction.
- IPUBWL (bits 25..20): Bits indicating specifically what is to be done by LDR/STR instruction.
    - I (bit 25): "0" indicates offset is immediate (#) rather than register with shift.
    - P (bit 24): "1" indicates pre-indexed, "0" indicates post-indexed.
    - U (bit 23): "1" indicates up (add the offset), "0" indicates subtract the offset.
    - B (bit 22): "1" indicates byte (LDRB), "0" indicates word (LDR).
    - W (bit 21): "1" indicates write back (!), i.e. update the base register.
    - L (bit 20): "1" indicates load (LDR), "0" indicates store (STR).

- Rn (bits 19..16): Register to contain result, R0 through R15
- Rd (bits 15..12): Base register, R0 through R15
- Offset (bits 11..0): Immediate range of +4095 to -4095. This can also be in a register with a shift.

Please recall that R13, R14, and R15 are the stack pointer SP, link register LR, and program counter PC, respectively. Although using any of these three registers as a base register or result register can provide some clever results, use them carefully. The exclamation point (!) indicates that the address generated by the sum of the base register and immediate offset is to be written back into the base register. This is usually done for stepping through arrays.

As you recall from the format of the ALU instructions, the second operand doesn't have to be an immediate constant, but instead can be a register with a shift. For example, the instruction "LDR R1,[R2,R3 LSR #2]" loads R1 from a word in an array whose beginning is pointed to by R2, and its "word position" is in R3. Since the ARM is byte-addressable, the word address must be multiplied by four, which is the same as shifting left two bit positions (LSR #2).

Figure 8.3 illustrates the machine code instruction generated for "LDR R5,[R6],#4" which is the post-index instruction we used in a previous lab for stepping through a table of words.

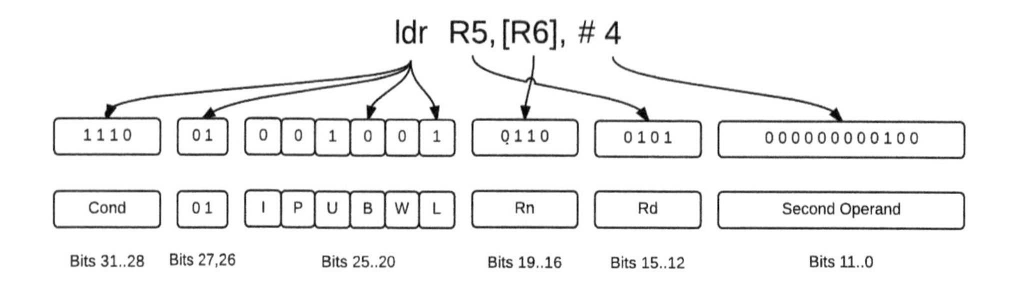

Figure 8.3: Machine code generated for post-indexed mode load word instruction.

In post-indexed mode, the write-back option (!) to update the base register is not an option. The exclamation symbol is not allowed since it is always implied. If an offset is present, such as either an immediate value or a register with a shift, it will be used to update the base register. It really would not make much sense to calculate a new index offset and not ever use it.

Big and Little Endian

We usually load words using the LDR instruction from an array written by the STR instruction or generated by the .WORD assembler directive. What happens if we load words that were written by the STRB instruction or generated by the .ASCII directive? Let's use the debugger again to provide the answer. Dump the string at address 0x100B8 three ways: as a string (x/s 0x100b8), twelve bytes (x/12b 0x100b8), and three words (x/3w 0x100b8).

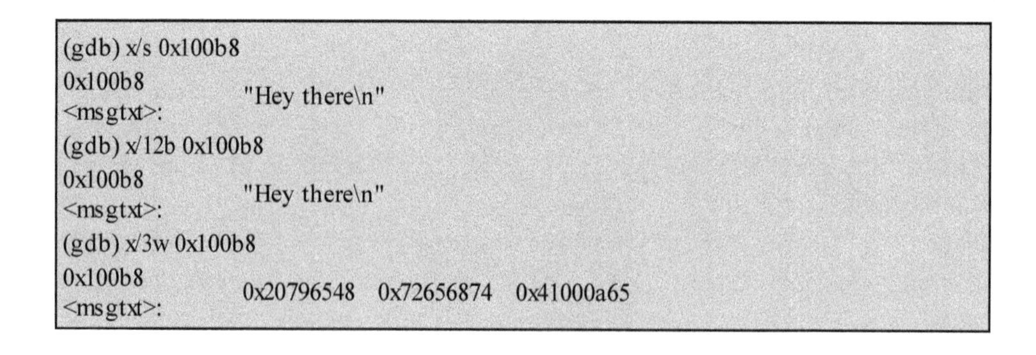

```
(gdb) x/s 0x100b8
0x100b8
<msgtxt>:        "Hey there\n"
(gdb) x/12b 0x100b8
0x100b8
<msgtxt>:        "Hey there\n"
(gdb) x/3w 0x100b8
0x100b8
<msgtxt>:        0x20796548   0x72656874   0x41000a65
```

Listing 8.8: Examine character string as bytes and words

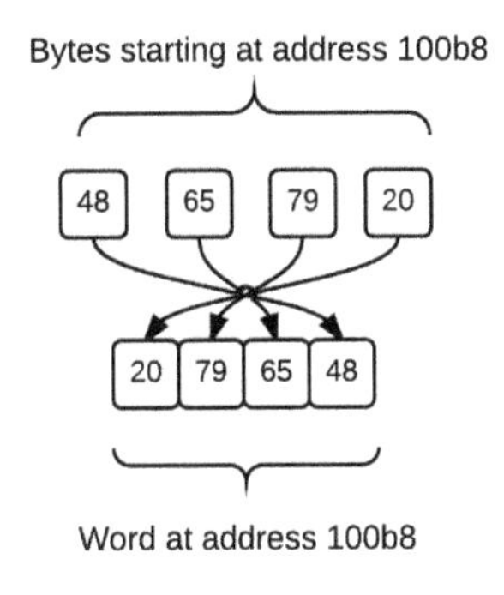

Figure 8.4: Load 4 consecutive bytes from memory

The first two byte-oriented dumps appear as expected, but the word-oriented dump looks backward to many people. The first word is 0x20796548 which in ASCII is " yeH" while the second is "reht" which have the words in the right order, but the bytes within each word appear to be reversed. This is what is referred to as "little endian" format meaning that the first byte from memory is loaded into the "little end" of the register (i.e., low order or rightmost side). The ARM processor within the Raspberry Pi uses little endian format by default which is consistent with most CPU architectures.

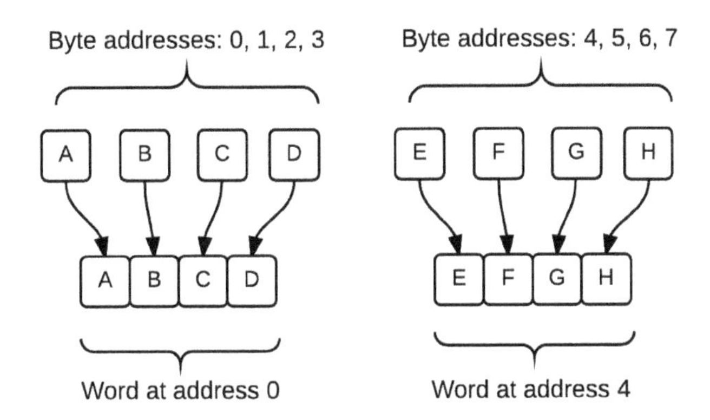

Figure 8.5: Big endian: The first byte goes to the "Big End" of the register.

Figure 8.5 illustrates the string "ABCDEFGH" being loaded as words in big endian format while Figure 8.6 shows the same string being loaded in little endian format. Note: The ARM 7 processor can be switched to big endian format by setting CPSR bit 9.

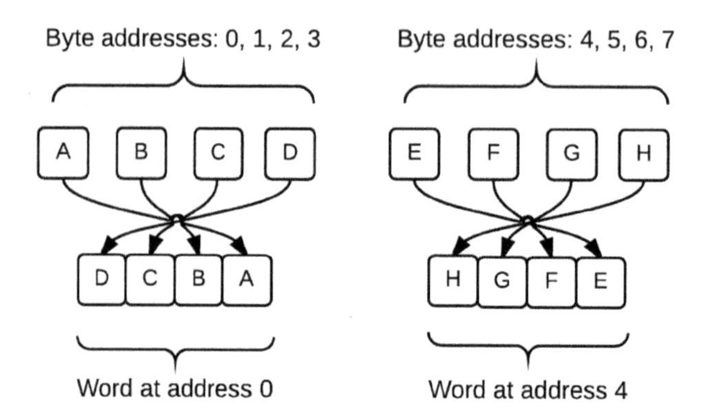

Figure 8.6: Little endian: The first byte goes to the "Little End" of the register.

Big endian format seems very natural to many people, while little endian seems so backward. Why would anyone prefer little endian? One advantage is in casting (i.e., conversion) among types of variables. In the example in Figure 8.7, memory address 1620 can represent an integer of various sizes, all with the same value of 25.

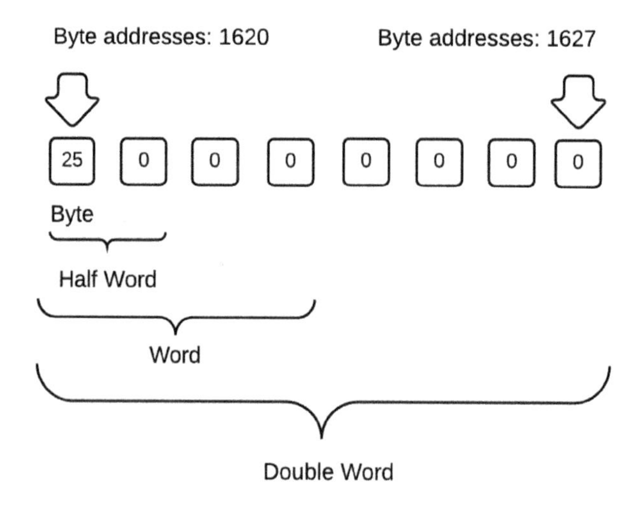

Figure 8.7: "Automatic" casting of a constant into different variable types

Alignment

Most byte-oriented CPU architectures prefer multi-byte memory accesses to be on aligned boundaries. A word should be loaded from an address that is a multiple of 4 bytes. Likewise a double-word and half-word should be on addresses that are multiples of 8 and 2, respectively. Words that are not aligned can lead to performance degradation or even a memory failure interrupt, depending on processor version and manufacturer. Listing 8.9 shows the use of the .ALIGN assembler directive that should be used to guarantee correct data memory assignment.

```
dig:     .ascii   "0123456789"   @ ASCII string of digits 0 through 9
         .ascii   "ABCDEF"       @ ASCII string of digits A through F
         .align   2              @ Force following address to be multiple of 4.
pow10:   .word    1              @ 10^0
         .word    10             @ 10^1
         .word    100            @ 10^2
```

Listing 8.9: Examples of character and word arrays

Store and Load Multiple Registers

The contents of as many as sixteen registers can be moved to or from memory with one "block transfer" instruction. The STM (Store Multiple) and LDM (Load Multiple) instructions move data between memory and registers. We've actually used a subset of these before disguised as PUSH and POP. An example instruction is STMIA R8,{R0-R7} where R8 is the base register providing the indirect memory location, and the set of registers is identified within a pair of braces like {R0-R7}.

- All 32 bits of each register are always moved (i.e., no single bytes).
- A register must be specified as a "base register" which points to the memory address
- Obviously, not all registers will be moved within one memory clock cycle, but at least only one instruction needs to be executed.
- Any combination of from one to sixteen registers can be specified.
- The order of registers doesn't matter as each of the following examples selects the same set of registers: R1,R2,R3,R5,R6, and R7.
    - {R1-R3,R5-R7}
    - {R6,R2,R7,R3,R5,R1}
    - {R6-R7,R5,R1-R3}

Block Transfers

The term "block transfer" in the ARM architecture is different than its meaning in most CPUs, where it refers to copying mutiple bytes of data from one location in memory to another. In the ARM, it refers to the group of eight load and store multiple instructions. Each instruction's name is built from three parts shown in Figure 8.8.

Load Multiple (ldm)	Increment (i)	Before (b)
Store Multiple (stm)	Decrement (d)	After (a)

Figure 8.8: Three parts of the block transfer instructions

- ldmib: Load Multiple Increment Before
- ldmia: Load Multiple Increment After
- ldmdb: Load Multiple Decrement Before
- ldmda: Load Multiple Decrement After
- stmib: Store Multiple Increment Before
- stmia: Store Multiple Increment After
- stmdb: Store Multiple Decrement Before
- stmda: Store Multiple Decrement After

Let's use the sample data in Figure 8.9 to describe how these instructions work. We have five words of data starting at byte address 40100. Register R8 will be used as the base register currently pointing to memory address 40108 which contains data value C0000. The four load multiple combinations are shown in Table 8.1. In the first line in the example, the address in base register R8 will be incremented after (IA) it is used to load (LDM) the data value C000 into register R0. Then the same instruction loads data value D000 into register R1, and the address is incremented after that also. Note: The address is being incremented, but not directly in R8, but in a CPU internal memory address register. In other words, the contents of register R8 are unchanged by these examples. As you might expect, there is a write-back option allowing R8 to be updated, and that will be shown soon.

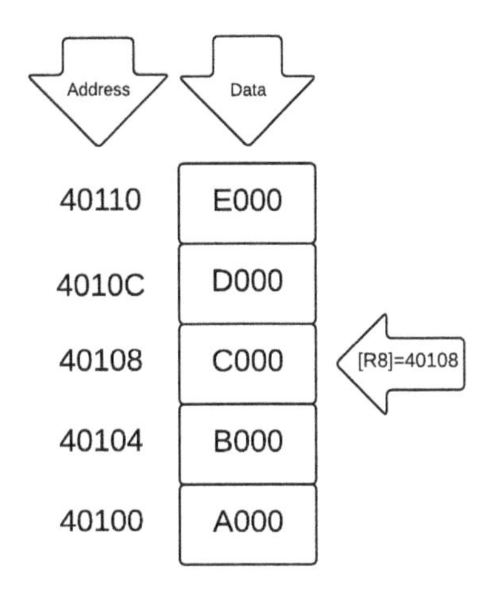

Figure 8.9: Sample data to describe block transfer

	R0	R1
ldmia R8,{R0,R1}	C000	D000
ldmib R8,{R0,R1}	D000	E000
ldmda R8,{R0,R1}	C000	B000
ldmdb R8,{R0,R1}	B000	A000

Table 8.1: Examples of loading two words from memory

Figure 8.10 provides an example of how the assembler constructs the machine language instruction for one of these block transfer instructions.

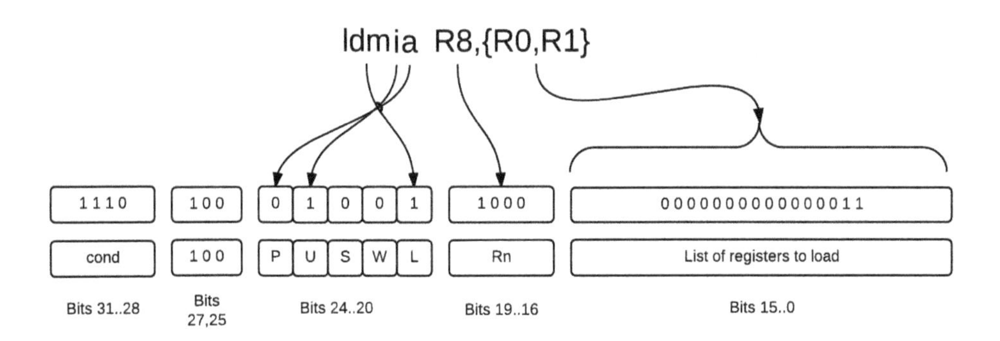

Figure 8.10: How the assembler builds the LDMIA R8,{R0,R1} instruction machine code

The load and store multiple instructions are similar to previous ones we've used, but has a register list:

- Cond (bits 31..28): Status indicating whether instruction should be executed (1110 indicates always).
- 1 0 0 (bits 27..25): Opcode indicating that this is a LDM/STM instruction.
- PUSWL (bits 24..20): Bits indicating specifically what is to be done by LDM/STM instruction.
  - P (bit 24): "1" indicates pre-indexed (before), "0" indicates post-indexed (after).
  - U (bit 23): "1" indicates increment address, "0" indicates decrement address.

- S (bit 22): special use only by operating system.
- W (bit 21): "1" indicates write-back, i.e., update the base register, "0" indicates no update.
- L (bit 20): "1" indicates load from memory (LDM), "0" indicates store (STM).

- Rn (bits 19..16): Indicates which register (R0 through R15) is used as base register.
- List (bits 15..0): Indicates which registers (R0 through R15) are loaded/stored.

As you might expect, all the multiple load/store instructions can be conditionally executed as the example in Listing 8.10 shows. The other thing you probably suspect is that the base register can be updated using the exclamation point, and that will be shown in Figure 8.11.

```
stmeqdb  R8,{R0-R7}                @ Store R0-R7 only if Z-flag is set
ldmgtdb  R8,{R0-R7}                @ Load R0-R7 only if !Z and (N = V)
```

Listing 8.10: Combining conditional execution and multiple register load/store

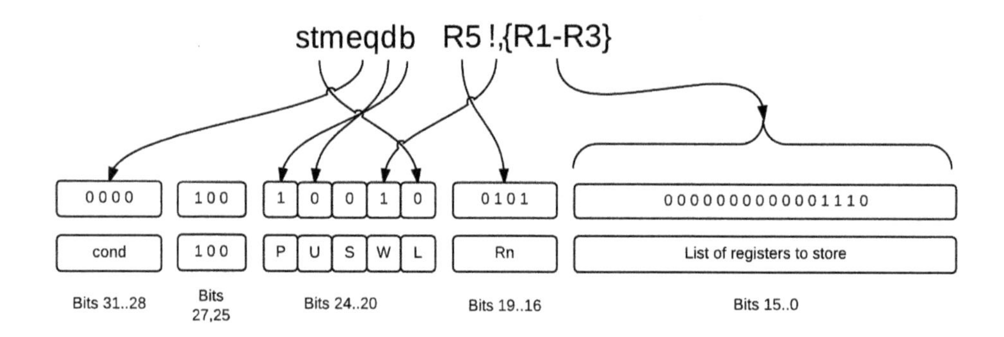

Figure 8.11: Block transfer instruction with conditional execution and write back into base register.

Top of the Stack

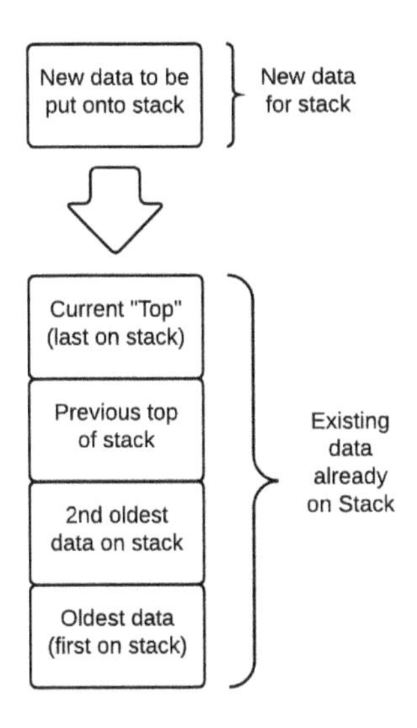

Figure 8.12: Stack "concept"

What do we mean by "pushing data on top of the stack" or "popping data from the top of the stack"? In other words, how do we implement the concept of a stack using program code? We've been saving and retrieving register contents on the stack using PUSH and POP, and as I alluded earlier, the assembler really uses the LDM and STM machine code instructions to implement the stack. A stack is referred to as a last-in first-out (LIFO) data structure, but we still need some type of pointer to where the last element of data is stored in memory.

Ascending or Descending, Full or Empty

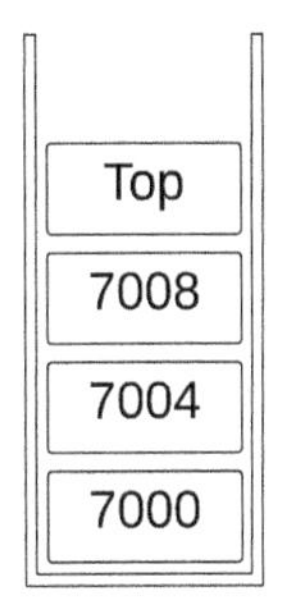

Figure 8.13: Ascending stack

When implementing a stack, there are two factors to consider regarding the stack pointer contents:

- Full or Empty: Does the stack pointer point to the last data pushed onto the stack (i.e., the location is full) or to the next location to receive data (i.e., the location is empty)?
- Ascending or Descending: When new data is pushed onto the stack, is the stack pointer incremented (ascending) or decremented (descending)?

Figure 8.13 illustrates an ascending stack which corresponds to the analogy of stacks of plates and trays in a cafeteria. This example shows the stack starting at address 7000 and incrementing for each word that is pushed onto the stack

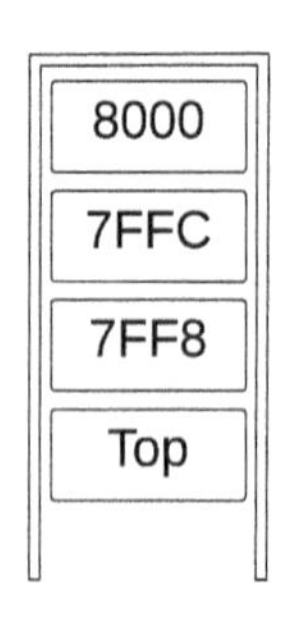

Figure 8.14: Descending stack

Figure 8.14 illustrates a descending stack, where the stack pointer is decremented when data is pushed onto the stack. This illustration shows the stack starting at address 8000 and descending for each word pushed onto the stack. An analogy for this approach is a stack of helium balloons put into a small elevator shaft.

So far in this discussion, I've used the term "stack pointer" to refer to the SP register, R13, but it could also refer to any of the other registers that could be pointing to other stacks separate from the system stack.

The block transfer instructions with names like LDMIA are very descriptive from a hardware viewpoint. They describe exactly what the instruction is to perform: Increment After, for example. On the other hand, names like PUSH and POP are more relevant to the software data structure. At this point, we really don't know exactly how PUSH and POP are implemented by the assembler. There is also a middle ground of names for block transfer instructions that refers to a stack's full/empty ascending/descending structure.

Load Multiple (ldm)	Empty (e)	Ascending (a)
Store Multiple (stm)	Full (f)	Descending (d)

Figure 8.15: Three parts of the block transfer instructions

- ldmea: Load Multiple Empty Ascending
- ldmed: Load Multiple Empty Descending
- ldmfa: Load Multiple Full Ascending
- ldmfd: Load Multiple Full Descending
- stmea: Store Multiple Empty Ascending
- stmed: Store Multiple Empty Descending
- stmfa: Store Multiple Full Ascending
- stmfd: Store Multiple Full Descending

Table 8.2 equates the hardware-descriptive instruction names to the data-structure-descriptive name for each of the eight block transfer instructions. It may seem like a small point, but which of the following would be less confusing for programming an Empty/Ascending stack? Push data onto the stack with a STMEA, and pop data with a

LDMEA? Or would you prefer pushing data with STMIA and popping data with LDMDB? The assembler doesn't care, and I think fewer programming errors would be made using STMEA/LDMEA than STMIA/LDMDB.

Stack type	Save on stack	Reload from stack
Empty Ascending	stmea (stmia)	ldmea (ldmdb)
Empty Descending	stmed (stmda)	ldmed (ldmib)
Full Ascending	stmfa (stmib)	ldmfa (ldmda)
Full Descending	stmfd (stmdb)	ldmfd (ldmia)

Table 8.2: "Stack relevant names" for the block transfer instructions

Each of the above four stack implementations is used in computer applications. We still don't know which one is used for the system stack in Linux. The assembler knows what type of stack is used, and that's how it can generate the proper block transfer instruction to be consistent with Linux.

Push and Pop

So what kind of stack are we really using for the system stack provided by Linux, and what instructions did the assembler really generate for the PUSH and POP? Again, let's use the debugger to provide the answers. Subroutine v_ascz begins with a PUSH, and that is at address 8088. We can use the x/3wi gdb command in Listing 8.11 to dump the first three instructions, but that really doesn't provide any new information. Using the x/3w does provide the hexadecimal value of E92D00FF which is what we need to determine the real machine code.

```
(gdb) x/3wi 0x8088
=>  0x8088 <v_ascz>:     push   {r0, r1, r2, r3, r4, r5, r6, r7}
    0x808c <v_ascz+4>:   sub    r2, r1, #1
    0x8090 <hunt4z>:     ldrb   r0, [r2, #1]!
(gdb) x/3w 0x8088
0x8088 <v_ascz>:   0xe92d00ff  0xe2412001  0xe5f20001
```

Listing 8.11: First three instructions of v_ascz displayed in machine code

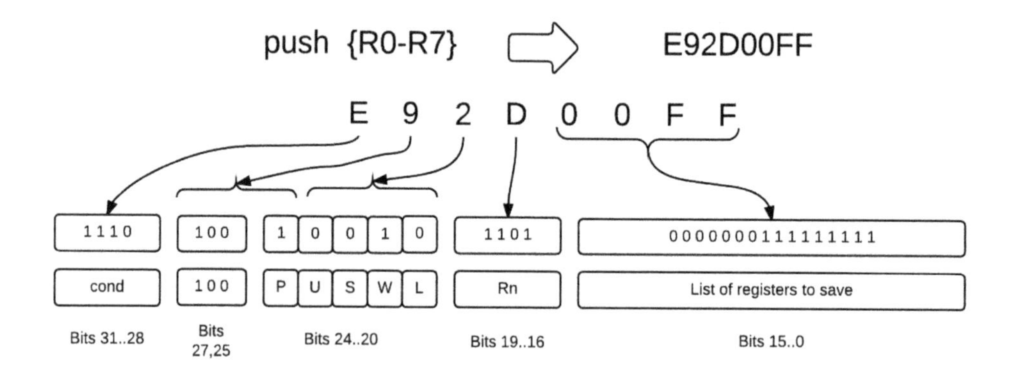

Figure 8.16: Machine code generated by assembler for PUSH {R0-R7}

Let's examine the bits in the machine code to see what instruction the assembler really generated for the PUSH.

- Cond (bits 31..28): Status indicating instruction should be executed (1110 indicates always).
- 1 0 0 (bits 27..25): Opcode indicating that this is a LDM/STM instruction.
- PUSWL (bits 24..20): Bits indicating specifically what is to be done by LDM/STM instruction.
    - P (bit 24): "1" indicates pre-indexed (before)
    - U (bit 23): "0" indicates decrement address.
    - S (bit 22): "0" indicates don't update the CPSR status bits.
    - W (bit 21): "1" indicates write-back, i.e., update the base register
    - L (bit 20): "0" indicates store instruction(STM).

- Rn (bits 19..16): Indicates register R13 (i.e., the SP) is used as the base register.
- List (bits 15..0): Indicates which registers (R0 through R7 in this example) are to be stored.

Therefore, we find that the assembler generated a STMDB SP!,{R0-R8} instruction, which from table 8.2 is the same instruction as STMFD SP!,{R0-R8}. If we do the same thing for the POP instruction, we get the following in Figure 8.17.

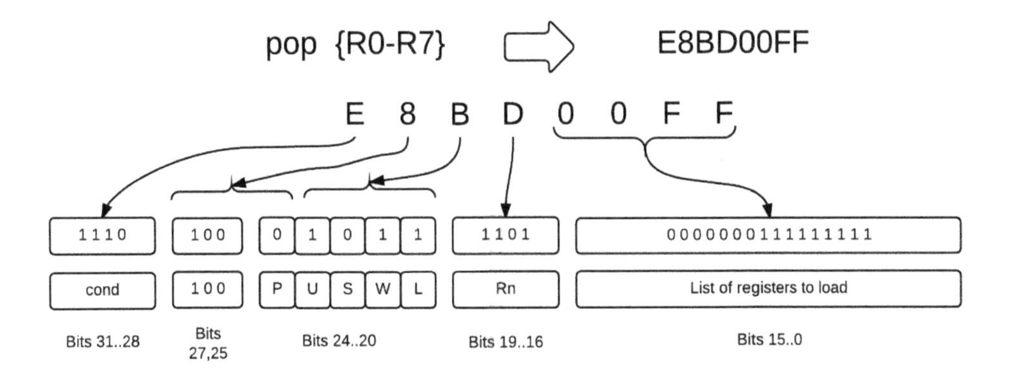

Figure 8.17: Machine code generated by assembler for POP {R0-R7}

Let's examine the bits in the machine code to see what instruction the assembler really generated for the POP.

- Cond (bits 31..28): Status indicating instruction should be executed (1110 indicates always).
- 1 0 0 (bits 27..25): Opcode indicating that this is a LDM/STM instruction.
- PUSWL (bits 24..20): Bits indicating specifically what is to be done by LDM/STM instruction.
    - P (bit 24): "0" indicates post-indexed (after)
    - U (bit 23): "1" indicates increment address.
    - S (bit 22): "0" indicates don't update the CPSR status bits.
    - W (bit 21): "1" indicates write-back, i.e., update the base register
    - L (bit 20): "1" indicates load instruction (LDM).

- Rn (bits 19..16): Indicates register R13 (i.e., the SP) is used as base register.
- List (bits 15..0): Indicates which registers (R0 through R7 in this example) are loaded.

We find that the assembler generated a LDMIA SP!,{R0-R8} instruction for POP {R0-R8} which from Table 8.2 has the same machine code as LDMFD SP!,{R0-R8}.

Since PUSH {R0-R8} is STMFD SP!,{R0-R8} and POP {R0-R8} is LDMFD SP!,{R0-R8}, the type of stack provided by Linux that we've been using is descending (D), and the top of the stack is full (F). This is a Linux convention and has been the most popular implementation used by most computer operating systems. It allows program code and local variables to grow upward from low memory addresses, and stack data to grow downward from high memory, and hopefully the two pointers never meet and run out of memory. Could it have been implemented differently? Sure. Programs can even have multiple stacks within them. PUSH and POP use the SP register, but other stacks could use other registers.

Are the block transfer instructions limited to just LIFO stacks? No, of course not.

First-in first-out (FIFO) queues can also be easily implemented, but you'll need two base registers: one pointing to "in" and one pointing to "out."

Coding and Debugging

We can learn a lot about something by taking it apart, that is, disassembling it. In order to begin to learn how an automobile engine works, I recommend finding a discarded lawn mower and taking it apart.

The preceding section ended with our disassembling of the PUSH and POP instructions "by hand." In this section we will use a program to disassemble the LDR and STR instructions that were introduced earlier in this lab. Figure 8.18 shows the assembly of a load byte instruction into machine code and its disassembly back into assembly language format.

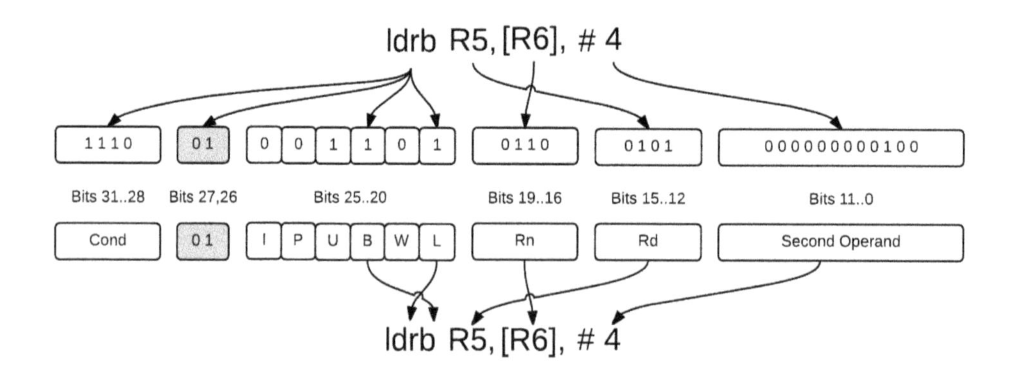

Figure 8.18: Assembly and disassembly of load byte instruction

Listing 8.12 contains subroutine dis_ls that disassembles and displays a load/store instruction using the following steps:

1. Output a tab character.
2. Output either "LDR" or "STR" depending on the value of bit 20.
3. Output a "B" if bit 22 indicates this instruction loads/stores a byte rather than a word.
4. Output the destination register, Rd:
    a. Output a tab character followed by "R"
    b. Call subroutine v_dec to output the register ID number.
5. Output the source register, Rs:
    a. Output a comma and left bracket followed by "R"
    b. Call subroutine v_dec to output the register ID number.

6. Output the offset to destination register value:
    a. Output a right bracket followed by a comma and hash.
    b. Call subroutine v_dec to output the immediate offset value.
7. Close out the disassembly display with a "new line" character.

```
1.@         Subroutine das_ls will disassemble and display "load/store" instructions.
2.@                    R0: contains the instruction to be disassembled and displayed
3.@                    LR: Contains the return address
4.@                    All register contents will be preserved
5.
6.          .global    das_ls          @ Provide program starting address to linker
7.
8.das_ls:   push       {R0-R4,LR}      @ Save contents of registers R0 -> R4 and LR.
9.          mov        R3,R0           @ Move "instruction" to R3.
10.
11.@        The opcode mnemonic display depends on Land B bits in instruction
12.@                   20: L(oad):     0 "STR" Store Register contents
13.@                                   1 "LDR" Load Register contents
14.@                   22: B(yte):     0 Word
15.@                                   1 "B" Byte
16.
17.         tst        R3,#1<<20       @ Test if this is a load or store instruction.
18.         ldrne      R1,=ldr_op      @ Load pointer to string "LDR"
19.         ldreq      R1,=str_op      @ Load pointer to string "STR"
20.         bl         v_ascz          @ Output the load or store opcode name.
21.         ldr        R1,=b_op        @ Load pointer to string "B"
22.         tst        R3,#1<<22       @ Test if this is a byte or word instruction.
23.         blne       v_ascz          @ Output the "B" for byte instructions.
24.
25.@        Display the destination and source registers
26.@                   Bits 12-15:     Destination register Rd
27.@                   Bits 16_19:     Source register Rn
28.
29.         ldr        R1,=tabr        @ Pointer to horizontal tab and "R" string
30.         bl         v_ascz          @ Tab and "R" precede destination register.
31.         mov        R0,R3,lsl #16   @ Left justify the destination register ID
32.         lsr        R0,#28          @ Right justify register ID in R0
33.         bl         v_dec           @ Display a number (0 through 15)
34.
35.         ldr        R1,=comlb       @ Pointer to ",[R" string.
36.         bl         v_ascz          @ ",[R" precedes source register.
37.         mov        R0,R3,lsl #12   @ Remove 12 bits left of the source ID
38.         lsr        R0,#28          @ Right justify register ID in R0
39.         bl         v_dec           @ Display a number (0 through 15)
40.
41.@        Finish the instruction with the post-indexed increment.
42.
```

```
43.          ldr     R1,=rbch        @ Pointer to "],#" string.
44.          bl      v_ascz          @ "],#" precedes post indexed increment.
45.          ldr     R1,=minus       @ Load pointer to string "-"
46.          tst     R3,#1<<23       @ Test if increment is Up or Down.
47.          bleq    v_ascz          @ Output the minus sign if "Down".
48.          mov     R0,R3,lsl #20   @ Left justify lower 12 bits.
49.          lsr     R0,#20          @ Offset is now right justified.
50.          bl      v_dec           @ Display the number 0 through 4095.
51.          ldr     R1,=lf          @ Pointer to line feed string.
52.          bl      v_ascz          @ End the instruction with a new line.
53.
54.          pop     {R0-R4,LR}      @ Restore saved register contents.
55.          bx      LR              @ Return to the calling program
56.
57. @        Strings used to format the instruction disassembly
58.
59. ldr_op:  .asciz  "\tLDR"         @ Opcode mnemonic for load register
60. str_op:  .asciz  "\tSTR"         @ Opcode mnemonic for store register
61. b_op:    .asciz  "B"             @ Opcode suffix for byte instruction
62. tabr:    .asciz  "\tR"           @ Precedes destination register
63. comlb:   .asciz  ",[R"           @ Precedes source register
64. rbch:    .asciz  "],#"           @ Precedes post-indexed increment value.
65. minus:   .asciz  "-"             @ Negative sign for decrementing index.
66. lf:      .asciz  "\n"            @ End of line character
67.          .end
```

Listing 8.12: Subroutine das_ls disassembles load/store instructions

Assumptions made while coding above version of subroutine das_ls:

1. The input "instruction" received in register R0 is really a load/store instruction: The program calling das_ls should have known that by checking bits 26 and 27. Subroutine das_ls could have double-checked that it is true, but then it would have to somehow flag the error.
2. No conditional execution or setting of CPSR status bits: This will be done in the examples in the Lab 9.
3. Post-indexed instruction format: This will be corrected in the next example in this section where das_ls is modified to also work with pre-indexed instructions.
4. Immediate mode for second operand: The disassembly of data processing instructions in Lab 9 will handle this case as well.

Listing 8.13 provides a main program to test subroutine das_ls. Notice that its list of test instruction contains both pre-indexed as well as post-indexed base registers. Obviously, the pre-indexed instructions will not disassemble properly, but it will be interesting to see what happens.

```
1.              .global   _start           @ Program starting address for linker
2._start:       ldr       R6,=tstlst       @ Point to list of test values.
3.sample:       ldr       R0,[R6],#4       @ Load next "instruction" to disassemble.
4.              bl        das_ls           @ Disassemble load/store instruction.
5.              cmp       R0,#0            @ Test if end of list of test values.
6.              bne       sample           @ Stop test loop after a zero value.
7.              mov       R0,#0            @ Exit Status code 0 for "normal completion"
8.              mov       R7,#1            @ Service code 1 terminates program.
9.              svc       0                @ Issue Linux command to stop program
10.
11.@            List of load/store sample instructions to disassemble
12.
13.tstlst:      ldr       R0,[R2],#15      @ Post-indexed instructions
14.             str       R5,[R6],#3000
15.             strb      R8,[R10],#500
16.             ldr       SP,[R0],#20      @ Stack pointer being loaded
17.             str       LR,[R11]         @ No post increment value provided
18.             ldr       R1,[R2,#3]       @ Pre-indexed instructions
19.             strb      R3,[R4,#5]!      @ Pre-indexed (up) with write-back.
20.             ldrb      R0,[R2,#-128]!   @ Pre-indexed (down) with write-back.
21.             .word     0                @ End of list.
22.             .end
```

Listing 8.13: Main program calling das_ls with test values

In addition to the main program in file model.s and subroutine das_ls in file das_ls.s, we will need to provide the linker with compiled object files v_ascz and v_dec as shown in Listing 8.14.

```
~$ nano model.s
~$ as -o model.o model.s
~$ nano das_ls.s
~$ as -o das_ls.o das_ls.s
~$ ld -o model model.o das_ls.o v_ascz.o v_dec.o
~$ ./model
```

Listing 8.14: Compile and link to test subroutine das_ls

Go ahead and run the new code. Table 8.3 shows the original “test instructions” on the left and the output from the disassembler program on the right.

LDR	R0,[R2],#15	LDR	R0,[R2],#15
STR	R5,[R6],#3000	STR	R5,[R6],#3000
STRB	R8,[R10],#500	STRB	R8,[R10],#500
LDR	SP,[R0],#20	LDR	R13,[R0],#20
STR	LR,[R11]	STR	R14,[R11],#0
LDR	R1,[R2,#3]	LDR	R1,[R2],#3
STRB	R3,[R4,#5]!	STRB	R3,[R4],#5
LDRB	R0,[R2,#-128]!	LDRB	R0,[R2],#-128
.WORD	0	STR	R0,[R0],#-0

Table 8.3: Compare output from das_ls disassembly to assembler source code

The first five test instructions disassembled and displayed correctly. The next three did not because they had pre-indexed base registers. The last line shows how a word of all zero bits is disassembled assuming it is a post-indexed load/store instruction.

Pre-Indexed Base Register

It's now time to properly handle the pre-indexed format as well. Figure 8.19 not only illustrates the pre-indexed mode (bit 24), but also the immediate, up/down, and write-back flag bits.

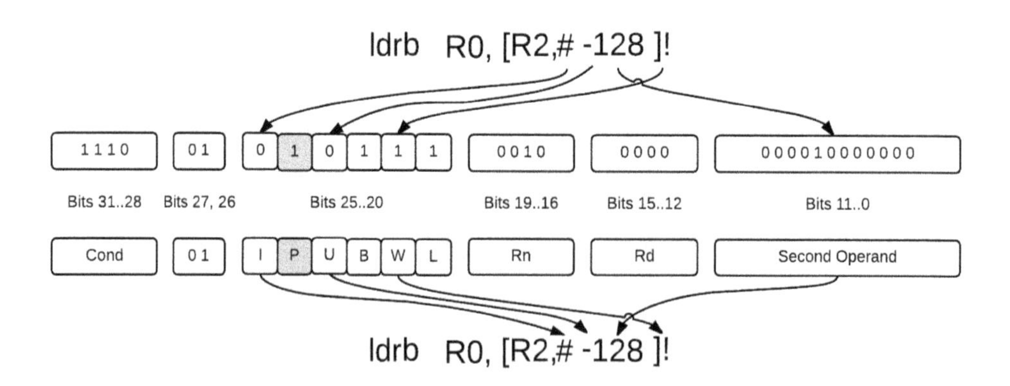

Figure 8.19: Pre-indexed format "load byte" instruction

Listing 8.15 shows the completed das_ls subroutine with modifications to accommodate both pre-indexed and post-indexed formats. The first 40 lines are exactly the same as the previous version in Listing 8.12.

```
1. @        Subroutine das_ls will disassemble and display "load/store" instructions.
2. @                    R0: contains the instruction to be disassembled and displayed
3. @                    LR: Contains the return address
4. @                    All register contents will be preserved
5.
6.          .global     das_ls          @ Provide program starting address to linker
7.
8. das_ls:  push        {R0-R4,LR}      @ Save contents of registers R0 -> R4 and LR.
9.          mov         R3,R0           @ Move "instruction" to R3.
10.
11. @       The opcode mnemonic display depends on L and B bits in instruction
12. @                   20: L(oad):     0 "STR" Store Register contents
13. @                                   1 "LDR" Load Register contents
14. @                   22: B(yte):     0 Word
15. @                                   1 "B" Byte
16.
17.         tst         R3,#1<<20       @ Test if this is a load or store instruction.
18.         ldrne       R1,=ldr_op      @ Load pointer to string "LDR"
19.         ldreq       R1,=str_op      @ Load pointer to string "STR"
20.         bl          v_ascz          @ Output the load or store opcode name.
21.         ldr         R1,=b_op        @ Load pointer to string "B"
22.         tst         R3,#1<<22       @ Test if this is a byte or word instruction.
23.         blne        v_ascz          @ Output the "B" for byte instructions.
24.
25. @       Display the destination and source registers
26. @                   Bits 12-15:     Destination register Rd
27. @                   Bits 16_19:     Source register Rn
28.
29.         ldr         R1,=tabr        @ Pointer to horizontal tab and "R" string
30.         bl          v_ascz          @ Tab and "R" precede destination register.
31.         mov         R0,R3,lsl #16   @ Left justify the destination register ID
32.         lsr         R0,#28          @ Right justify register ID in R0
33.         bl          v_dec           @ Display a number (0 through 15)
34.
35.         ldr         R1,=comlb       @ Pointer to ",[R" string.
36.         bl          v_ascz          @ ",[R" precedes source register.
37.         mov         R0,R3,lsl #12   @ Remove 12 bits left of the source ID
38.         lsr         R0,#28          @ Right justify register ID in R0
39.         bl          v_dec           @ Display a number (0 through 15)
40.
41. @       Output the increment value for both pre and post indexed modes
42.
43.         ldr         R1,=rbr         @ Load pointer to string "]"
44.         tst         R3,#1<<24       @ Test if this is pre-indexed instruction.
45.         bleq        v_ascz          @ Output "]" to close post-indexed format.
46.         ldr         R1,=comh        @ Pointer to ",#" string.
47.         bl          v_ascz          @ ",#" precedes increment value.
48.         ldr         R1,=minus       @ Load pointer to string "-"
49.         tst         R3,#1<<23       @ Test if increment is Up or Down.
```

```
50.          bleq      v_ascz          @ Output the minus sign if "Down".
51.          mov       R0,R3,lsl #20   @ Left justify lower 12 bits.
52.          lsr       R0,#20          @ Offset is now right justified.
53.          bl        v_dec           @ Display the number 0 through 4095
54.          ldr       R1,=rbr         @ Load pointer to string "]"
55.          tst       R3,#1<<24       @ Test if this is pre-indexed instruction.
56.          blne      v_ascz          @ Output "]" to close pre-indexed format.
57.
58. @        Write the exclamation point if the write-back bit is set.
59.
60.          ldr       R1,=wrex        @ Load pointer to string "!"
61.          tst       R3,#1<<21       @ Test if write-back bit set.
62.          blne      v_ascz          @ Output the "!" for index update.
63.
64. @        Finish the instruction with an end of line character and return.
65.
66.          ldr       R1,=lf          @ Pointer to line feed string.
67.          bl        v_ascz          @ End the instruction with a new line.
68.          pop       {R0-R4,LR}      @ Restore saved register contents.
69.          bx        LR              @ Return to the calling program
70.
71. @        Strings used to format the instruction disassembly
72.
73. ldr_op:  .asciz    "\tLDR"         @ Opcode mnemonic for load register
74. str_op:  .asciz    "\tSTR"         @ Opcode mnemonic for store register
75. b_op:    .asciz    "B"             @ Opcode suffix for byte instruction
76. tabr:    .asciz    "\tR"           @ Precedes destination register
77. comlb:   .asciz    ",[R"           @ Precedes source register
78. comh:    .asciz    ",#"            @ Precedes pre-indexed increment value.
79. rbr:     .asciz    "]"             @ Closes pre-indexed increment value.
80. wrex:    .asciz    "!"             @ Indicates pre-indexed register updated.
81. minus:   .asciz    "-"             @ Negative sign for decrementing index.
82. lf:      .asciz    "\n"            @ End of line character
83.          .end
```

Listing 8.15: Subroutine das_ls with support for pre-indexed instructions

Notice the changes beginning on line 41 to accommodate both index formats as well as the up/down flag providing the minus sign. If the write-back bit is set, then line 62 will display it. Go ahead and run this version. Only the das_ls subroutine needs to be recompiled, since the main program in model.s remains the same. The side-by-side test instructions displayed with their corresponding outputs are shown in Table 8.4.

LDR	R0,[R2],#15	LDR	R0,[R2],#15
STR	R5,[R6],#3000	STR	R5,[R6],#3000
STRB	R8,[R10],#500	STRB	R8,[R10],#500
LDR	SP,[R0],#20	LDR	R13,[R0],#20
STR	LR,[R11]	STR	R14,[R11],#0
LDR	R1,[R2,#3]	LDR	R1,[R2],#3
STRB	R3,[R4,#5]!	STRB	R3,[R4,#5]!
LDRB	R0,[R2,#-128]!	LDRB	R0,[R2,#-128]!
.WORD	0	STR	R0,[R0],#-0

Table 8.4: Compare output from improved das_ls disassembly to assembler source code

It now looks like all of the instructions are being disassembled correctly. An interesting observation from Table 8.4 is that the write-back bit is not set for any of the post-indexed instructions, even though all post-indexed instructions update the base register. This is a case where the CPU hardware knows what must be done, and the instruction doesn't have to remind it.

So what's still missing to properly disassemble these instructions? Lab 9 will provide the following enhancements for data processing instructions, which can also be retrofitted here:

1. The second operand can also be in registers. It's not just an immediate constant value.
2. The setting of the CPSR status bits and conditional execution must be included.
3. Registers R13, R14, and R15 are best displayed as the SP, LR, and PC, respectively.

Maintenance

Review Questions

1. The PUSH instruction places new data onto the stack and POP removes that data. Do you think there is a safeguard to prevent the stack from being overfilled? What do you think will happen if the stack's maximum size is exceeded?
2. Why doesn't the GNU "as" assembler allow the exclamation point (!) on the post-indexed mode instructions?
3. Can the order of registers stored in memory with the block transfer instructions be specified? If not, why is it not possible?
4. * The LDR and STR instructions allow a negative offset. Is this negative direction set with two's complement or sign and magnitude format?

Programming Exercises

1. Write a program to load registers R0, R1, and R2 from an array stored in memory in big endian format. Assume the CPSR bit to change between big and little endian is not available and only the ARM's default little endian format is possible.
2. Using the load and store multiple block instructions, write a program to move the contents of registers R1, R2, R3,and R4 to R4, R3, R2, and R1, respectively.

Lab 9
Disassembler

At this point, some of my students are very comfortable with the ARM instruction format and others are not. In order to gain a better understanding of a subject, sometimes it is helpful to look at it from another angle, such as taking something apart rather than assembling it. With that idea in mind, we will continue the disassembly procedure begun in Lab 8 for the load/store instructions, but continue it for the set of data processing instructions. Macros and object libraries will be introduced in Lab 9 to improve the style of assembler coding and linking, respectively.

Prototype

In this Prototype section, we'll construct a preliminary version of subroutine das_dp which will disassemble (i.e., build assembler source code from ARM machine code) data processing instructions, such as ADD, SUB, AND, etc. The following two sections in this lab will enhance das_dp with the complete handling of the second operand and include macros to facilitate coding and maintenance. The main program used to test subroutine das_dp is presented in Listing 9.0. It has a list of various data processing instructions, including those with CPSR processing.

```
1.              .global    _start          @ Program starting address for linker
2. _start:      ldr        R6,=tstlst      @ Point to list of test values.
3. sample:      ldr        R0,[R6],#4      @ Load next "instruction" to disassemble.
4.              bl         das_dp          @ Display "data processing" instruction.
5.              cmp        R0,#0           @ Test if end of list of test values.
6.              bne        sample          @ Stop test loop after a zero value.
7.              mov        R0,#0           @ Exit Status code 0 for "normal completion"
8.              mov        R7,#1           @ Service code 1 terminates program.
9.              svc        0               @ Issue Linux command to stop program
10.
11. @           List of data processing type instructions to test disassembly.
12.
13. tstlst:     eor        R5,R6,R7,lsr R8 @ Using four different registers
14.             ands       R0,R3,#0xFF     @ Immediate value with setting CPSR bits
15.             andeqs     R2,#0b11110     @ Immediate with CPSR bits used twice
16.             mov        R4,#512         @ Move immediate value
```

```
17.        addne    R1,R5        @ Add with conditional execution
18.        lsr      R2,R3,#7     @ Move instruction with shift.
19.        ror      R5,R2        @ Simple move with rotate
20.        .word    0            @ End of list of instructions.
21.        .end
```

Listing 9.0: Main program in model.s calls subroutine das_dp to test disassembly of instructions.

The set of ARM data processing instructions are characterized by the typical RISC architecture format of two operands and one result. The first operand and the result are always in individual registers. The second operand has various formats that can be as "simple" as an immediate value or as complicated as two registers with a shift count. Listing 9.1 contains a first version of subroutine das_dp that uses the same style of programming found in the disassembly of the load/store instructions in Lab 8. We will not perform the disassembly of the second operand at this time, but will include everything else by implementing the following steps:

1. Output the opcode name:
    a. Look up the 4-bit opcode value in the table of opcode names.
    b. Display the name preceded by a tab character.
2. Output the condition codes:
    a. Look up the 4-bit condition code value in table of code names.
    b. Display the condition code immediately adjacent to the opcode.
3. Output the "update status" "S" flag if bit set.
4. Output the destination register, Rd:
    a. Output a tab character followed by "R"
    b. Call subroutine v_dec to output the register ID number.
5. Output the source register, Rs:
    a. Output a comma followed by "R"
    b. Call subroutine v_dec to output the register ID number.
6. Without examining the second operand, close the disassembly by displaying a "new line" character.

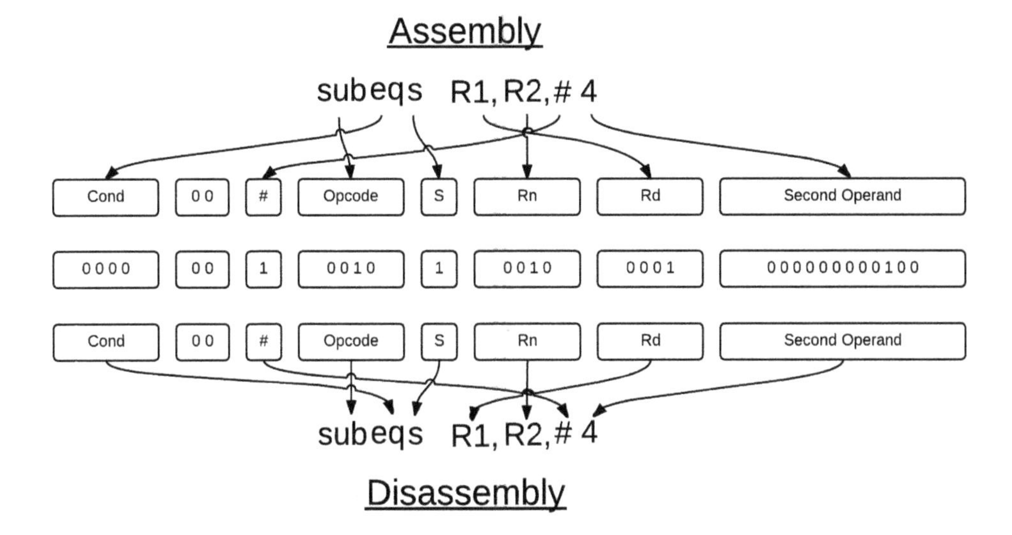

Figure 9.0: Assembly and disassembly of machine code instruction.

```
1. @         Subroutine das_dp will disassemble and display "data processing"
             instructions.
2. @                   R0: contains the instruction to be disassembled and displayed
3. @                   LR: Contains the return address
4. @                   All register contents will be preserved
5.
6.           .global   das_dp          @ Provide program starting address to linker
7.
8. das_dp:   push      {R0-R4,LR}      @ Save contents of registers R0 through R4
                                       and LR.
9.           mov       R3,R0           @ Move "instruction" to R3 where macro will
                                       use it.
10.
11. @        Display opcode field of the data processing instruction: opcode{cond}
             {s}
12. @                  Opcode          Opcode is in bits 24 down to 21
13. @                  Cond            Condition is in bits 31 down to 28
14. @                  S               "Set status codes" flag is in bit 20
15.
16.          ldr       R1,=tab         @ Pointer to horizontal tab string
17.          bl        v_ascz          @ Tab precedes Opcode.
18.          ldr       R1,=opcode      @ Pointer to table of Opcode names.
19.          mov       R2,R3,lsl #7    @ "Shake off" unwanted bits on the left.
20.          lsr       R2,#28          @ Right justify Opcode in R2.
21.          add       R1,R1,R2,lsl #2 @ Set R1 pointing to specific Opcode name.
22.          bl        v_ascz          @ Display Opcode name as text string.
23.
24.          ldr       R1,=cond        @ Pointer to table of condition codes.
```

```
25.            mov      R2,R3,lsr #28   @ Right justify condition code in R2.
26.            add      R1,R1,R2,lsl #2 @ Set R1 pointing to specific condition code.
27.            bl       v_ascz          @ Display condition name as text string.
28.
29.            ldr      R1,=s_suf       @ Load pointer to string "s"
30.            tst      R3,#1<<20       @ Test if instruction updates status flags.
31.            blne     v_ascz          @ The "s" indicates status to be updated.
32.
33. @          Display the destination and source registers.
34. @                   D-reg           Destination register in bits 15 down to 12.
35. @                   S-reg           Source register is in bits 19 down to 16.
36.
37.            ldr      R1,=tabr        @ Pointer to horizontal tab and "R" string
38.            bl       v_ascz          @ Tab and "R" precede destination register.
39.            mov      R0,R3,lsl #16   @ Remove 16 bits left of the destination ID
40.            lsr      R0,#28          @ Right justify register ID in R0
41.            bl       v_dec           @ Display the number 0 through 15
42.
43.            ldr      R1,=comr        @ Pointer to comma and "R" string
44.            bl       v_ascz          @ Comma and "R" precede source register.
45.            mov      R0,R3,lsl #12   @ Remove 12 bits left of the source ID
46.            lsr      R0,#28          @ Right justify register ID in R0
47.            bl       v_dec           @ Display the number 0 through 15
48.
49. @          Close the instruction for now with a line feed.
50.
51.            ldr      R1,=nl          @ Pointer to line feed character string
52.            bl       v_ascz          @ Line feed terminates line.
53.
54.            pop      {R0-R4,LR}      @ Restore saved register contents.
55.            bx       LR              @ Return to the calling program
56.
57. @          Lists of mnemonic names used in various fields with ARM instruction
               format
58.
59.            .align   2               @ Start on 4-byte boundary.
60. opcode:    .asciz   "and"           @ 0000 [Rd] = [Rn] AND (2nd operand)
61.            .asciz   "eor"           @ 0001 [Rd] = [Rn] ExclusiveOr (2nd operand)
62.            .asciz   "sub"           @ 0010 [Rd] = [Rn] - (2nd operand)
63.            .asciz   "rsb"           @ 0011 [Rd] = (2nd operand) - [Rn]
64.            .asciz   "add"           @ 0100 [Rd] = [Rn] + (2nd operand)
65.            .asciz   "adc"           @ 0101 [Rd] = [Rn] + (2nd operand) + C
66.            .asciz   "sbc"           @ 0110 [Rd] = [Rn] (2nd operand) + C - 1
67.            .asciz   "rsc"           @ 0111 [Rd] = (2nd operand) - [Rn] + C - 1
68.            .asciz   "tst"           @ 1000 [Rn] AND (2nd operand) => status
                                        bits
69.            .asciz   "teq"           @ 1001 [Rn] ExclusiveOr (2nd operand) =>
                                        status
70.            .asciz   "cmp"           @ 1010 [Rn] + (2nd operand) => status bits
```

```
71.             .asciz    "cmn"      @ 1011 [Rn] - (2nd operand) => status bits
72.             .asciz    "orr"      @ 1100 [Rd] = [Rn] InclusiveOR (2nd operand)
73.             .asciz    "mov"      @ 1101 [Rd] = [Rn]
74.             .asciz    "bic"      @ 1110 [Rd] = [Rn] AND NOT (2nd operand)
75.             .asciz    "mvn"      @ 1111 [Rd] = NOT [Rn]
76.
77. cond:       .asciz    "eq\0"     @ 0000 Equal (zero); Z set
78.             .asciz    "ne\0"     @ 0001 Not equal (non-zero); Z clear
79.             .asciz    "hs\0"     @ 0010 Unsigned hiher or same; C set -- also
                                     "cs"
80.             .asciz    "lo\0"     @ 0011 Unsigned lower; C clear --also "cc"
81.             .asciz    "mi\0"     @ 0100 Minus or negative; N set
82.             .asciz    "pl\0"     @ 0101 Plus or positive; N clear
83.             .asciz    "vs\0"     @ 0110 Overflow; V set
84.             .asciz    "vc\0"     @ 0111 No overflow; V clear
85.             .asciz    "hi\0"     @ 1000 Unsigned higher; C set and Z clear
86.             .asciz    "ls\0"     @ 1001 Unsigned lower or same; C clear or Z
                                     set
87.             .asciz    "ge\0"     @ 1010 Signed greater than or equal to; N = V
88.             .asciz    "lt\0"     @ 1011 Signed less than; N not same as V
89.             .asciz    "gt\0"     @ 1100 Signed greater than; Z clear and N
                                     equals V
90.             .asciz    "le\0"     @ 1101 Signed less than or equal; Z set or N
                                     != V
91.             .asciz    "\0 \0"    @ 1110 Always; any status bits OK -- also
                                     "al"
92.             .asciz    "??\0"     @ 1111 Never (reserved) -- also "nv"
93.
94.
95. tab:        .asciz    "\t"       @ Strings to be displayed in disassembled
                                     instructions.
96. tabr:       .asciz    "\tR"      @ Precedes destination register
97. comr:       .asciz    ",R"
98. s_suf:      .asciz    "s"
99. nl:         .asciz    "\n"
100.            .end
```

Listing 9.1: Subroutine das_dp disassembles “data processing” instructions

Go ahead and run this new code. Listing 9.2 shows the linking of the four object modules that are needed.

```
~$ nano model.s
~$ as -o model.o model.s
~$ nano das_dp.s
~$ as -o das_dp.o das_dp.s
~$ ld -o model model.o das_dp.o v_dec.o v_ascz.o
~$ ./model
```

Listing 9.2: Sequence of edit, compile, link, and execute

Table 9.0 is a side-by-side comparison of the program's output with the original assembly language source code for each instruction in the test set. Go ahead and add some more test cases of your choice and rerun the program. For variety, be sure to use sample instructions containing registers LR, SP, and PC, as well as different combinations of CPSR status codes. The coding does not have to make sense. We're only using it as "data" and trying to catch errors in the programming of the disassembler.

EOR	R5,R6,R7,LSR R8	EOR	R5,R6
ANDS	R0,R3,#0xFF	ANDS	R0,R3
ANDEQS	R2,#0b11110	ANDEQS	R2,R2
MOV	R4,#512	MOV	R4,R0
ADDNE	R1,R5	ADDNE	R1,R5
LSR	R2,R3,#7	MOV	R2,R3
ROR	R5,R2	MOV	R5,R2
.WORD	0	AND	R0,R0

Table 9.0: Compare assembler source code to output from das_dp disassembly

In the event you want to use the debugger, the "as" command lines in Listing 9.2 must contain the -g option so that the assembler passes diagnostic information onto the linker and debugger in the object module files. The linker command line remains the same, and of course, you start the debugger with a "gdb model" command line.

Introductions

Nine new assembler directives will be introduced in the Principles section as part of using macros to generate tables of data and sequences of code. The Archiver utility program will be introduced in the Coding and Debugging section to simplify and organize the method for including multiple precompiled object files while linking a program.

	List of assembler directives introduced in Lab 9
.align	Set memory address to half-word, word, double-word, ... border
.endif	Marks end of .ifnc and ifc conditional assembly
.endm	Marks end of .macro
.endr	Marks end of repeat (.irp) block of code
.ifc	Assemble the following code if two string arguments have same value
.ifnc	Assemble the following code if two string arguments have different values
.irp	Repeat the following block of code for a list of parameters
.macro	Define assembly-time function
.set	Assign value to assembly time variable

	List of command lines introduced in Lab 9
ar	Archiver: Build a library of object modules

Principles

As previously described in Lab 5, the set of data processing instructions has the general format of $\mathbf{D = N - M{\times}2^S}$ where the minus sign indicating subtraction is just one of the sixteen possible "data processing" operation codes.

$\mathbf{D = N - M{\times}2^S}$ where 2^S is not really an exponent, but a shift count.

- **D**: Destination register to hold the 32-bit result
- **N**: First operand register: In subtraction, it is the minuend (quantity from which another quantity is to be subtracted)
- $\mathbf{M{\times}2^S}$: Second operand: In subtraction, it is the subtrahend (quantity to subtract). There are three possible formats for M and S:
    1. **M** and **S** are both in registers: The contents of register M are shifted (logical, algebraic, or circular) by the value in register S.

2. **M** is a register and **S** is a constant: The contents of register M are shifted (logical, algebraic, or circular) by a constant (range of 0 through 31).
3. **M** and **S** are both constants: This is somewhat like scientific notation where M is the base (constant 0 through 255) and S is the exponent (shift count 0 through 30, even integers).

The disassembler will have to reconstruct the following assembly language fields for each instruction:

1. Opcode: AND,EOR, SUB, ...
2. Condition code: EQ, NE, HI, ..
3. Update CPSR "S" flag
4. Destination register: R0, R1, R2, ...
5. First operand (register): R0, R1, R2, ...
6. Second operand:
    - Immediate constant
    - Base register with shift count

In the Prototype section, the opcodes and condition codes were generated by indexing into tables: one a list of 16 opcodes and the other a list of 16 condition codes. The names in each list were different and the locations where the index values are found within the instruction are different, but the assembler coding for displaying the opcode is basically the same as that for displaying the condition code. This similarity hints at using a subroutine, doesn't it? However, we have another more appropriate option: *macros*. Listing 9.3 shows one macro call line that will replace the seven lines of assembler code in the proprotype program that picks out the 4-bit opcode from the data processing instruction and outputs its mnemonic name.

```
        d_fld   opcode,24,21,tab    @ This one macro replaces the following 7
                                    lines

16.     ldr     R1,=tab             @ Pointer to horizontal tab string
17.     bl      v_ascz              @ Tab precedes Opcode.
18.     ldr     R1,=opcode          @ Pointer to table of Opcode names.
19.     mov     R2,R3,lsl #7        @ "Shake off" unwanted bits on the left.
20.     lsr     R2,#28              @ Right justify Opcode in R2.
21.     add     R1,R1,R2,lsl #2     @ Set R1 pointing to specific Opcode name.
22.     bl      v_ascz              @ Display Opcode name as text string.
```

Listing 9.3: Macro d_fld replaces 7 lines of asembler code

Actually, the same macro will be able to reconstruct the first five fields of the instruction being disassembled. We will, of course, need a third table containing a list of register names, but that will even enable us to output LR, SP, and PC instead of R13, R14, and R15, respectively. Only the second operand in the data processing instruction will need special coding that is different from that of the other five fields as shown in Figure 9.1. The second operand, is more complex having three possible configurations containing both mnemonics as well as numbers.

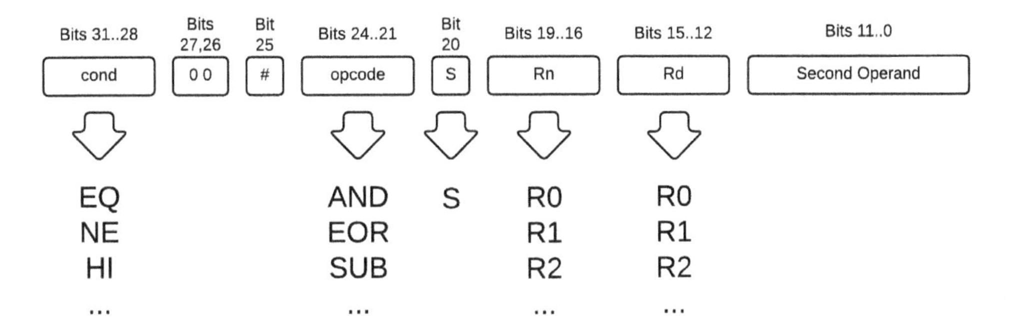

Figure 9.1: Disassembly of five fields into mnemonics

We will need three index tables: opcodes, conditionals, and register names, as follows.

```
opcode:   ascz8   and   @ 0000 [Rd] = [Rn] AND (2nd operand)
          ascz8   eor   @ 0001 [Rd] = [Rn] ExclusiveOr (2nd operand)
          ascz8   sub   @ 0010 [Rd] = [Rn] - (2nd operand)
          ascz8   rsb   @ 0011 [Rd] = (2nd operand) - [Rn]
          ascz8   add   @ 0100 [Rd] = [Rn] + (2nd operand)
          ascz8   adc   @ 0101 [Rd] = [Rn] + (2nd operand) + C
          ascz8   sbc   @ 0110 [Rd] = [Rn] (2nd operand) + C - 1
          ascz8   rsc   @ 0111 [Rd] = (2nd operand) - [Rn] + C - 1
          ascz8   tst   @ 1000 [Rn] AND (2nd operand) => status bits
          ascz8   teq   @ 1001 [Rn] ExclusiveOr (2nd operand) => status bits
          ascz8   cmp   @ 1010 [Rn] + (2nd operand) => status bits
          ascz8   cmn   @ 1011 [Rn] - (2nd operand) => status bits
          ascz8   orr   @ 1100 [Rd] = [Rn] Inclusive OR (2nd operand)
          ascz8   mov   @ 1101 [Rd] = [Rn]
          ascz8   bic   @ 1110 [Rd] = [Rn] AND NOT (2nd operand)
          ascz8   mvn   @ 1111 [Rd] = NOT [Rn]
```

Table 9.1: List of ARM data processing operation codes

```
cond:   ascz8   eq   @ 0000 Equal (zero); Z set
        ascz8   ne   @ 0001 Not equal (non-zero); Z clear
        ascz8   hs   @ 0010 Unsigned hiher or same; C set -- also "cs"
```

```
        ascz8    lo     @ 0011 Unsigned lower; C clear --also "cc"
        ascz8    mi     @ 0100 Minus or negative; N set
        ascz8    pl     @ 0101 Plus or positive; N clear
        ascz8    vs     @ 0110 Overflow; V set
        ascz8    vc     @ 0111 No overflow; V clear
        ascz8    hi     @ 1000 Unsigned higher; C set and Z clear
        ascz8    ls     @ 1001 Unsigned lower or same; C clear or Z set
        ascz8    ge     @ 1010 Signed greater than or equal to; N equals V
        ascz8    lt     @ 1011 Signed less than; N not same as V
        ascz8    gt     @ 1100 Signed greater than; Z clear and N equals V
        ascz8    le     @ 1101 Signed less than or equal; Z set or N not same as
                        V
        ascz8    \0     @ 1110 Always; any status bits OK -- also "al"
        ascz8    ??     @ 1111 Never (reserved) -- also "nv"
```

Table 9.2: List of ARM conditional execution codes

```
reg:    ascz8    R0     @ 0000 General purpose register set names
        ascz8    R1     @ 0001
        ascz8    R2     @ 0010
        ascz8    R3     @ 0011
        ascz8    R4     @ 0100
        ascz8    R5     @ 0101
        ascz8    R6     @ 0110
        ascz8    R7     @ 0111
        ascz8    R8     @ 1000
        ascz8    R9     @ 1001
        ascz8    R10    @ 1010
        ascz8    R11    @ 1011
        ascz8    R12    @ 1100
        ascz8    LR     @ 1101 Could also be "R13"
        ascz8    SP     @ 1110 Could also be "R14"
        ascz8    PC     @ 1111 Could also be "R15"
```

Table 9.3: List of ARM register names

Notice how I didn't use .ascii or .asciz to generate the strings in the above tables. The reason is that I wanted all of the strings to be the same length so that I could easily index into these tables. For example, the strings "R9" and "R10" have lengths of two bytes and three bytes, respectively. Naturally, I could have manually appended additional nulls (e.a., "R9\0") to the end of all the two byte register names, but that would be somewhat messy. Before we make the macro that will reconstruct the fields of the instruction being disassembled, we will make a simpler one, named ascz8, that will construct the preceding tables.

Macros

Almost all assemblers for all CPU designs have some form of "macro" capability, and the GNU "as" assembler is no exception. Macros provide a means for generating custom data formats and custom sequences of instructions. They not only provide for quicker initial program development, but also provide better documentation and maintenance. As shown in Listing 9.4, a macro begins with the .macro statement and ends with .endm. Macro processors for other ARM assemblers may begin with macro and end with mend, so macro syntax is not portable among all assemblers, even for the same CPU architecture.

```
        .macro    ascz8      item
        .asciz    "\item"    @ Name of item in list
        .align    3          @ Move to next 8-byte alignment boundary.
        .endm
```

Listing 9.4: Macro ascz8 generate an 8-byte name.

Macros typically generate more than one line of assembler code for each macro call line. Macro processing is performed during a first pass through the assembler source file that expands macro statements before the assembler converts them to machine code. You can think of a macro as a subroutine that is called at assembly time. Instead of performing a calculation using the contents of the registers, it generates lines of assembly language code that will be included in the program. Arguments appear after the macro name and are used to provide differences in the assembly language code that is generated.

For the example in Listing 9.4, the ascz8 macro is called with one argument named "item." Wherever "item" appears in the text of the macro following a back slash, the value from the macro call line will be substituted. For example, the line "ascz8 add" will generate ".asciz "add"" on one line followed by ".align 3" on the next. The purpose of the ".align 3" line is to start the following instruction or data on the next 8-byte boundary so that all elements in the table are the same length of 8, or 2^3. If it was necessary to make all table elements 16 bytes long (2^4), then ".align 4" would be used.

We not only use macros to generate table data, but program instructions as well. As you examine the current dis_dp subroutine, you notice that the program segments displaying each of the mnemonic fields in the instruction being disassembled will be very similar and will perform the following:

1. Call subroutine v_ascz to display a "field separation character" such as a tab or blank.
2. Select the desired field within the machine code instruction by removing unwanted bits left of the field.
3. Finish selecting the field by removing unwanted bits right of the field.

4. Set R1 pointing to the name to be displayed by adding the address of the beginning of the list of field names to 8 times (i.e., 8 bytes per name) the value found in the field.
5. Call subroutine v_ascz to display the contents of the field.

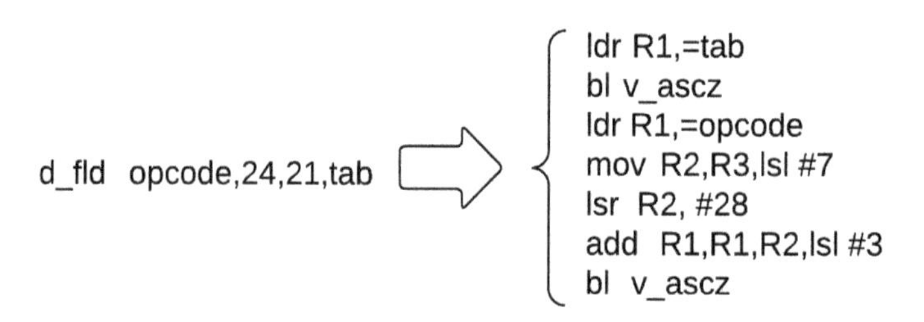

Figure 9.2: Macro expansion generates assembly language instructions.

For example to display the opcode as show in Figure 9.2, we would first display a tab character by calling subroutine v_ascz, then pick out the 4-bit opcode field from bits 24 down to 21, multiply that value by 8 (i.e., shift it left 3 bit positions), add it to the address of the beginning of the table of opcode names, and finally call v_ascz again to display the name of the opcode. We will be using this same set of instructions for the other mnemonic fields within the instruction being disassembled.

```
.macro  d_fld            field, leftbit, rightbit, prefix=0
.ifnc   \prefix,0
ldr     R1,=\prefix      @ Pointer to string to output before op code
bl      v_ascz           @ Display text string terminated by a null.
.endif
.set    left, 31-\leftbit
.set    right, left + \rightbit
ldr     R1,=\field       @ Pointer to beginning of table of fields to display
mov     R2,R3,lsl #left  @ "Shake off" unwanted bits on the left.
lsr     R2, #right       @ "Shake off" unwanted bits on the right.
add     R1,R1,R2,lsl #3  @ Set R1 pointing to specific field name to display.
bl      v_ascz           @ Display field name as text string ending with a
                         null.
.endm
```

Listing 9.5: Macro d_fld generates instructions to display an “instruction field” for the disassembler.

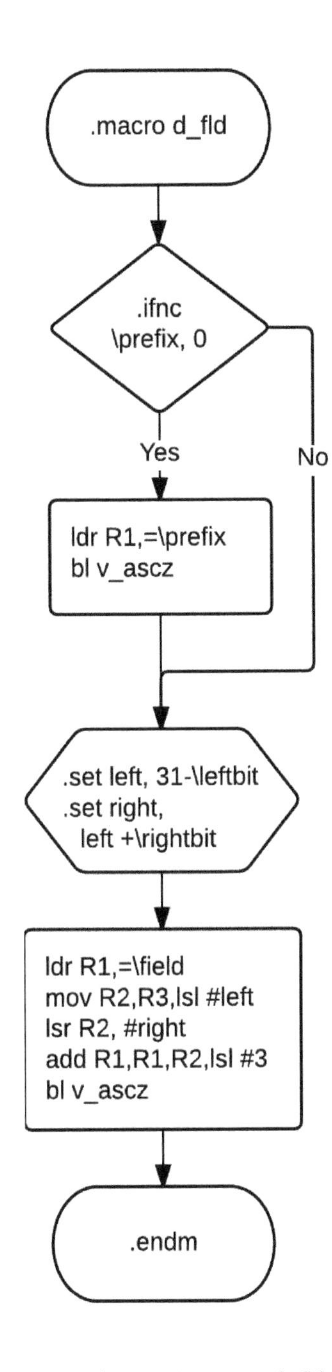

Figure 9.3: Macro d_fld

The flowchart in Figure 9.3 illustrates macro d_fld as if it were a subroutine. In its process blocks, the macro is not performing calculations, but is generating lines of assembler code that will later be compiled (i.e., assembled) by the assembler. The conditional diamond containing the assembler directive .ifnc provides an alternate sequence of code to be generated depending on a value received in the macro's input arguments. In the setup block, assembler directive .set assigns values to "assembly time" variables.

Macro d_fld is called with four arguments as listed on the first line of Listing 9.5:

- **field:** Address of the table containing list of field names
- **leftbit:** Bit position of left side of field
- **rightbit:** Bit position of right side of field
- **prefix:** Address of the string to be output before field

Notice how the argument "prefix=0" assigns a default value of "0" to argument "prefix" in the event no value is provided on the call to macro d_fld. Argument "prefix" is a memory address (i.e., an assembly language label identifying the beginning of a table or string), and I used "0" as the default because that would not be a legitimate label name. In the flowchart, the .ifnc diamond shows that no assembly language code will be generated to display a prefix string if argument "prefix" has a value of "0" (i.e., no prefix is present on the macro call).

Each of the fields to be displayed is located within a particular range of bit positions. For example, the opcode field is within bit positions 21 through 24. The macro must generate code to shake off the unwanted bits and right justify the field value so that it can be used as an index into a table of names. This is done by two shifts: a left logical shift to get rid of the unwanted bits on the left and a right logical shift to get rid on bits on the right as well as right justify the value. In order to make the code more descriptive and easier to program, the following two assembly time variables are used:

- **left:** How far to shift to the left (31 - position of leftmost bit)
- **right:** How far to shift to the right to right justify the field.

This macro generates a sequence of assembly language instructions that operate on register contents:

- **R1:** Will point to the name to be displayed by subroutine v_ascz
- **R2:** Will contain the designated field removed from the instruction
- **R3:** Contains the instruction to be disassembled
- **LR:** Will be used for return addresses for calls to v_ascz

If this were a macro that would be used in multiple places within a large program, I'd probably make the following adjustments to it:

1. Instead of assuming that the instruction to be disassembled is always found in register R3, I'd have its location as a fifth argument. Not only could the instruction be located in a different register, it could even be located in memory. The reason I didn't do it here is that the source instruction is always in the same location (R3), and who needs to make the macro look even more complicated than necessary?
2. The macro is using and thereby destroying the contents of three "hidden" registers. Without looking at the macro code, a programmer calling macro d_fld would not know that the contents of registers R1, R2, and LR would be changed. Of course, I could have saved those registers onto the stack like what is done in subroutines, but that would have added an extra eight bytes to each macro call.
3. I would put it into a separate file and use the assembler .include directive to copy it to each assembler source file in which it is needed. In fact, the d_fld macro would be very useful in the das_ls subroutine in Lab 8 for disassembling load/store machine code instructions.

ARM data processing instructions have three general formats, but they all have the same beginning which contains the opcode name with condition status bits followed by the destination register and source register. Listing 9.6 shows the d_fld macro being called to generate the entire instruction except for the second operand.

bits	name	Contents
31..28	Cond	Only execute this instruction on condition of the value in the NZCF flags
27..26	00	These two bits are always zero for "data processing" instructions
25	#	Immediate operand flag
24..21	Opcode	Which operation (add, sub, and, orr, eor, ...)
20	S	Indicates that this instruction will modify the condition codes
19..16	Rn	ID number of register containing the first operand
15..12	Rd	ID number of register to receive the result
11..0	op2	Three formats possible for second operand

Table 9.4: General bit layout for ARM "data processing" instructions such as SUB

```
@       Display part of the data processing instruction: opcode{cond}{s} Rd,Rn

        d_fld   opcode,24,21,tab     @ Opcode is in bits 24 down to 21
        d_fld   cond,31,28           @ Condition is in bits 31 down to 28
        d_fld   s,20,20              @ "Set status codes" flag is in bit 20
        d_fld   reg,15,12,tab        @ Destination register is in bits 15 down to 12
        d_fld   reg,19,16,comma      @ Source register is in bits 19 down to 16
```

Listing 9.6: Build first part of disassembled instruction.

Second Operand

Thanks to the macro, the code in Listing 9.6 will actually reconstruct over 75% of the assembly language instruction from the machine code instruction. However, as in many projects where the last 10% of the project consumes 90% of the effort, the more complicated coding is yet to come. As previously noted, the second operand is of the form $\mathbf{M\times2^S}$ where M and S can both be registers, both be constants, or M is a register with S being a constant.

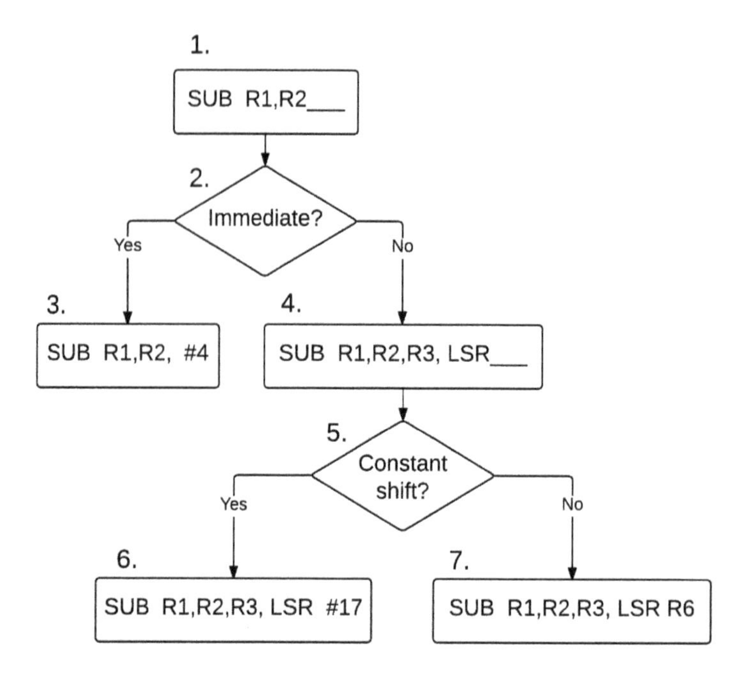

Figure 9.4: Stages of reconstruction of the disassembled instruction

Figure 9.4 shows the seven possible stages of reconstructing an ARM data processing instruction. The code we have so far is at stage 1 of Figure 9.4 where the blank at the end of the instruction indicates work yet to be done.

1. SUB R1,R2 _____: Everything but second operand is done.
2. Test: Check if the second operand is in registers or immediate.
3. SUB R1,R2, #4: Finished instruction with immediate constant
4. SUB R1,R2,R3, LSR __: Second operand is a register being shifted.
5. Test: Check if shift count is in register or immediate.
6. SUB R1,R2,R3, LSR #17: Finished instruction with constant shift count
7. SUB R1,R2,R3, LSR R6: Finished instruction with shift count in register

Second Operand is an Immediate Constant

Let's continue programming the disassembler with the case where the second operand is an immediate value as shown in Figure 9.5. If the immediate bit is set, then the second operand located in the lower 12 bits of the instruction word is a constant. This is stage 3 in the above list and figure.

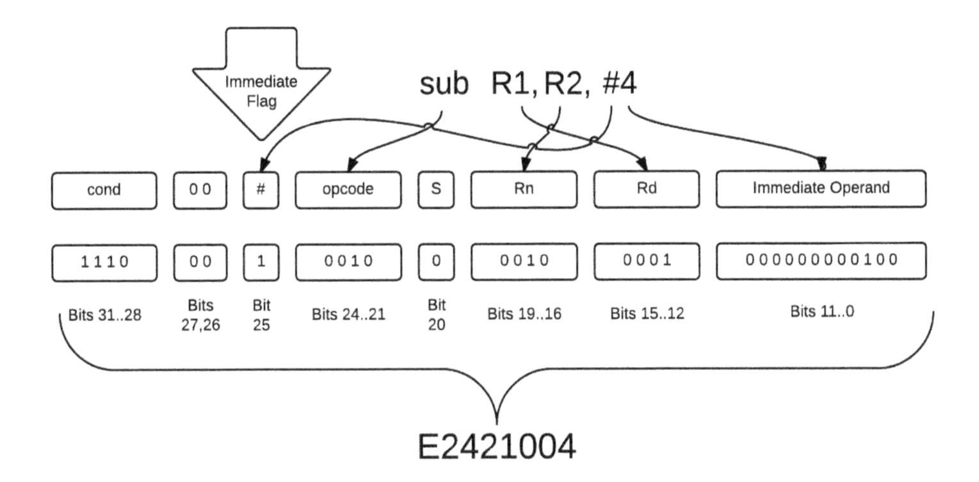

Figure 9.5: "Assemble" SUB with immediate value to ARM machine code 0xE2421004

Figure 9.5 has the correct bit pattern for the second operand having a value of 4, but it has oversimplified the immediate value field. Rather than being a single 12 bit binary number having a decimal range of 0 through 4095, the ARM implementation enables a 32-bit number to be generated by combining an 8-bit base with a 4-bit shift. It sacrifices precision in order to provide a greater range.

Any 8-bit pattern can be moved to any even bit-position. Below are four 9-bit numbers, where one will work in the second operand format, but three cannot.

- 0x101 (binary 100000001) cannot work because it is more than an 8-bit pattern.
- 0x102 (binary 100000010) cannot work because even though it is an 8-bit pattern, it cannot be shifted to an odd bit position.
- 0x103 (binary 100000011) cannot work because it is more than an 8-bit pattern.
- 0x104 (binary 100000100) works. Not only is it a 7-bit pattern, but it shifts to an even bit position.

Basically, any 8-bit pattern can be moved to any even bit-position. Figure 9.6 shows several examples of an immediate constant being converted into an 8-bit value along with a rotate value. Note that the rotate is only to the right, and its range of 0 through 15 (4-bit field) must be multiplied by 2 in order to make the full 30 bit rotation possible.

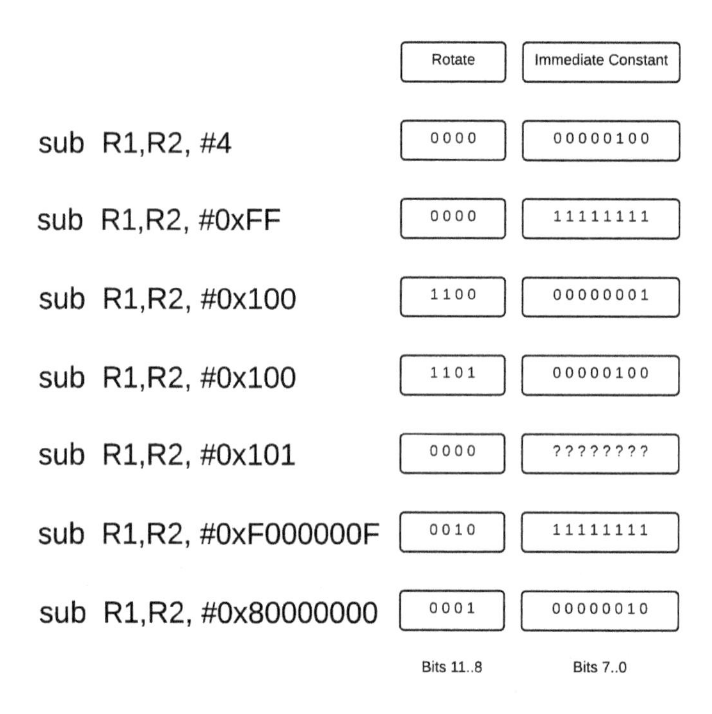

Figure 9.6: Six immediate value examples and one impossible value (0x101)

The code segment in Listing 9.7 combines the constants in the base and shift count to produce a single integer. On lines 103 and 104, a third macro (d_int) is called to display it in both a decimal and hexadecimal format, respectively. This completes stage 3 of Figure 9.4, the instruction having an immediate constant for the second operand.

```
93.
94. @      There are two formats for the second operand: immediate value or
           register.
95. @      If the immediate bit is set, display the 2nd operand in decimal and hex.
96.
97.        tst    R3,#0x02000000   @ Check if "immediate" bit is set for 2nd
                                   operand.
98.        beq    shreg            @ If no immediate bit, 2nd operand uses
                                   register.
99.        and    R0,R3,#0xFF      @ Move base value of immediate operand
                                   ino R0.
100.       lsr    R2,R3,#7         @ "Almost" right justify shift count (times 2)
101.       and    R2,#0b11110      @ Total shift count is double the value.
102.       ror    r0,R2            @ Finish immediate value with a rotate.
103.       d_int  dec,hstcom       @ Display the immediate value in decimal.
```

```
104.            d_int     hex,hstat      @ Display the immediate value in
                                         hexadecimal.
105.            b         dpfin          @ Go finish the disassembly with a line feed.

195. hstcom:    .asciz    ", #"
196. hstat:     .asciz    "\t@ #0x"
```

Listing 9.7: Disassembly of immediate constant second operand.

The d_int macro has two arguments:

1. **format:** Values of "dec" and "hex" will result in subroutine calls to v_dec and v_hex, respectively. Although it's not used, a value of bin would result in a call like BL v_bin.
2. **prefix:** A character string to be displayed before the integer. In this code, it puts a comma and hash symbol before the decimal integer, and it puts the hexadecimal value into the comments section following an @ symbol.

```
.macro    d_int          format, prefix
ldr       R1,=\prefix    @ Pointer to string to output before integer
bl        v_ascz         @ Display text string terminated by a null.
.ifc      \format,hex
mov       R2,#0          @ Code to print no leading zeroes
.endif
bl        v_\format      @ Call one of the display subroutines in view.o.
.endm
```

Listing 9.8: Macro d_int generates formatted integers for the disassembler.

Second Operand is a Register with a Shift

We are now at stage 4 of Figure 9.4. The register to be shifted is located in bits 3 down to 0, and the type of shift is indicated in bits 6 and 5. The mnemonics for the contents of both of these fields can be displayed with the d_fld macro. After that, bit 4 determines whether the shift count is an immediate constant or another register.

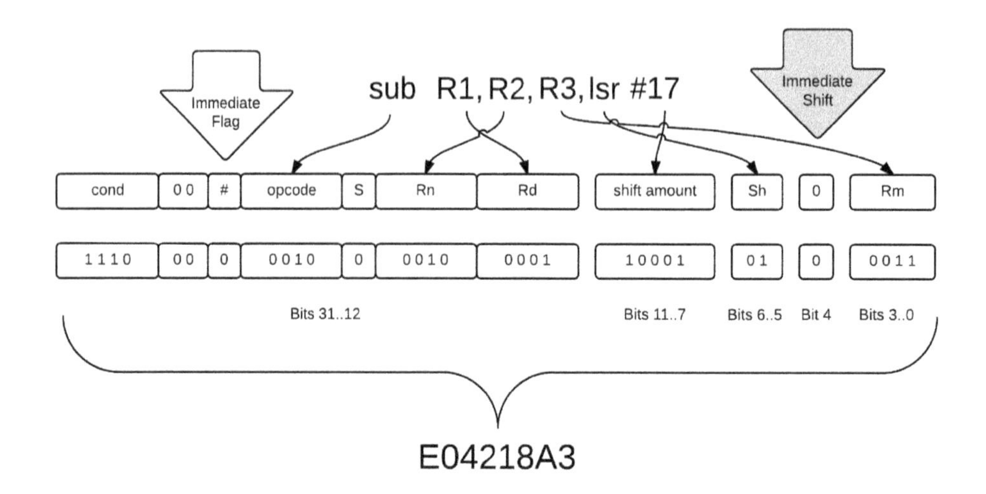

Figure 9.7: "Assemble" SUB with immediate constant shift instruction to ARM machine code

```
183. shift:    ascz8    lsl    @ 00 Logical Shift Left
184.           ascz8    lsr    @ 01 Logical Shift Right
185.           ascz8    asr    @ 10 Algebraic Shift Right
186.           ascz8    ror    @ 11 Rotate Right (Circular shift)
```

Listing 9.9: Table of four shifts available in second operand.

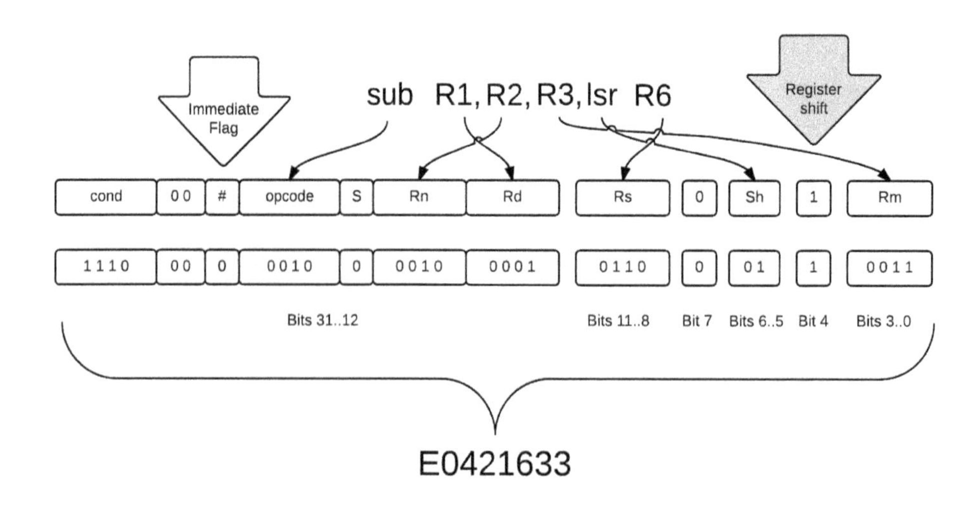

Figure 9.8: "Assemble" SUB with shift count in register to ARM machine code

Listing 9.10 provides the code which finishes the disassembly of instructions with a shift (stages 6 and 7 of figure 9.4).

```
107.@        The 2nd operand is a shift. Display the register to be shifted and shift
             type.
108.
109.shreg:    d_fld   reg,3,0,comma     @ Base register is in bits 3 down to 0
110.          d_fld   shift,6,5,comma   @ Shift type is in bits 6 down to 5.
111.          tst     R3,#0b010000      @ Check if "constant" shift or register value.
112.          bne     shrr              @ If bit set, shift count is in register.
113.          mov     R0,R3, lsl #20    @ Copy and left justify immediate shift count.
114.          lsr     R0,#27            @ Right justify shift count to "shake off" bits.
115.          d_int   dec,hst           @ Display the immediate shift in decimal.
116.          b       dpfin             @ Go finish the disassembly with a line feed.
117.
118.@        The 2nd operand is a register shifted by the contents of a register
119.
120.shrr:     d_fld   reg,11,8,blank    @ Shift register is in bits 11 down to 8
```

Listing 9.10: Finish the disassembler by processing the shift (stages 6 and 7)

Coding and Debugging

We're ready to complete the das_dp subroutine and test it. We have the main program from the Prototype section, and the new das_dp subroutine will also contain a call to subroutine v_hex, so Listing 9.11 provides the new work flow.

```
~$ nano das_dp.s
~$ as -o das_dp.o das_dp.s
~$ ld -o model model.o das_dp.o v_dec.o v_hex.o v_ascz.o
~$ ./model
```

Listing 9.11: Sequence of edit, compile, link, and execute

Notice how the list of object files on the linker command line is getting pretty long. Is there a better way? There is, and it's a resource that has been available on almost every computer operating system in the past 60 years: the object library. Rather than list each file that is needed, we can put all the subroutines into a library, and let the linker find those object files that it needs.

Object Library

The "ar" archiver program can combine a variety of files into a single file. At one time, it was a popular utility for building a set of several different file types including source code files, graphic files, and data files. However, today it is primarily used to generate a library of object files. Figure 9.9 shows a program being linked using a library. The names of two of the object files will be provided on the linker command line while the other two object files will be found in the library by the linker.

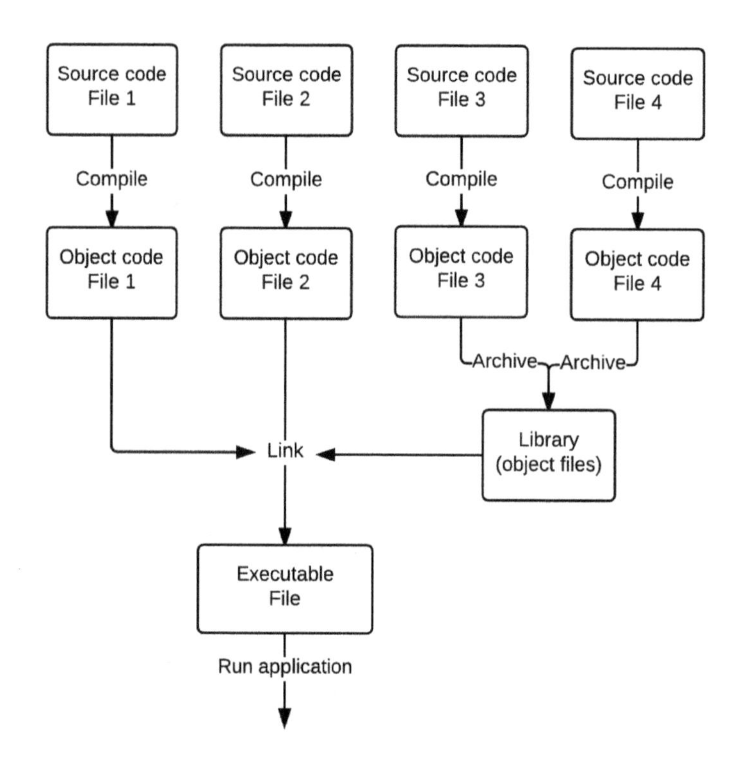

Figure 9.9: Linking with a library

Listing 9.12 shows three subroutines being added to a library named view.a. This "library" of subroutines can then be called from a variety of "main" programs. Of course, in a large application, it would not be practical to put all of the display subroutines into a single file, but for now, it is a start in the right direction.

```
~$ as -o model.o model.s
~$ as -o v_dec.o v_dec.s
~$ as -o v_hex.o v_hex.s
~$ as -o v_ascz.o v_ascz.s
~$ ar rvs view.a v_ascz.o v_hex.o v_dec.o
~$ ld -o model model.o view.a
```

Listing 9.12: Using an object library to link a program

The options commonly used to add object files to a library are rvs, and options t, d, and x are useful for accessing existing libraries.

- r: Replace object file if one with same file name is already in the library
- v: Verbose tells the ar program to display details of what it has completed
- s: Symbol table is to be updated with the subroutine ".global" name
- t: Table of contents of files in the library are listed
- d: Delete a file from the library
- x: Extract files (copy, not remove) from the library

The above options can be performed for each file listed on the "ar" command, or if no files listed, then all files will be extracted or appear in the table of contents. If the "v" option also appears with the above options, then more information will be displayed.

Listing 9.13 follows which contains the completed das_dp subroutine.

```
1.@     Subroutine das_dp will disassemble and display ARM "data processing"
        instructions.
2.@              R0: contains the instruction to be disassembled and displayed
3.@              LR: Contains the return address
4.@              All register contents will be preserved
5.
6.      .global  das_dp        @ Provide program starting address to linker
7.
8.@     Macro "ascz8 item" appends an item to list of field names in an instruction.
9.
10.     .macro   ascz8         item
11.     .asciz   "\item"       @ Name of item in list
12.     .align   3             @ Move to next 8-byte alignment boundary.
13.     .endm
14.
15.@    Macro "d_int format,prefix" displays formatted integer value.
16.@             format:       Format: dec, hex, bin (decimal, hexadecimal,
                               binary)
17.@             prefix:       String to display before integer.
18.
19.     .macro   d_int         format,prefix
20.     ldr      R1,=\prefix   @ Pointer to string to output before integer
21.     bl       v_ascz        @ Display text string terminated by a null.
22.     .ifc     \format,hex
23.     mov      R2,#0         @ Code to print no leading zeroes
24.     .endif
25.     bl       v_\format     @ Call one of the display subroutines in view.o.
26.     .endm
27.
```

Listing 9.13a: Subroutine das_dp entry point and two macros

28. @	Macro "d_fld field,leftbit,rightbit,prefix" displays field in instructions.		
29. @		field:	Beginning of table containing list of field names
30. @		leftbit:	Bit position of left side of field
31. @		rightbit:	Bit position of right side of field
32. @		prefix:	String to be output before field
33.			
34. @	Examples to generate the name of following fields from register R3:		
35. @		d_fld	reg,15,12,comma @ Register name in bits 15..12
36. @		d_fld	opcode,24,21,tab @ Opcode name in bits 24..21
37.			
38.			
39. @	Registers used by macro "d_fld" (R1, R2, LR used and not preserved):		
40. @		R3:	Input containing the instruction to be disassembled.
41. @		R1:	Used as pointer into list of mnemonic names.
42. @		R2:	Used to "pull out" desired segment of instruction.
43. @		LR:	Used to call subroutines.
44.			
45. @	Parameters generated in macro:		
46. @		left:	Left shift count to get rid of extra bits on left
47. @		right:	Right shift count to right justify selected "field"
48.			
49.	.macro	d_fld	field,leftbit,rightbit,prefix=0
50.	.ifnc	\prefix,0	
51.	ldr	R1,=\prefix	@ Pointer to string to output before op code
52.	bl	v_ascz	@ Display text string terminated by a null.
53.	.endif		
54.	.set	left, 31-\leftbit	
55.	.set	right, left + \rightbit	
56.	ldr	R1,=\field	@ Pointer to beginning of table of fields to display
57.	mov	R2,R3,lsl #left	@ "Shake off" unwanted bits on the left.
58.	lsr	R2, #right	@ "Shake off" unwanted bits on the right.
59.	add	R1,R1,R2,lsl #3	@ Set R1 pointing to specific field name to display.
60.	bl	v_ascz	@ Display field name as text string ending with a null.
61.	.endm		

Listing 9.13b: Macro d_fld

62.		
63. @	Subroutine das_dp will disassemble and display "data processing" instructions.	
64. @		R0: contains the instruction to be disassembled and displayed
65. @		LR: Contains the return address

```
66. @                 All register contents will be preserved
67.
68. das_dp:  push  {R0-R4,LR}         @ Save contents of registers R0 through R4
                                      and LR.
69.          mov   R3,R0              @ Move "instruction" to R3 where macro
                                      will use it.
70.
71. @        Display beginning of the data processing instruction: opcode{cond}{s}
             Rd,Rn
72.
73.          d_fld opcode,24,21,tab   @ Opcode is in bits 24 down to 21
74.          d_fld cond,31,28         @ Condition is in bits 31 down to 28
75.          d_fld s,20,20            @ "Set status codes" flag is in bit 20
76.          d_fld reg,15,12,tab      @ Destination register is in bits 15 down to
                                      12
77.          d_fld reg,19,16,comma    @ Source register is in bits 19 down to 16
78.
```

Listing 9.13c: Generate entire instruction except for second operand

```
79. @     There are two formats for the second operand: immediate value or register.
80. @     If the immediate bit is set, display the 2nd operand in decimal and hex.
81.
82.       tst    R3,#0x02000000    @ Check if "immediate" bit is set for 2nd
                                   operand.
83.       beq    shreg             @ If no immediate bit, 2nd operand uses
                                   register.
84.       and    R0,R3,#0xFF       @ Move base value of immediate operand ino
                                   R0.
85.       lsr    R2,R3,#7          @ "Almost" right justify shift count (times 2)
86.       and    R2,#0b11110       @ Total shift count is double the value in field.
87.       ror    r0,R2             @ Complete the immediate value using a rotate.
88.       d_int  dec,hstcom        @ Display the immediate value in decimal.
89.       d_int  hex,hstat         @ Display the immediate value in hexadecimal.
90.       b      dpfin             @ Go finish the disassembly with a line feed.
91.
```

Listing 9.13d: Second operand is immediate value

```
92. @        The 2nd operand is a shift. Display the register to be shifted and shift
             type.
93.
94. shreg:   d_fld reg,3,0,comma      @ Base register is in bits 3 down to 0
95.          d_fld shift,6,5,comma    @ Shift type is in bits 6 down to 5.
96.          tst   R3,#0b010000       @ Check if "constant" shift or register value.
97.          bne   shrr               @ If bit set, shift count is in register.
```

```
98.             mov     R0,R3, lsl #20      @ Copy and left justify immediate shift count.
99.             lsr     R0,#27              @ Right justify shift count to "shake off" bits.
100.            d_int   dec,hst             @ Display the immediate shift in decimal.
101.            b       dpfin               @ Go finish the disassembly with a line feed.
102.
103. @          The 2nd operand is a register shifted by the contents of a register
104.
105. shrr:      d_fld   reg,11,8,blank      @ Shift register is in bits 11 down to 8
106.
107. @          Finish the disassembled instruction with a line feed and then return.
108.
109. dpfin:     ldr     R1,=nl              @ Line feed to finish the disassembled
                                            instruction.
110.            bl      v_ascz
111.            pop     {R0-R4,LR}          @ Restore saved register contents.
112.            bx      LR                  @ Return to the calling program
```

Listing 9.13e Second operand is a register shift

```
113.
114. @          Lists of mnemonic names used in various fields with ARM instruction
                format
115.
116.            .align  3           @ Start on 8-byte boundary.
117. opcode:    ascz8   and         @ 0000 [Rd] = [Rn] AND (2nd operand)
118.            ascz8   eor         @ 0001 [Rd] = [Rn] ExclusiveOr (2nd operand)
119.            ascz8   sub         @ 0010 [Rd] = [Rn] - (2nd operand)
120.            ascz8   rsb         @ 0011 [Rd] = (2nd operand) - [Rn]
121.            ascz8   add         @ 0100 [Rd] = [Rn] + (2nd operand)
122.            ascz8   adc         @ 0101 [Rd] = [Rn] + (2nd operand) + C
123.            ascz8   sbc         @ 0110 [Rd] = [Rn] (2nd operand) + C - 1
124.            ascz8   rsc         @ 0111 [Rd] = (2nd operand) - [Rn] + C - 1
125.            ascz8   tst         @ 1000 [Rn] AND (2nd operand) => status bits
126.            ascz8   teq         @ 1001 [Rn] ExOr (2nd operand) => status bits
127.            ascz8   cmp         @ 1010 [Rn] + (2nd operand) => status bits
128.            ascz8   cmn         @ 1011 [Rn] - (2nd operand) => status bits
129.            ascz8   orr         @ 1100 [Rd] = [Rn] Inclusive OR (2nd operand)
130.            ascz8   mov         @ 1101 [Rd] = [Rn]
131.            ascz8   bic         @ 1110 [Rd] = [Rn] AND NOT (2nd operand)
132.            ascz8   mvn         @ 1111 [Rd] = NOT [Rn]
133.
134. cond:      ascz8   eq          @ 0000 Equal (zero); Z set
135.            ascz8   ne          @ 0001 Not equal (non-zero); Z clear
136.            ascz8   hs          @ 0010 Unsigned hiher or same; C set -- also "cs"
137.            ascz8   lo          @ 0011 Unsigned lower; C clear --also "cc"
138.            ascz8   mi          @ 0100 Minus or negative; N set
139.            ascz8   pl          @ 0101 Plus or positive; N clear
```

```
140.              ascz8    vs         @ 0110 Overflow; V set
141.              ascz8    vc         @ 0111 No overflow; V clear
142.              ascz8    hi         @ 1000 Unsigned higher; C set and Z clear
143.              ascz8    ls         @ 1001 Unsigned lower or same; C clear or Z set
144.              ascz8    ge         @ 1010 Signed greater than or equal to; N = V
145.              ascz8    lt         @ 1011 Signed less than; N not same as V
146.              ascz8    gt         @ 1100 Signed greater than; Z clear and N = V
147.              ascz8    le         @ 1101 Signed less than or equal; Z set or N <> V
148.              ascz8    \0         @ 1110 Always; any status bits OK -- also "al"
149.              ascz8    ??         @ 1111 Never (reserved) -- also "nv"
150.
151. reg:         ascz8    R0         @ 0000 General purpose Registers
152.              ascz8    R1         @ 0001
153.              ascz8    R2         @ 0010
154.              ascz8    R3         @ 0011
155.              ascz8    R4         @ 0100
156.              ascz8    R5         @ 0101
157.              ascz8    R6         @ 0110
158.              ascz8    R7         @ 0111
159.              ascz8    R8         @ 1000
160.              ascz8    R9         @ 1001
161.              ascz8    R10        @ 1010
162.              ascz8    R11        @ 1011
163.              ascz8    R12        @ 1100
164.              ascz8    LR         @ 1101 Could also be "R13"
165.              ascz8    SP         @ 1110 Could also be "R14"
166.              ascz8    PC         @ 1111 Could also be "R15"
167.
168. shift:       ascz8    lsl        @ 00 Logical Shift Left
169.              ascz8    lsr        @ 01 Logical Shift Right
170.              ascz8    asr        @ 10 Algebraic Shift Right
171.              ascz8    ror        @ 11 Rotate Right (Circular shift)
172.
173. s:           ascz8    \0         @ 0 Set status bit "s" is zero
174.              ascz8    s          @ 1 Set status bit "s" is one
175.
176. tab:         .asciz   "\t"       @ Strings to be displayed in instructions.
177. comma:       .asciz   ","
178. blank:       .asciz   " "
179. hst:         .asciz   " #"
180. hstcom:      .asciz   ", #"
181. hstat:       .asciz   "\t@ #0x"
182. nl:          .asciz   "\n"
183.              .end
```

Listing 9.13f: Tables used by das_dp subroutine

Now that we have a completed version of subroutine das_dp, let's rerun the program from the Prototype section to disassemble some data processing instructions. The command lines are provided in Listing 9.14:

1. We do not need to recompile the main program, model.s, appearing in Listing 9.0 since it has not been changed.
2. The v_ascz, v_hex, and v_dec subroutines have already been added to the view.a object library as shown in Listing 9.12.
3. The "nano das_dp" command line represents the modifications just made to subroutine das_dp, which of course, could have been done using any editor. Substituting "cp RPi_Asm_Bridge/Listing_9_13.txt das_dp.s" would also have worked (see Appendices A and K).
4. The linker command line combines the main program with the das_dp subroutine and the other subroutines needed from the view.a object library.

```
~$ nano das_dp.s
~$ as -o das_dp.o das_dp.s
~$ ld -o model model.o das_dp.o view.a
~$ ./model
```

Listing 9.14: Sequence of edit, compile, link, and execute

Listing 9.15 shows the program's output which is a completed version of the output presented in the Prototype section..

```
eor       R5,R6,R7,lsr R8
ands      R0,R3, #255      @ #0xFF
andeqs    R2,R2, #30       @ #0x1E
mov       R4,R0, #512      @ #0x200
addne     R1,R1,R5,lsl #0
mov       R2,R0,R3,lsr #7
mov       R5,R0,R5,ror R2
andeq     R0,R0,R0,lsl #0
```

Listing 9.15: Output from disassembler program

Listing 9.16 pairs each of the sample instructions from Listing 9.0 lines 13 through 20 with its disassembled output appearing in Listing 9.15.

```
13.    eor      R5,R6,R7,lsr R8  @ Using four different registers
       eor      R5,R6,R7,lsr R8
14.    ands     R0,R3,#0xFF      @ Immediate value with setting CPSR bits
       ands     R0,R3, #255      @ #0xFF
15.    andeqs   R2,#0b11110      @ Immediate with CPSR bits used twice
       andeqs   R2,R2, #30       @ #0x1E
16.    mov      R4,#512          @ Move immediate value
       mov      R4,R0, #512      @ #0x200
17.    addne    R1,R5            @ Add instruction with conditional execution
       addne    R1,R1,R5,lsl #0
18.    lsr      R2,R3,#7         @ Move instruction with shift.
       mov      R2,R0,R3,lsr #7
19.    ror      R5,R2            @ Simple move with rotate
       mov      R5,R0,R5,ror R2
20.    .word    0                @ End of list of instructions to disassemble.
       andeq    R0,R0,R0,lsl #0
```

Listing 9.16: Comparison of input and output from das_dp subroutine.

Some interesting things we discover by comparing the disassembled machine code to the original assembler code:

1. Line 16 shows that the operand register field Rn is zero when an immediate value is loaded into the destination register.
2. Line 17 shows that all register to register data moves involve a shift, even if the shift doesn't do anything (shift 0).
3. Lines 18 and 19 show that all shifts are really move instructions.
4. Line 20 shows that almost every binary value can be interpreted as a machine code instruction, even binary zero.

Emulator

CPU designs vary by their instruction sets as well as the size and number of registers. A CPU emulator is a software program that enables machine code instructions designed for one CPU to be run on another. For example, an executable program written in Z80 machine code can be run on an ARM processor using an emulator program. An emulator takes the disassembler concept one step further: it not only decodes each instruction, but actually executes the program.

Maintenance

Review Questions

1. How is a macro similar to a subroutine?
2. * How is a macro different from a subroutine?
3. * Give an example of a useful macro that generates neither any instructions nor any data.
4. Why do you think the 12-bit immediate constant was divided into an 8-bit base with a 4-bit shift count? Would a 7-bit base with a 5-bit shift be any better or worse?
5. * An emulator doesn't have to be programmed in assembly language. What would be the advantage of writing one in C or another higher level language?

Programming Exercises

1. Go back and rewrite subroutine das_ls from Lab 8 using the macros developed in this lab.
2. Finish building the view.a library by assembling and then "archiving" the object modules for the display subroutines in the previous labs.

Lab 10
Fixed Point

We've worked with integers in the previous labs, and now we begin to work with the floating point format used to represent fractions as well as very large and very small real numbers needed in science and engineering. However, not all numbers with a decimal point should be represented in floating point.

Prototype

The main program in Listing 10.0 takes the sum of a list of integers and displays them in decimal. The comments beginning with line 23 state that there is an implied decimal point in each of the numbers. The two decimal digits enable this example to represent dollars and cents or meters and centimeters.

1.	.global	_start	@ Provide program starting address to linker
2. _start:	ldr	R6,=tstlst	@ Point to list of test values.
3.	mov	R3,#0	@ Initialize running total to zero.
4. sample:	ldr	R0,[R6],#4	@ Load next 32-bit value from list.
5.	bl	v_dec	@ Display the next number in the list.
6.	ldr	R1,=nl	@ Point to memory buffer containing Line Feed character.
7.	bl	v_ascz	@ Put each number in list on a separate list.
8.	add	R3,R0	@ Add new value to running total.
9.	cmp	R0,#0	@ Test if we've reached end of list of test values.
10.	bne	sample	@ Stop test loop after processing a zero value.
11.	ldr	R1,=dashes	@ Point to memory buffer containing dashes.
12.	bl	v_ascz	@ Separate the list from the total.
13.	mov	R0,R3	@ Move running total to R0 for display.
14.	bl	v_dec	@ Display final sum of all values in list.
15.	ldr	R1,=nl	@ Point to memory buffer containing Line Feed character.
16.	bl	v_ascz	@ Put each number in list on a separate list.
17.	mov	R0,#0	@ Exit Status code set to 0 indicates "normal completion"
18.	mov	R7,#1	@ Service command code 1 terminates this program.

```
19.          svc      0           @ Issue Linux command to terminate program
20.
21.          .data
22. tstlst:                       @ List of integers representing fixed point
                                  values.
23.          .word    25          @ Represents 0.25
24.          .word    111         @ Represents 1.11
25.          .word    -267        @ Represents -2.67
26.          .word    1234        @ Represents 12.34
27.          .word    -7          @ Represents -0.07
28.          .word    0           @ End of list
29. dashes:  .asciz   "-----\n"   @ Line to separate list from sum
30. nl:      .asciz   "\n"        @ Go to new line
31.          .end
```

Listing 10.0: Main program to calculate sum of list of integers

Go ahead and run the sample program. For the link command, you'll need the view.a library from the previous lab for the v_dec and v_ascz subroutines. Listing 10.1 shows the output which provides the correct answer, but it certainly would be helpful to actually see the decimal point. This example could have represented meters and millimeters if there were three decimal places, but again where's the decimal point? That will be solved by adapting a copy of v_dec to become v_fix2 in the Coding and Debugging section

```
25
111
-267
1234
-7
0
-----
1096
```

Listing 10.1: Output from program in Listing 10.0

Introductions

Lab 10 introduces no new commands, instructions, or services, but does provide additional programming practice.

Principles

The main reason for including this rather simple lab about "fixed point" is to dispel the idea that all numbers appearing with a decimal point are internally stored as floating point. The second reason is to assist you in understanding the floating point format. I know I've not defined floating point yet, so here's a quick definition: Floating point format is a container for real numbers, like those from measurements and those required in scientific notation such as 6.02×10^{23}.

Whole Numbers and Integers

Most people count: one, two, three, etc. which provides a system known as the natural or counting numbers. When we include zero, we now have the set of whole numbers. Including negative numbers in the set such as …,-3, -2, -1, 0, 1, 2, 3, … gives us the set of integers (where the … extends to infinity in the negative and positive directions). Because binary computers store numbers in a finite number of bits, we are limited to a range of values that can be represented:

Range of numbers represented by an eight-bit byte:

- Whole numbers: 0 to 255
- Integers: Negative 128 to positive 127 using two's complement

Range of numbers represented by a sixteen-bit half-word:

- Whole numbers: 0 to 65,535
- Integers: -32,268 to +32,267 using two's complement

Range of numbers represented by a 32-bit word:

- Whole numbers: 0 to 4,294,967,295
- Integers: -2,147,483,648 to +2,147,483,647 using two's complement

Range of numbers represented by a 64-bit double-word:

- Whole numbers: 0 to 18,446,744,073,709,551,616
- Integers: -9,223,372,036,854,775,808 to +9,223,372,036,854,775,807

There are two problems with the above:

1. The range is too small. Common scientific values like the number of molecules in a gram of a substance would require about 25 decimal digits, which is far beyond that available in even the 64-bit double precision format.
2. No fractions or mixed numbers are available, not even ½ or ¼ let alone something like 360.40.

Floating point addresses both of the above problems, but introduces a couple new problems of its own: loss of precision and reduced performance. The proper use and understanding of floating point is somewhat complicated, so I will be introducing it over the next few labs while also presenting alternatives.

Fixed Point

You can bank on it. Although monetary systems using dollars and cents may look like floating point numbers because they have a decimal point, they're generally stored as fixed point integers. The main reason is precision, and a second reason is performance.

- Precision: Even a value like 1.10_{10} cannot be accurately stored in floating point format (See Lab 12 for details).
- Performance: Floating point format is more complicated than binary integer format resulting in slower computations as well as slower conversion between text format and floating point format.

Packed decimal format, which is a cross between binary and character formats, was once very popular for business and accounting applications. Some mainframe computers and even early microcomputers had special machine code instructions for performing arithmetic on what looked somewhat like a string of base 10 digits.

Coding and Debugging

Subroutine v_dec has been modified slightly to become subroutine v_fix2 which still displays a number as decimal digits, but also displays a decimal point. This format could represent dollars and cents or meters and centimeters. The position of the decimal point is assumed and is fixed at two in this modification, and there will be a minimum of three digits displayed. I put the v_fix2 subroutine into a separate file, v_fix2, that will have to be linked with the main program.

```
1.              .global    v_fix2            @ Make entry point external for linker access.
2.
3. @            Subroutine v_fix2 will display a fixed point number with 2 decimal places
4. @                       R0: contains a number to be displayed in decimal
5. @
6. @                       LR: Contains the return address
7. @                       All register contents will be preserved
8.
9. v_fix2:      push       {R0-R8,LR}        @ Save contents of registers R0 through R8, LR
10.
11.             mov        R3,R0             @ R3 will hold a copy of input word to be displayed.
12.             mov        R2,#1             @ Number of characters to be displayed at a time.
13.             mov        R0,#1             @ Code for stdout (standard output, i.e., monitor)
14.             mov        R7,#4             @ Linux service command code to write string.
```

Listing 10.2a: Subroutine entry point and initialization of registers

Lines 1 and 9 in Listing 10.2a show the subroutine name change from v_dec to v_fix2.

```
15.
16. @          If bit-31 is set, then register contains a negative number and "-" should be output.
17.
18.            cmp    R3,#0          @ Determine if minus sign is needed.
19.            bge    absval         @ If positive number, then just display it.
20.            ldr    R1,=msign      @ Address of minus sign in memory
21.            svc    0              @ Service call to write string to stdout device
22.            rsb    R3,R3,#0       @ Get absolute value (negative of negative).
23.
24. @          Get highest power of ten this number will use (i.e., is it greater than 10?, 100?, ...)
```

```
25.
26. absval:   ldr     R6,=pow10+12    @ Point to 1000's column of power of ten table.
27.           sub     R8,R6,#8        @ Mark position for decimal point.
28. high10:   ldr     R5,[R6],#4      @ Load next higher power of ten.
29.           cmp     R3,R5           @ Test if at the highest power of ten needed.
30.           bge     high10          @ Continue search for larger power of ten.
31.           sub     R6,#8           @ We stepped two integers too far.
```

Listing 10.2b: Display minus sign and determine number of decimal digits

Line 26 was modified so that a minimum of three digits will be displayed (one's column and two decimal places). Line 27 is new and loads register R8 with the address to identify where the decimal point will be needed. See line 49 for its use.

```
32.
33. @         Loop through powers of 10 and output each to the standard output
              (stdout) monitor display.
34.
35. nxtdec:   ldr     R1,=dig-1       @ Point to 1 byte before "0123456789" string
36.           ldr     R5,[R6],#-4     @ Load next lower power of 10 (move right 1 dec
                                      column)
37.
38. @         Loop through the next base ten digit to be displayed (i.e., thousands,
              hundreds, ...)
39.
40. mod10:    add     R1,#1           @ Set R1 pointing to the next higher digit '0'
                                      through '9'.
41.           subs    R3,R5           @ Do a count down to find the correct digit.
42.           bge     mod10           @ Keep subtracting current decimal column
                                      value
43.           addlt   R3,R5           @ We counted one too many (went negative)
44.           svc     0               @ Write the next digit to display
45.
46. @         Insert decimal point between hundred's and ten's columns
47.
48.           ldr     R1,=dpoint      @ Load memory address of decimal point
                                      character.
49.           cmp     R6,R8           @ Test if R6 is pointing to thousand's column
50.           svceq   0               @ Display the decimal point between digits.
51.           cmp     R5,#10          @ Test if we've gone all the way to the one's
                                      column.
52.           bgt     nxtdec          @ If 1's column, go output rightmost digit and
                                      return.
```

Listing 10.2c: Loop to display each decimal digit except one's column

The decimal point is displayed using the code on lines 45 through 50 which are new.

```
53.
54.@        Finish decimal display by calculating the one's digit.
55.
56.          ldr      R1,=dig           @ Pointer to "0123456789"
57.          add      R1,R3             @ Generate offset into "0123456789" for one's
                                        digit.
58.          svc      0                 @ Write out the final digit.
59.
60.          pop      {R0-R8,LR}        @ Restore saved register contents
61.          bx       LR                @ Return to the calling program
62.
63.          .data
64.pow10:    .word    1                 @ 10^0
65.          .word    10                @ 10^1
66.          .word    100               @ 10^2
67.          .word    1000              @ 10^3 (thousand)
68.          .word    10000             @ 10^4
69.          .word    100000            @ 10^5
70.          .word    1000000           @ 10^6 (million)
71.          .word    10000000          @ 10^7
72.          .word    100000000         @ 10^8
73.          .word    1000000000        @ 10^9 (billion)
74.          .word    0x7FFFFFFF        @ Largest integer in 31 bits (2,147,483,647)
75.dig:      .ascii   "0123456789"      @ ASCII string of digits 0 through 9.
76.msign:    .ascii   "-"               @ Needed for negative decimal numbers.
77.dpoint:   .ascii   "."               @ Decimal point.
78.          .end
```

Listing 10.2d: Display one’s column and the data tables

Line 77 is new and simply contains the decimal point character. The main program from the Prototype section will only have to be modified to call subroutine v_fix2 instead of v_dec as shown on lines 5 and 14 in Listing 10.3.

```
1.           .global  _start            @ Provide program starting address to linker
2._start:    ldr      R6,=tstlst        @ Point to list of test values.
3.           mov      R3,#0             @ Initialize running total to zero.
4.sample:    ldr      R0,[R6],#4        @ Load next 32-bit value from list.
5.           bl       v_fix2            @ Display the next number in the list.
6.           ldr      R1,=nl            @ Point to memory buffer containing Line Feed
                                        character.
7.           bl       v_ascz            @ Put each number in list on a separate list.
8.           add      R3,R0             @ Add new value to running total.
9.           cmp      R0,#0             @ Test if we've reached end of list of test
                                        values.
10.          bne      sample            @ Stop test loop after processing a zero value.
11.          ldr      R1,=dashes        @ Point to memory buffer containing dashes.
```

12.	bl	v_ascz	@ Separate the list from the total.
13.	mov	R0,R3	@ Move running total to R0 for display.
14.	bl	v_fix2	@ Display final sum of all values in list.
15.	ldr	R1,=nl	@ Point to memory buffer containing Line Feed character.
16.	bl	v_ascz	@ Put each number in list on a separate list.
17.	mov	R0,#0	@ Exit Status code set to 0 indicates "normal completion"
18.	mov	R7,#1	@ Service command code 1 terminates this program.
19.	svc	0	@ Issue Linux command to terminate program
20.			
21.	.data		
22. tstlst:			@ List of integers representing fixed point values.
23.	.word	25	@ Represents 0.25
24.	.word	111	@ Represents 1.11
25.	.word	-267	@ Represents -2.67
26.	.word	1234	@ Represents 12.34
27.	.word	-7	@ Represents -0.07
28.	.word	0	@ End of list
29. dashes:	.asciz	"-----\n"	@ Line to separate list from sum
30. nl:	.asciz	"\n"	@ Go to new line
31.	.end		

Listing 10:3: Two small modifications to main program in model.s

You have two options for linking: include v_fix2.o on the linker command line (Figure 10.4) or include the v_fix2 subroutine in the view.a library along with the other display subroutines (Figure 10.5). The r (replace), v (verbose), and s (symbol table) options are typically used on the ar (archive) command line when adding (or replacing) an object in a library.

```
~$ nano model.s
~$ as -o model.o model.s
~$ nano v_fix2.s
~$ as -o v_fix2.o v_fix2.s
~$ ld -o model model.o v_fix2.o view.a
~$ ./model
```

Listing 10.4: Edit, compile, link, and execute with v_fix2 separate

```
~$ nano model.s
~$ as -o model.o model.s
~$ nano v_fix2.s
~$ as -o v_fix2.o v_fix2.s
~$ ar rvs view.a v_fix2.o
~$ ld -o model model.o view.a
~$ ./model
```

Listing 10.5: Alternate approach where v_fix2 is added to view.a library

Go ahead and run the new code. Listing 10.6 shows the new output including the two decimal places.

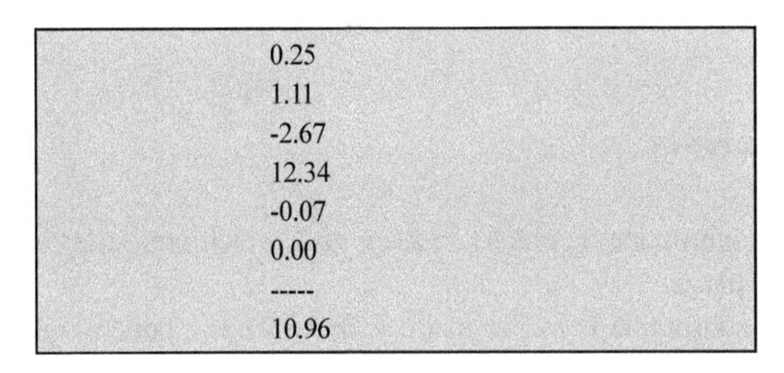

```
0.25
1.11
-2.67
12.34
-0.07
0.00
-----
10.96
```

Listing 10.6: Fixed point display showing "implied" decimal points

Maintenance

Review Questions

1. * What are the advantages of fixed point over floating point?
2. What limitation does fixed point have that would require the use of floating point to remedy?
3. Briefly describe "packed decimal" arithmetic as described in an article on the Internet?

Programming Exercises

1. Modify subroutine v_fix2 to make v_fix3 which supports the display of three decimal places.
2. Modify subroutine v_fix2 to make v_fix which is a generalization of subroutines v_fix2 and v_fix 3 where the number of decimal places is provided in register R2.

Lab 11
Binary Point

A decimal point separates a decimal whole number from a decimal fraction. Likewise, a binary point separates a binary whole number from a binary fraction. How can we represent fractions in binary? In lab 11, we will continue to ease ourselves into floating point format by first working with binary fractions and converting them to decimal for display. We learned in the previous lab that due to performance and precision concerns, many applications use integers to represent what appears to be fractions. Likewise, the technique presented in this lab not only helps to introduce floating point format, but also can be chosen as an alternative to floating point in many applications.

Prototype

The main program in Listing 11.0 loops through a list of binary fractions and calls subroutine v_flt0 to display them in decimal. In each of the binary fractions, the most significant bit (i.e., bit 31 in each word) has the value of ½ followed by a bit with a value of ¼, etc.

```
1.              .global  _start        @ Provide program starting address to linker
2. _start:      ldr      R6,=tstlst    @ Point to list of test values.
3. sample:      ldr      R0,[R6],#4    @ Load next 32-bit "fraction" test value.
4.              bl       v_flt0        @ Call subroutine to display "fraction" in R0 as
                                       ASCII.
5.              cmp      R0,#0         @ Test if we've reached end of list of test
                                       values.
6.              bne      sample        @ Stop test loop after processing a zero value.
7.
8.              mov      R0,#0         @ Exit Status code set to 0 indicates "normal
                                       completion"
9.              mov      R7,#1         @ Service command code (1) will terminate this
                                       program
10.             svc      0             @ Issue Linux command to terminate program
11.
12.             .data
13. tstlst:                            @ List of test fractions.
14. F1_2:       .word    0x80000000    @ .5 (1/2)
```

```
15. F1_4:      .word     0x40000000    @ .25 (1/4)
16. F7_8:      .word     0xE0000000    @ .875 (7/8)
17. F9_16:     .word     0x90000000    @ .5625 (9/16)
18. F1_32:     .word     0x08000000    @ .03125 (1/32)
19. F63_64:    .word     0xFC000000    @ .984375 (63/64)
20.            .word     0             @ End of list
21.            .end
```

Listing 11.0: Main test program for subroutine v_flt0

The assembler .word directive assumes whole numbers are being input which are right justified in a 32-bit word. The fractions that we plan to use are left justified. We could, of course, write a macro for fractions to left justify them, but we'll only need it in this one example, and I want to emphasize that fractions are left justified. For the sake of compactness, the above list of examples of fractions is entered in hexadecimal on lines 14 through 19:

- ½ => $10000000000000000000000000000000_2 = 80000000_{16}$
- ¼ => $01000000000000000000000000000000_2 = 40000000_{16}$
- ¾ => $11000000000000000000000000000000_2 = C0000000_{16}$
- 7/8 => $11100000000000000000000000000000_2 = E0000000_{16}$
- 9/16 => $10010000000000000000000000000000_2 = 90000000_{16}$
- 1/32 => $00001000000000000000000000000000_2 = 08000000_{16}$
- 63/64 => $11111100000000000000000000000000_2 = FC000000_{16}$

As shown in Listing 11.1, an unsigned multiply instruction in a loop converts the binary fraction to decimal digits for display. The UMULL instruction multiplies the contents of two 32-bit registers resulting in a 64-bit product that is stored in two 32-bit registers. For example, multiplying 0.314 times 10 would provide the whole number 3 in one register and the fraction 0.14 in another. The "3" would be displayed in the current pass thought the loop, while the "1" and "4" would be displayed in the next two passes through the loop. The CPU hardware really doesn't "know" that the fraction is a "fraction," but works with it as if it was a very large 32-bit positive number.

```
1. @          Subroutine v_flt0 will display a "fraction" in decimal digits
2. @                       R0: contains the number to be displayed
3. @                                       Bit 31 = 1/2, Bit 30 = 1/4, Bit 29 = 1/8, Bit 28 =
                                           1/16, ...
4. @                       LR: Contains the return address
5. @                       All register contents will be preserved
6.
7.            .global      v_flt0          @ Provide entry point address to linker
8.
9. v_flt0:    push         {R0-R8,LR}      @ Save contents of registers R0 through R8
                                           and LR.
10.           mov          R3,R0           @ R3 will hold a copy of input word to be
                                           displayed.
11.           mov          R4,#10          @ Base 10 used to "shift" over each digit.
12.           ldr          R5,=dig         @ Set R5 pointing to "0123456789" string
13.           mov          R2,#1           @ Number of characters to be displayed at a
                                           time.
14.           mov          R0,#1           @ Code for stdout (standard output, i.e.,
                                           monitor display)
15.           mov          R7,#4           @ Linux service command code to write string.
16.
17. @         Loop through powers of 10 and display each digit on the standard output
              (stdout).
18.
19. nxtdfd:   umull        R3,R1,R4,R3     @ "Shift" next decimal digit into register R1.
20.           add          R1,R5           @ Set pointer to digit in "0123456789"
21.           svc          0               @ Write out one digit.
22.           cmp          R3,#0           @ Set Z flag if mantissa is now zero.
23.           bne          nxtdfd          @ Go display next decimal digit.
24.
25.           pop          {R0-R8,LR}      @ Restore saved register contents.
26.           bx           LR              @ return to the calling program
27.
28. dig:      .ascii       "0123456789"    @ ASCII string of digits 0 through 9.
29.           .end
```

Listing 11.1: Subroutine v_flt0 displays binary fraction.

Go ahead and run the sample code. You won't need the other display routines from the view library, but it won't hurt to include view.a when you link the program. Try some other test values, such as 0b11000000000000000000000000000000. It makes a better example in binary, but be sure to include all 32 bits of the word.

Introductions

The only new instruction introduced in Lab 11 is the unsigned multiply.

List of ARM instructions introduced in Lab 11			
11.1.19.	umull	R3,R1,R4,R3	@ Unsigned 64-bit product: {R3},[R1] = [R4] * [R3].

Principles

So why should the high order (leftmost) bit represent ½? Why is a whole number right justified in a register, yet a fraction is left justified?

Shifting decimal digits to the left relative to the decimal point multiplies a number by 10, while shifting to the right divides by 10. Likewise, shifting binary digits (i.e., bits) to the left relative to the binary point multiplies a number by 2, while shifting to the right divides by 2. In Figure 11.0, the whole number is in register R1, and the fraction is in register R0, but any two registers would have worked. In this example, we start out with the whole number 5 in register R1 with a fraction value of 0 in register R0. Dividing by 2 (i.e., shifting right one bit position) gives us a value of 2.5, where R1 contains the number 2 and register R0 has the high order bit set indicating ½. Dividing by 2 again gives us 1.25, where R1 contains the whole number 1, and R0 contains the fraction of .25.

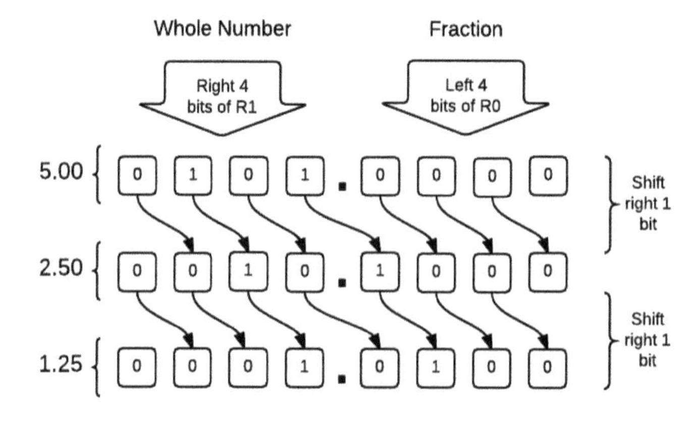

Figure 11.0: Fraction is left justified in register.

We could have even "logically divided" a single 32-bit register in two, where the upper 16 bits would be the whole number and the lower 16 bits the fraction. The register would not even have to be divided in half. It could be 12 bits and 20 bits, for example.

A decimal number is really a short notation for a polynomial of powers of 10. For example: 137_{10} is $1\times10^2 + 3\times10^1 + 7\times10^0$ which is 100 + 30 + 7. Fractions continue that pattern where .036 is $0\times10^{-1} + 3\times10^{-2} + 6\times10^{-3}$. Likewise, a binary number is really a short notation for a polynomial of powers of 2. For example: 10101_2 is $1\times2^4 + 0\times2^3 + 1\times2^2 + 0\times2^1 + 1\times2^0$. The binary fraction $.1101_2$ is $1\times2^{-1} + 1\times2^{-2} + 0\times2^{-3} + 1\times2^{-4}$.

In the example below, it seems reasonable to make the first bit to the right of the binary point equal to 2^{-1} which is ½. The second bit to the right of the binary point is 2^{-2} which is $½^2$ which is ¼.

- $1111._2 = 1\times2^3 + 1\times2^2 + 1\times2^1 + 1\times2^0 = 8 + 4 + 2 + 1$
- $111.1_2 = 1\times2^2 + 1\times2^1 + 1\times2^0 + 1\times2^{-1} = 4 + 2 + 1 + ½$
- $11.11_2 = 1\times2^1 + 1\times2^0 + 1\times2^{-1} + 1\times2^{-2} = 2 + 1 + ½ + ¼$

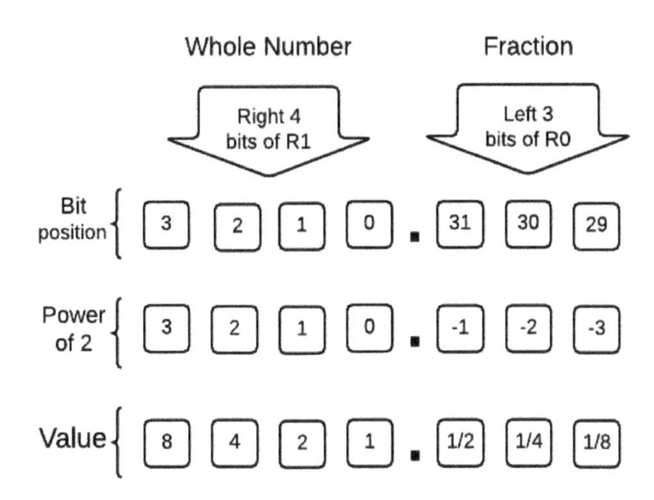

Figure 11.1: Whole number in register R1 and fraction in register R0

Coding and Debugging

We're not going to enhance the v_flt0 subroutine here, but modify it a couple of times to gain a deeper appreciation for representing fractions on binary computers. However for some applications where floating point is unavailable or performance issues arise, binary point is the best option.

Other Bases

Subroutine v_flt0 displays the fraction in base ten, but that can easily be changed to another base by changing the constant on line 11 of Listing 11.1. Try changing the instruction to **mov R4,#7** for base 7 or **mov R4,#3** for base 3. What did you notice about the outputs, especially simple values like ½? Try it again for base 8 and base 4 where ½ should display as .4 and .2, respectively.

One of the things we learn about fractions is that we can't represent all of them accurately. We can come close, depending on how many bits we have available, but some numbers will not be exact, and every time we use them in a calculation, the error will grow in the result. This is something we've become accustomed to in base ten where 1/3 = 0.33333..., 1/7 = 0.142857142857..., and 1/11 = 0.090909... are among an infinite number of fractions that lose precision when representing them in certain bases.

When Is Enough Enough?

For whole numbers, we know how many digits to display, but for fractions, it can vary. Sometimes the v_flt0 subroutine will display a value as 0.400000012 or .399999941 when it really should have been simply 0.4.

The number of decimal places displayed by subroutine v_flt0 is determined by the CMP compare instruction on line 22 of Listing 11.1. The CMP is actually a subtract instruction that doesn't store the results, but does set the status bits. Perhaps, instead of comparing to zero, try comparing to a number like 1000 to see if the loop exits sooner. Of course, the BNE instruction on line 23 would have to be changed to a BGE. Another approach would be to use a TST instruction instead of the CMP instruction. It sets the same status bits as the AND, but the data result is not stored anywhere.

In higher level languages like C, the number of digits displayed to the right of the decimal point can be specified. This not only accommodates the inaccuracy of many fractions, but also acknowledges that floating point numbers generally represent measured data that have a limited precision due to the measurement itself. It does, however, transfer the responsibility of how many digits to display to the application programmer who would also be knowledgeable of the precision of the measurements.

Maintenance

Review Questions

1. * "By hand, without a computer," convert the following decimal fractions into binary and provide the answers in hexadecimal.
    a. 0.5
    b. 0.625
    c. 0.25
    d. 0.03125
    e. 0.0078125
2. * "By hand, without a computer," convert the following binary fractions from hexadecimal back into real numbers in base 10.
    a. .C0000000
    b. .E0000000
    c. .10000000
    d. .50000000
3. * Using a calculator for division by powers of 2, convert the following binary fractions from hexadecimal into the real number that each is "approaching" in base 10.
    a. 33333333
    b. 66666666
    c. CCCCCCCC
    d. E6666666

Programming Exercises

1. Modify the v_flt0 subroutine to display the fractions in hexadecimal. Only two lines of code need to be modified.
2. A couple of modifications to subroutine v_flt0 were suggested in the Coding and Debugging section to reduce situations like .4 being displayed as .400000012, but they would not correct .399999941. What modification would work for both situations?

Lab 12
Floating Point Display

Floating point format sacrifices precision for range. In Lab 12, we will examine the IEEE 754 standard floating point format as well as develop subroutine v_flt for displaying floating point numbers. We've previously seen a somewhat similar structure while examining the "immediate" value for the second operand of ARM data processing instructions.

Prototype

Figure 12.0 has the correct machine code for moving a value of 48 (hexadecimal 30) into a register, but it has oversimplified the immediate value field. Rather than being a single 12 bit binary number having a decimal range of 0 through 4095, the ARM implementation enables a 32-bit number to be generated by combining an 8-bit base value with a 4-bit shift count. It sacrifices four bits of precision in order to provide a greater range. This technique enables any 8-bit pattern to be moved to any even bit-position of a 32-bit register. The other 24 bits of the register will be zeros.

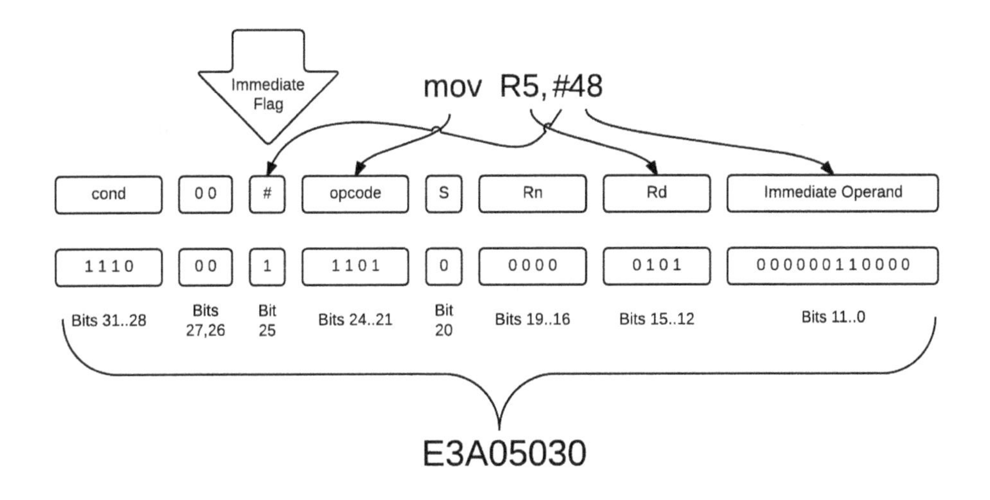

Figure 12.0: "Assemble" MOV with immediate value of 48 (hex 30) to ARM machine code.

As previously discussed in Lab 9 and illustrated in figures 9.3 and 9.4, the immediate value is of the form $\mathbf{M}\times\mathbf{2}^{-\mathbf{S}}$ where **M** is an 8-bit constant and **S** is actually a rotation count to the right, which is somewhat similar to a base 2 exponent. The 4-bit **S** field, having a range of 0 through 15, is doubled to accommodate the complete rotation of a 32-bit register. One thing that is interesting with this approach, and it could have also been true with floating point format, is that some "numbers" can be generated by different combinations of **M** and **S**. Figure 12.1 shows four different ways of loading the integer value 48 (hexadecimal 30) into a register. The first two lines are the same, of course, just being expressed in decimal and hexadecimal, but the others are unique.

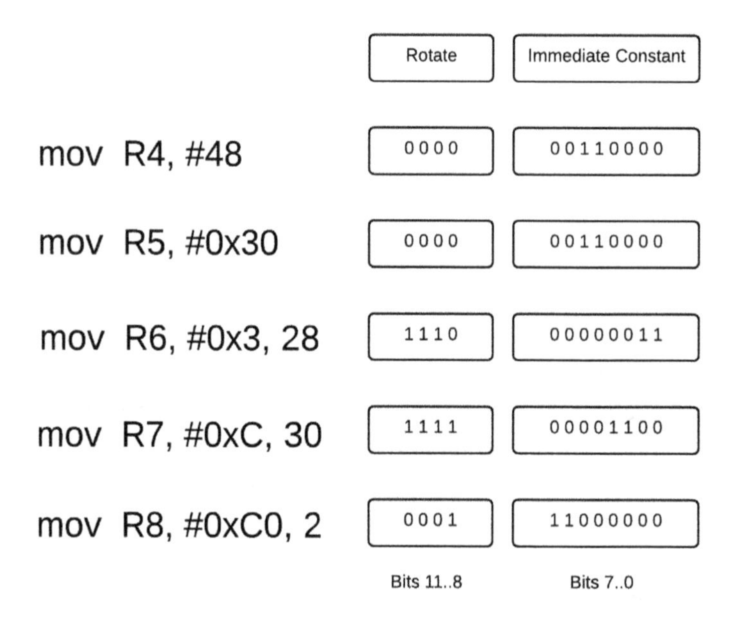

Figure 12.1: Four ways to load an immediate value of 48.

1.		.global	_start	@ Provide program starting address to linker
2.	_start:	mov	R4,#48	@ Load immediate 0x30 into register R4.
3.		.word	0xE3A05030	@ Load R5 with 0x30 rotated 0 bits to the right
4.		.word	0xE3A06E03	@ Load R6 with 0x3 rotated 28 bits to the right
5.		.word	0xE3A07F0C	@ Load R7 with 0xC rotated 30 bits to the right
6.		.word	0xE3A081C0	@ Load R8 with 0xC0 rotated 2 bits to the right
7.		mov	R0,#0	@ Exit Status code set to 0 indicates "normal completion"
8.		mov	R7,#1	@ Service command code 1 terminates this program.
9.		svc	0	@ Issue Linux command to terminate program
10.		.end		

Listing 12.0: Assembly code to load an immediate value of 48

The prototype program for Lab 12 appearing in Listing 12.0 is a short one used to demonstrate four ways of loading the value 48 into registers. Lines 2 through 6 correspond to the five lines of code in Figure 12.1. I normally would not recommend entering instructions directly as machine code using the .word directive, but I wanted to force various combinations of a base value and shift count.

This program is complete in itself, but does require assembly with the "g" option because the debugger will be needed. Compile and link the sample program. Start the program with the gdb debugger, issue a "b 7" breakpoint command, and start the program with the "run" command. The "x/5i _start" dump command will disassemble the first five instructions of the program, including the ones entered as machine code.

```
(gdb) b 7
Breakpoint 1 at 0x8068: file model.s, line 7.
(gdb) run
Starting program: /home/pi/model

Breakpoint 1, _start () at model.s:7
7          mov        R0,#0       @ Exit Status code set to 0 indicates "normal
                                  completion"
(gdb) x/5i _start
0x8054 <_start>:        mov     r4, #48 ; 0x30
0x8058 <_start+4>:      mov     r5, #48 ; 0x30
0x805c <_start+8>:      mov     r6, #3, 28 ; 0x30
0x8060 <_start+12>:     mov     r7, #12, 30 ; 0x30
0x8064 <_start+16>:     mov     r8, #192, 2 ; 0x30
```

Listing 12.1: Immediate values demonstrating base with shift count

The register dump in Listing 12.2 shows that registers R4 through R8 were loaded with the same immediate value, even though four different combinations of base and rotate were used.

```
(gdb) i r
r0     0x0     0
r1     0x0     0
r2     0x0     0
r3     0x0     0
r4     0x30    48
r5     0x30    48
r6     0x30    48
r7     0x30    48
r8     0x30    48
r9     0x0     0
r10    0x0     0
r11    0x0     0
```

r12	0x0	0
sp	0x7efff860	0x7efff860
lr	0x0	0
pc	0x8068	0x8068 <_start+20>
cpsr	0x10	16
(gdb)		

Listing 12.2: Identical results from different immediate source code

Introductions

The assembler .float directive will generate IEEE 754 floating point numbers, which we'll take apart and display using a new subroutine v_flt. The inclusive "or" instruction ORR has been described in previous labs, but not used in a sample program until now.

	List of ARM instructions introduced in Lab 12		
12.3.19.	orr	R3,#0x80000000	@ Set the "assumed" high order bit.

	List of assembler directives introduced in Lab 12
.float	Generate floating point number

Principles

History credits the first working binary floating point computer to Konrad Zuse and his Z3 from 1941. Almost all of the mainframe computers of the 1960s were available with floating point hardware. Many of the minicomputers of the 1970s did as well depending on the manufacturer and model. When the microcomputers emerged in the 1970s and 1980s, they did not initially support floating point hardware. The early ones were eight bit processors like the Motorola 6800 and the Intel 8080. These microprocessors did support 16-bit (double byte) arithmetic and addressing but no floating point hardware. Floating point could be supported in software, of course, but it was extremely slow. Floating point coprocessors like the Intel 8087 and 80287 were developed as options to accompany the Intel 8086 and 80286 CPUs, respectively. Many of the complex instruction set microcomputers (CISC) that followed actually contained floating point arithmetic on the same chip.

Moving real number data (i.e., floating point) from one computer system to another was not impossible, but certainly more difficult than necessary, and it was even prone to error. The problem with the floating point formats present in the mainframes and minicomputers was that although they were almost identical in concept, their implementations were incompatible. In the 1960s, even the size of a floating point number varied: 32 bits, 36 bits, 48 bits, and 60 bits were common, and double precision added another four sizes. Some computers used one's complement; some used two's complement. Most had the exponent in base 2, while one used base 16 and another base 8.

In the 1960s, ASCII was defined to address incompatibility among character sets in different computers. Likewise in 1985, the Institute of Electrical and Electronics Engineers (IEEE) standard 754 was defined to address the incompatibility among floating point formats used by various computer manufacturers. This standard was later refined in 2008, as well as becoming standard ISO/IEC/IEEE 60559:2011.

Floating Point Implements Scientific Notation

When we look at real numbers expressed in scientific notation such as 6.0221409×10^{23}, 9.10938356×10^{-31}, and $-1.60217662\times10^{-19}$ used in science, we observe the following:

1. The number is positive or negative
2. The significant (left of the ×10)
    - Is in base 10
    - Contains a decimal point
    - Has a precision related to the number of digits

3. The exponent (right of the ×10)
    - Can be negative or positive
    - Is in base 10
    - Is a whole number (i.e., although exponents like 5.23 are certainly allowed in mathematics, we only use integers in scientific notation)

So how are these base 10 real numbers with a wide range of values implemented in floating point? Figure 12.2 illustrates the floating point components and their locations within IEEE standard 754's single precision format.

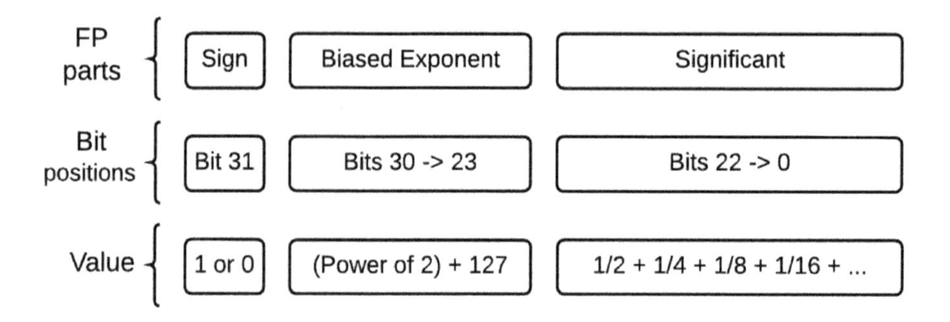

Figure 12.2: Single precision floating point fields in IEEE 754 format

Normalization

In the Prototype section, we were able to represent the same number four different ways by having different combinations of base value and shift count. The floating point fields described above would allow the same flexibility, but to a much greater extent, if it were not for a requirement called "normalization."

Does 220 equal 2.2×10^2 and equal 2200×10^{-1}? Of course. What about binary? Is 110_2 equal to $1.10_2\times2^2$ and equal to $1100_2\times2^{-1}$? That is also true. In scientific notation, a number is expressed in "normalized" form when it has exactly one non-zero digit left of the decimal point. When a floating point value is "normalized," it has exactly one non-zero digit to the left of the binary point. This restriction leads to the following three advantages:

- Each real number is represented by a unique floating point value. Of the above three decimal choices, only 2.2×10^2 is in scientific notation. Of the above three binary numbers, only $1.10_2\times2^2$ is eligible for floating point format.
- Since there are only two binary symbols, and the digit left of the binary point cannot be "0," it must therefore be a "1." For this reason, the IEEE 754 format doesn't include this bit in the 32-bit format, and thereby "gains" an extra bit of precision.
- In "normalized" floating point format, the number of significant digits will be consistent and maximized. Note: I didn't say that the precision of all floating point numbers is equal.

The requirement of normalization was not new when it appeared in the IEEE 754 standard, but was present on all floating point hardware of the 1960s. Of course, it varied somewhat from one manufacturer's implementation to another.

Conversion to IEEE 754 Floating Point

Let's look at a couple of examples to see how a floating point number in IEEE 754 format is constructed. The first example will be the easier one to convert from base 10, since it is only a whole number and requires only the steps listed below:

1. Convert the number to base 2.
2. Normalize it.
3. Bias the base 2 exponent by adding 127, and store it into bits 23 through 30.
4. Store the fractional part of the normalized binary number into bits 0 through 22. Note: Nothing is done with the "1" that is to the left of the decimal point.

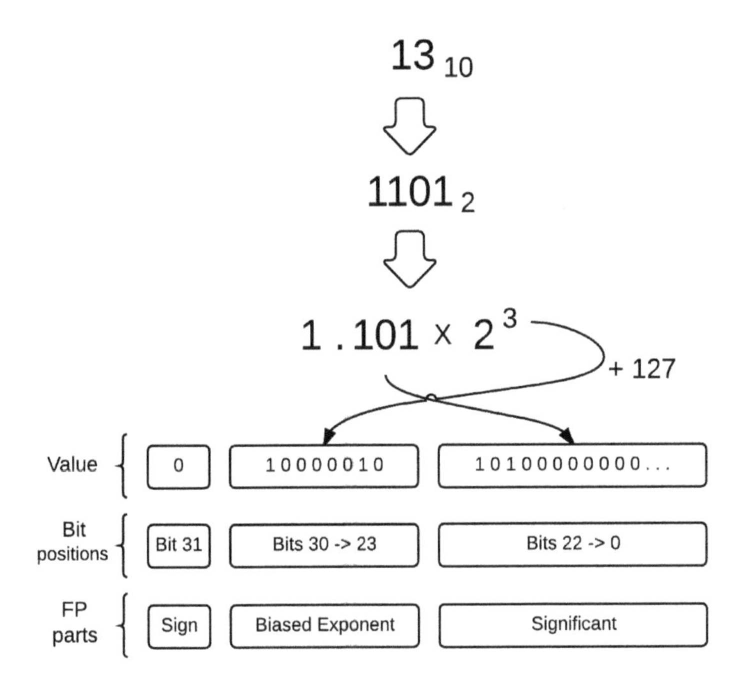

Figure 12.3: Pack 13.0 into single precision floating point fields in IEEE 754 format

Let's take a more thorough examination of the construction of floating point representation using the more complicated example shown in Figure 12.4. Although several different programing approaches can be taken, the following the steps are pretty common:

1. Set the sign bit: 1 if negative, 0 if positive.
2. Convert the base 10 exponent into a base 2 exponent
3. Convert the fraction to base 2 as the significant
4. Normalize the significant
5. Bias the base 2 exponent by adding 127, and store it into bits 23 through 30.
6. Store the significant of the normalized binary number into bits 0 through 22. Note: Nothing is done with the 1-bit that is to the left of the decimal point.

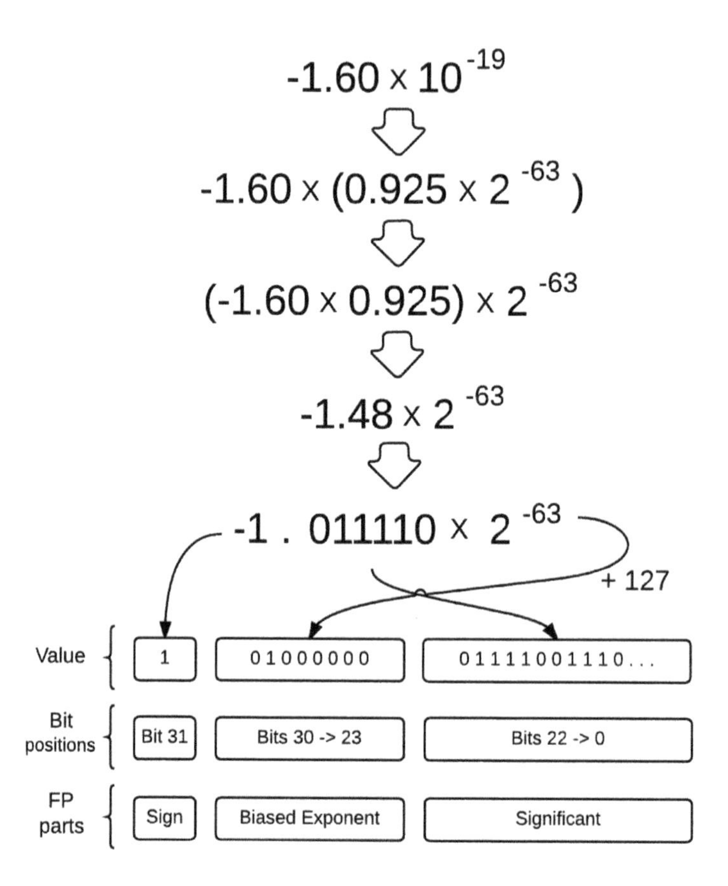

Figure 12.4: Convert scientific notation into floating point format

Why Bias the Exponent?

The obvious answer is that floating point must support a range of both positive and negative exponents. Appendix B describes four ways of indicating negative numbers: sign/magnitude, bias, one's complement, and two's complement. They all work, but why not just pick one and use it consistently? One would think that even though differences arise among different computer manufactures, at least there would be consistency within a single machine. The Raspberry Pi uses three of the four techniques, while some computers of the past, such as the CDC 6600, have employed all four techniques for representing negative numbers.

Both one's and two's complement use the same arithmetic unit for signed and unsigned numbers. They can also extend to virtually any size "word" by combining multiple bytes using the carry flag. The bias format, on the other hand, used in the exponent, enables the integer compare instruction (CMP) to work on floating point numbers. One way to look at it is that floating point is a package, and integer format is homogeneous.

Where Did the Most Significant "1" Bit Go?

We put the sign bit into bit position 31. We biased and put the base 2 exponent into bits 23 through 30. We put the fractional part of the normalized binary number into bits 0 through 22, but we discarded what seems to be the most significant bit of all. Every additional bit included in a binary number doubles its range, so if a bit is "always" going to be the same in the floating point format, why not allow that bit position in the 32-bit word to either extend the range of the exponent or the precision of the significant? Secondly, if the bit is not there in the format, it is nearly impossible to make a non-normalized floating point number.

A Note on Normalization

I have to admit that I took a bit of liberty in the above floating point description. I did so because that description has been the one that my students have found the easiest and quickest to accept. You may find other descriptions where the bias is 128 (hex 80) because that is one half of the 256 range provided in the 8-bit exponent field. The exponent is then decremented because normalization from a hardware viewpoint has the significant being less than one.

Traditionally, a normalized floating point number is defined as one where the significant is shifted until its high order bit is a 1 bit (i.e., the significant is greater than or equal to ½, but less than 1). Unlike IEEE 754, most floating point formats did not remove the high order bit even though it "always" had to be a 1. It was even possible to generate non-normalized floating point numbers, but their use in arithmetic usually produced undesirable results. A special non-normalized case is present in the IEEE 754 standard that will be described in the Lab 13.

Significant or Mantissa?

The terms "significant" and "mantissa" refer to the fractional part in the floating point format and are used somewhat interchangeably in the literature. The term "mantissa" has been used to describe floating point format for several decades beginning in the mid 1940s. The term "significant" is preferred in the IEEE 754 documentation apparently because "mantissa" has been associated with logarithms for centuries, and the "fraction" in the floating point format really isn't a logarithm.

Coding and Debugging

We will gain more understanding of the floating point format by converting examples from IEEE 754 format to base 10 display format. In this section of Lab 12, we will work with a limited range of values by only using "integer" arithmetic. Accommodating very large and very small numbers like $1.2345\times10^{+30}$ and 1.2345×10^{-30} will be left for Lab 18, when the floating point processor will be used. Listing 12.3 contains the v_flt floating point display subroutine that should be compiled and added to the view library.

```
1. @        Subroutine v_flt will display a floating point number in decimal digits.
2. @                  R0: contains the number to be displayed
3. @                  LR: Contains the return address
4. @                  All register contents will be preserved
5.
6.          .global   v_flt               @ Provide entry point address to linker
7.
8. v_flt:   push      {R0-R8,LR}          @ Save contents of registers R0 - R8, LR.
9.
10. @       Output a "minus sign" if number is negative
11.
12.         ldr       R1,=minus           @ Negative sign character "message"
13.         mov       R2,#1               @ Number of characters in message.
14.         movs      R6,R0,lsl #1        @ Move sign bit into "C" flag
15.         blcs      v_asc               @ Display the "-" if sign bit was set.
16.
17. @       Initialize the whole number in R0 and fraction in R3.
18.
19.         mov       R3, R0,lsl #8       @ Left justify normalized mantissa to R3.
20.         orr       R3,#0x80000000      @ Set the "assumed" high order bit.
21.         mov       R0,#0               @ Whole part = 0 (Number < .9999...)
22.         cmp       R6,#0               @ Test if both mantissa and exp = 0.
23.         beq       disp                @ Go display 0.0 if significant and exp = 0.
```

Listing 12.3a: Subroutine v_flt entry point and initialization

```
24.
25. @       Get the exponent and remove its bias
26.
27.         mov       R6,R6,lsr #24       @ Right justify biased exponent.
28.         subs      R6, #126            @ Remove the exponent bias.
29.         beq       disp                @ If exponent = 0, need no shifting.
30.         blt       shiftr              @ Values <.5 must be shifted right.
31.
```

```
32. @          Shift mantissa left: floating point number is greater than (or eq) 1.0
33.
34.            rsb      R5,R6,#32          @ Convert left shift to right shift count.
35.            mov      R0,R3,lsr R5       @ Get the whole number potion of the number.
36.            lsl      R3,R6              @ Get the fractional part of the number
37.            b        disp               @ Go display both whole number and fraction.
38.
39. @          Shift mantissa right (floating point number is less than .5).
40.
41. shiftr:    rsb      R6,R6,#0           @ Calculate positive shift count (to right).
42.            lsr      R3,R6              @ "Divide by 2" for each bit shifted.
43.
44. @          The floating point number is now divided into two registers:
45. @                   R0: Has the whole number (left of the decimal point)
46. @                   R3: Has the fraction (right of the decimal point)
47.
48. @          Display the whole number in base 10.
49.
50. disp:      bl       v_dec              @ Display the number in R0 in decimal digits.
51.
52. @          Display decimal point separating the whole number from the fraction.
53.
54.            ldr      R1,=point          @ Pointer to decimal point.
55.            bl       v_asc              @ Put decimal point into display.
56.
57. @          Display the fraction in base 10
58.
59.            mov      R4,#10             @ Base 10 used to "shift" over each digit.
60.            ldr      R5,=dig            @ Set R5 pointing to "0123456789" string
61.
62. @          Loop through powers of 10 and display each digit.
63.
64. nxtdfd:    umull    R3,R1,R4,R3        @ "Shift" next decimal digit into R1.
65.            add      R1,R5              @ Set pointer to digit in "0123456789"
66.            bl       v_asc              @ Write out one digit.
67.            cmp      R3,#0              @ Set Z flag if mantissa is now zero.
68.            bne      nxtdfd             @ Go display next decimal digit.
69.
70.            pop      {R0-R8,LR}         @ Restore saved register contents.
71.            bx       LR                 @ return to the calling program
72.
73. dig:       .ascii   "0123456789"       @ ASCII string of digits 0 through 9.
74. minus:     .ascii   "-"                @ Negative sign
75. point:     .ascii   "."                @ Decimal point
76.            .end
```

Listing 12.3b: Combine the significant and exponent into one “integer.”

Listing 12.4 contains the main program that will call subroutine v_flt to display a series of IEEE 754 format floating point numbers.

```
1.              .global   _start          @ Provide program starting address to linker
2. _start:      ldr       R6,=tstval      @ Point to list of test values.
3. sample:      ldr       R0,[R6],#4      @ Load next 32-bit floating point test value.
4.              bl        v_flt           @ Subroutine to display floating point number.
5.              ldr       R1,nl           @ Pointer to line ending characters.
6.              bl        v_ascz          @ Separate the test values with new lines.
7.              cmp       R0,#0           @ Test if end of list of test values.
8.              bne       sample          @ Stop test loop after processing a zero value.
9.
10.             mov       R0,#0           @ Exit Status of 0 indicates "normal completion"
11.             mov       R7,#1           @ Service command code 1 terminates program
12.             svc       0               @ Issue Linux command to terminate program
13.
14.             .data
15. tstval:     .float    0.5             @ .5 (1/2)
16.             .float    0.25            @ .25 (1/4)
17.             .float    -1.0
18.             .float    100.0
19.             .float    1234.567
20.             .float    -9876.543
21.             .float    7070.7070
22.             .float    3.3333
23.             .float    694.3e-9
24.             .float    6.0221e2
25.             .float    6.0221e23
26.             .word     0               @ End of list
27. nl:         .asciz    "/n"            @ Line ending characters.
28.             .end
```

Listing 12.4: Main test program for subroutine v_flt

As show in Listing 12.5, both the main program in file model.s and the subroutine in v_flt.s will have to be separately compiled. I recommend adding v_flt to the view library using the "ar" command line. Both object files will then be linked along with any other subroutines needed from the view library.

```
~ $ nano v_flt.s
~ $ as -o v_flt.o v_flt.s
~ $ ar rvs view.a v_flt.o
~ $ nano model.s
~ $ as -o model.o model.s
~ $ ld -o model model.o view.a
~ $ ./model
```

Listing 12.5: Build program to display floating point format numbers.

Go ahead and run the program using the 11 sample data examples included in the main program. Table 12.0 provides the output corresponding to each test value. The first thing we notice is that all of the output except for the last line is very close to the value generated using the assembler's .float directive. The second thing we notice is that some of the examples have too many many digits displayed, some of which are not correct. We'll look more into this in Lab 14 as well as a programming exercise in the Maintenance section, but can you think of a couple very easy ways to correct that now? Also, try some other test values of your own, such as 0.1 or 0.2.

Assembler text	**Output from v_flt**
.float 0.5	0.5
.float 0.25	0.25
.float -1.0	-1.0
.float 100.0	100.0
.float 1234.567	1234.5670166015625
.float -9876.543	-9876.54296875
.float 7070.7070	7070.70703125
.float 3.3333	3.333300113677978515625
.float 694.3e-9	0.000000694068148732185363769531 25
.float 6.0221e2	602.21002197265625
.float 6.0221e23	0.0

Table 12.0: Output from v_flt for 11 IEEE 754 values

Maintenance

Review Questions

1. * "By hand, without a computer," convert the following real numbers into single precision IEEE 754 floating point and provide the answers in hexadecimal.
    a. 128.0
    b. 9.25
    c. -9.25
    d. 0.03125
    e. 128.03125
    f. 0.0
    g. -0.0
2. * "By hand, without a computer," convert the following IEEE 754 floating point numbers from hexadecimal back into real numbers in base 10.
    a. 42a80000
    b. C1A80000
    c. 424C8000
    d. BF100000
    e. 3DCCCCCD
3. Which of the four ways to represent negative numbers described in Appendix B allows for both a positive and a negative zero?
4. In IEEE 754 floating point format, zero is represented by both the significant and the exponent being zero. If this was not the case, what value, expressed as a power of 2, would a word of all zero bits represent?
5. Why is it impossible to have a non-normalized IEEE 754 format value?

Programming Exercises

1. Lines 67 and 68 in Listing 12.3 control the loop through each of the digits that is displayed by the v_flt subroutine. Replace either or both of these instructions with something similar to eliminate the extra debris of digits that are incorrectly displayed.
2. Modify the v_flt subroutine so that register R2 contains an input argument providing the number of decimal digits to display. Be sure to accommodate rounding to the desired number of places, not just truncation.

Lab 13
Floating Point Coprocessor

Besides the ARM Central Processing Unit, the Raspberry Pi has other specialized processors for input/output, system control, and floating point operations. In Lab 13, we will examine features common to both the VFPv3 and NEON coprocessors using a program that multiplies two floating point numbers and displays the result.

Prototype

Floating point operations are performed in their own set of registers using their own set of instructions. In the prototype example in Listing 13.0, floating point registers S0 and S1 are loaded with floating point values of -3.5 and 10.5 from the ARM's memory. The two values are multiplied on line 5 leaving the result in register S0. This result is then moved back into the ARM's R0 register so that it can be displayed using subroutine v_flt.

```
1.          .global     _start        @ Provide program starting address to linker
2. _start:  ldr         R6,=tstval    @ Point to list of test values.
3. sample:  vldr        S0,[R6]       @ Load next 32-bit floating point test value.
4.          vldr        S1,[R6,#4]    @ Second operand for floating point operation.
5.          vmul.f32    S0,S0,S1      @ Multiply the two floating point operands
6.          vmov        R0,S0         @ Move product to ARM register for display.
7.          bl          v_flt         @ Subroutine to display floating point number.
8.          ldr         R1,=nl        @ Pointer to line ending characters.
9.          bl          v_ascz        @ Separate the floating point test values.
10.         mov         R0,#0         @ Exit Status code 0 indicates "normal
                                      completion"
11.         mov         R7,#1         @ Command code to terminate this program.
12.         svc         0             @ Issue Linux command to terminate program
13.         .data
14. tstval: .float      -3.5, 10.5    @ Floating point test values
15. nl:     .asciz      "\n"          @ Line ending characters.
16.         .end
```

Listing 13.0: Floating point multiplication of two numbers

You'll need the view library from previous labs for the v_flt and v_ascz subroutines. The edit, compile, link, and execute sequence of commands will be the same as before, but either the "-mfpu=vfp3" or "-mfpu=neon" option must be included on the assembler command as shown in Listing 13.1.

```
~$ nano model.s
~$ as -g -o model.o model.s -mfpu=vfp3
~$ ld -o model model.o view.a
~$ ./model
-36.75
```

Listing 13.1: Edit, compile, link, and execute for floating point

Go ahead and run the sample code. Substitute some other operand values on line 14, and rerun the test. You should also try other operations by replacing the VMUL on line 5 with VADD or VDIV. The ".F32" suffice must be on each of the preceding operations to indicate that 32-bit floating point values are used.

Introductions

All of the new instructions introduced in this lab involve floating point, including those instructions that move data into and out of the floating point registers.

	List of Floating Point instructions introduced in Lab 13		
13.0.3.	vldr	S0,[R6]	@ Load next 32-bit floating point test value.
13.0.5.	vmul.f32	S0,S1	@ Multiply the two floating point operands
13.0.6.	vmov	R0,S0	@ Move the product back into ARM register.
13.5.3.	vldm	R6!, {S0,S1}	@ Load next two 32-bit values.

Principles

The floating point coprocessor is an independent processor having its own set of registers and instructions. It does not contain its own memory for program and data storage, but "watches" all the instructions that the ARM processor is fetching. Basically, a floating point operation is performed by the following programming sequence:

1. One or more coprocessor instructions move source data into the desired floating point registers.
2. Perform the desired floating point operation using a coprocessor instruction.
3. Use another coprocessor instruction to move the floating point result into either one of the ARM's registers or memory.

There are 32 32-bit floating point data registers named S0 through S31, where the "S" implies single-precision. These registers can be grouped in various ways for vector operations or double precision, however in this lab, we will only use the registers in their simplest mode.

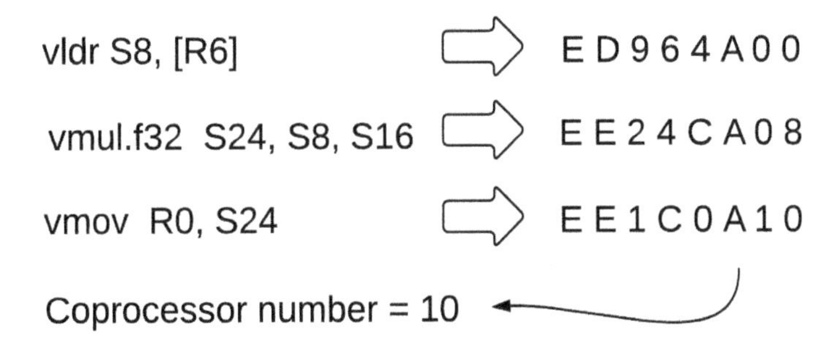

Figure 13.0: Coprocessor ID of 10 (hex 0xA) indicates single precision floating point instructions.

Figure 13.0 shows the three floating point instructions similar to those in Listing 13.0 converted into machine code. I won't be breaking these instructions apart like I did for the ARM instructions, but I do want you to notice one thing all three have in common: the coprocessor ID number. Keep in mind the following points regarding the execution of floating point operations:

1. The ARM is fetching the instructions from memory, but a coprocessor will actually perform the specified operation.
2. The ARM knows the ID number of all coprocessors available in its configuration and will set an error code if an instruction is fetched with an invalid ID number.
3. All coprocessors present on the chip will be watching all instructions loaded by the ARM and will process the instructions that match their unique ID numbers.

At first glance, it almost sounds like the ARM and its coprocessors are isolated entities, but the above points indicate that they are actually very tightly coupled.

Data can be moved between the ARM's registers and the floating point processor's registers with the VMOV instruction. Data can also be moved between memory and the floating point registers using the VLDR and VSTR instructions. It almost seems like these data load and store instructions are just like the corresponding ARM LDR and STR instructions, but they're not because the "write back" feature that updates ARM index registers is not always available. Note: There are additional instructions for "interleaved" and multiple-register to memory transfers that will be presented in the next two labs, and they will have index-register write-back capability.

```
vldr    S16,[R6,#4]   @ Pre-indexed mode is allowed.
vldr    S16,[R6,#4]!  @ Pre-indexed mode write-back is not allowed.
vldr    S16,[R6],#4   @ Post-indexed mode write-back is not allowed.
```

Listing 13.2: Automatic updating of index registers is not possible.

The other point to notice at this time is that the floating point operations have a suffix of ".F32" after the name of the operation. In the next two labs, we will have several variations of this suffice that indicate the type and size of the operands and operations performed within the coprocessors, but in this lab we will only work with single precision floating point.

Multiprocessors

Having coprocessors is not a new idea. During the 1960s, it was common for mainframes from Burroughs, Control Data, and Univac to have multiprocessors. During the 1980s, microcomputers, such as the Intel 8086, had coprocessors for I/O and floating point. However, thanks to today's fabrication techniques, the Raspberry Pi is able to have not only the ARM processor, but its coprocessors physically located on the same silicon "chip."

Not a Number (NaN)

The IEEE 754 format includes a "value" known as NaN (Not a Number) which results from operations like square root of a negative number or division by zero. There are other special cases where the exponent is either all one bits or all zero bits as shown in Figure 13.1. These cases can be generated by floating point instructions, and can also be used as operands in floating point instructions.

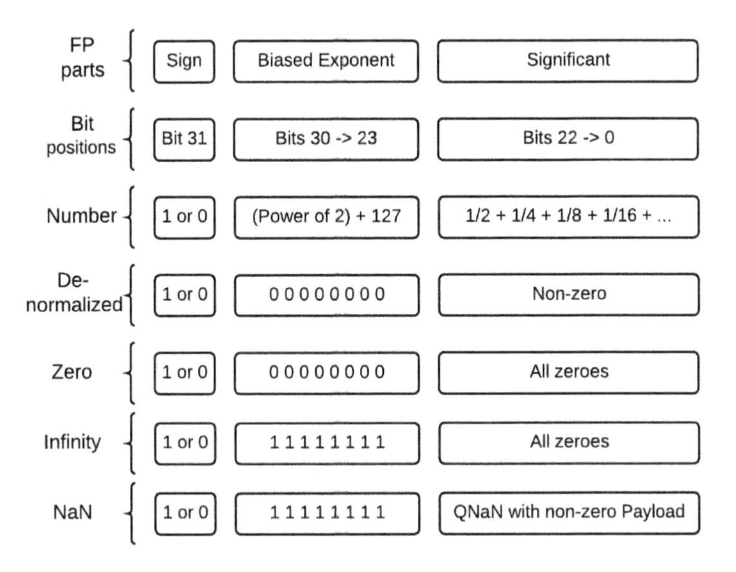

Figure 13.1: IEEE 754 floating point numbers and special cases.

The following observations can be made from Figure 13.1:

- It would be impossible to make a number that is not normalized if it were not for these de-normalized special cases. How could you normalize a floating point value of zero? You can't. That "assumed" high order 1-bit in the significant of the IEEE 754 would always get in the way. How small can a normalized number be? How small can it be if de-normalized, yet still not zero?
- All four of the special cases can be both positive and negative. Positive and negative infinity are certainly different, but positive and negative zero really are the same "value."
- Infinity is certainly "not a number" from a math perspective, but there are also other cases within the IEEE 754 standard as described below.

For an exponent of all one-bits, infinity is identified by its all-zero significant, while the two types of NaN are identified by a non-zero "Payload" (definition varies for bits 0 through 21) plus the QNaN quiet flag in bit 22:

- Signaling QNaN=0: If signaling NaN values are in one of the operands of a floating point operation, a CPU floating point interrupt will occur that requires immediate special handling by the operating system and application program.
- Quiet QNaN=1: These quiet NaN values will run through the floating point processor as smoothly as any normalized floating point number. The result will probably be another quiet NaN that will have to be examined later, but no immediate action is necessary.

Coding and Debugging

The purpose of this lab is to get the assembly language programmer comfortable using floating point operations without getting bogged down in the many details of vector operations and loss of precision that arise while using the NEON and VFPv3 coprocessors. The exercise in Listing 13.3 is therefore a rather simple one consisting of a loop that multiples several pairs of single precision floating point numbers.

1.	.global	start	@ Provide program starting address to linker
2. _start:	ldr	R6,=tstval	@ Point to list of test values.
3. sample:	vldr	S0,[R6]	@ Load next 32-bit floating point test value.
4.	vldr	S1,[R6,#4]	@ Load second operand for floating point operation.
5.	vmul.f32	S0,S1	@ Multiply the two floating point operands
6.	vmov	R0,S0	@ Move the product back into ARM register for display.
7.	bl	v_flt	@ Call subroutine to display floating point number.
8.	ldr	R1,=nl	@ Pointer to line ending characters.
9.	bl	v_ascz	@ Separate the floating point test values with new lines.
10.	add	R6,#8	@ Move to next pair of floating point operands.
11.	cmp	R1,R6	@ Test if end of list of test values.
12.	bgt	sample	@ Continue loop until all tests done.
13.	mov	R0,#0	@ Exit Status code set to 0 indicates "normal completion"
14.	mov	R7,#1	@ Service command code 1 terminates this program.
15.	svc	0	@ Issue Linux command to terminate program
16.			
17.	.data		
18. tstval:	.float	3.0, 10.0	@ Floating point test values
19.	.float	0.25, -1.0	
20.	.float	100.0, 1234.567	
21.	.float	-9876.543, 7070.7070	
22.	.float	3.3333, 694.3e-9	
23.	.float	6.0221e2, 6.0221e23	
24.	.float	0, 0	@ End of list
25. nl:	.asciz	"\n"	@ Line ending characters.
26.	.end		

Listing 13.3: Loop through a list of floating point multiplications

Listing 13.4 shows the output from the execution of the program in Listing 13.3. The first four answers seem reasonable, but the last three are strange. The fifth line, resulting from 3.3333 times 694.3×10^{-9}, has "debris" following the expected value of 0.000002314. This problem is not in the multiplication, but from the display provided by subroutine v_flt. Suggestions for correcting this were provided in the previous lab. The next strange output was the zero resulting from the multiplication of 6.0221×10^{2} and 6.0221×10^{23} which is clearly beyond the intended range of subroutine v_flt as it was programmed in the previous lab.

```
30.0
-0.25
123456.703125
-69834144.0
0.00000231410376727581024169921875
0.0
0.5
```

Listing 13.4: Output from loop of floating point multiplications

The value of 0.5 on the last line is also an error in v_flt. If I knew about it, why didn't I already correct it in the code? See question number 1 in the Maintenance section to gain some additional insight into the floating point format.

Square Root?

We know floating point operations have been available from the 1930s, and the floating point processors on commercial computers in the 1960s and 1970s were limited to add, subtract, multiply, and divide. Today's floating point processors following the IEEE 754 standard also include operations like square root and "fused" multiplication. In the next two labs, we will employ a few of these other instructions.

The program in Listing 13.5 produces exactly the same output as the previous example, but uses a Vector Load Multiple instruction on line 3. Not only can we load both floating point registers with one instruction, but we can also increment the base register, thereby replacing three instructions with one.

```
1.          .global     _start          @ Provide program starting address to linker
2. _start:  ldr         R6,=tstval      @ Point to list of test values.
3. sample:  vldm        R6!,{S0,S1}     @*Load next two 32-bit values.
4.          vmul.f32    S0,S0,S1        @ Multiply the two floating point operands
5.          vmov        R0,S0           @ Move the product into ARM register for
                                        display.
6.          bl          v_flt           @ Subroutine displays floating point number.
7.          ldr         R1,=nl          @ Pointer to line ending characters.
8.          bl          v_ascz          @ Separate the floating point test values.
9.          cmp         R1,R6           @ Test if end of list of test values.
10.         bgt         sample          @ Continue loop until all tests done.
11.         mov         R0,#0           @ Status code 0 indicates "normal completion"
12.         mov         R7,#1           @ Service command code to terminate program.
13.         svc         0               @ Issue Linux command to terminate program
14.
15.         .data
16. tstval: .float      3.0, 10.0       @ Floating point test values
17.         .float      0.25, -1.0      @ .25 (1/4)
18.         .float      100.0, 1234.567
19.         .float      -9876.543, 7070.7070
20.         .float      3.3333, 694.3e-9
21.         .float      11.08, 360.195
22.         .float      12.0e20, 12.0e-22
23.         .float      6.0221e2, 6.0221e23
24. nl:     .asciz      "\n"            @ Line ending character.
25.         .end
```

Listing 13.5: Load Multiple floating point registers with auto-increment

NEON or VFPv3

Are NEON and VFPv3 instruction sets the same? No, but there is an overlap for the simpler instructions. Both “processors” have unique definitions and are available with other CPUs besides the ARM.

Maintenance

Review Questions

1. * Why does the v_flt program display 0.5 when it should display 0 for a floating point number consisting of all 32 zero bits?
2. * By examining Figure 13.1, what is the smallest absolute value non-zero normalized number?
3. By examining Figure 13.1, what is the smallest absolute value non-zero de-normalized number?
4. By examining Figure 13.0, how many coprocessors do you think the ARM architecture can support?
5. What type of data processing would work best with quiet NaN values?
6. What type of data processing would work best with signaling NaN values?
7. What are the advantages of the Raspberry Pi being a monolithic SOC System On a Chip, where the ARM and the coprocessors are on the same silicon “chip"?

Programming Exercises

1. Modify the v_flt subroutine to handle the special cases of zero, infinity, and NaN.
2. Using the modified version of v_flt from above, use the .word directive to construct test value operands for the above special cases. Be sure to try both the quiet and signaling forms of NaN.

Lab 14
Precision

Errors creep into measured data while it's being measured, when it's put into floating point format, and during calculations. In Lab 14, we will examine ways to minimize the loss of precision of real numbers and begin to consider performance issues while we work with both the VFPv3 and NEON coprocessors.

Prototype

Listing 14.0 is nearly identical to Listing 13.5 from the previous lab, except it has been modified to use double precision. Each double precision value is defined as a 64-bit package rather than the 32-bit package defined for single precision. Because it has "double" the number of bits used for single precision, the following changes are needed in the program:

- Registers D0 and D1 are used instead of S0 and S1.
- The instruction suffix of ".F64" is used instead of ".F32."
- The 64-bit result is converted back into single precision using the VCVT.F32.F64 instruction because the v_flt display subroutine expects single precision.
- The assembler directive ".double" is used to generate the 64-bit test values.
- Each double precision value uses 8 bytes of storage instead of 4.

1.	.global	_start	@ Provide program starting address to linker
2. _start:	ldr	R6,=tstval	@ Point to list of test values.
3. sample:	vldm	R6!,{D0,D1}	@ Load next two 64-bit values.
4.	vmul.f64	D0,D1	@ Multiply the two floating point operands
5.	vcvt.f32.f64	S0,D0	@ Convert product back into single precision
6.	vmov	R0,S0	@ Move the product into ARM register for display.
7.	bl	v_flt	@ Call subroutine to display floating point number.
8.	ldr	R1,=nl	@ Pointer to line ending characters.
9.	bl	v_ascz	@ Separate the floating point test values.
10.	cmp	R0,#0	@ Test if end of list of test values.

11.	bne	sample	value.
12.	mov	R0,#0	@ Exit Status code 0 indicates "normal completion"
13.	mov	R7,#1	@ Service command code to terminate this program.
14.	svc	0	@ Issue Linux command to terminate program
15.	.data		
16. tstval:	.double	3.0, 10.0	@ Double precision floating point
17.	.double	0.25, -1.0	@ .25 (1/4)
18.	.double	100.0, 1234.567	
19.	.double	-9876.543, 7070.7070	
20.	.double	3.3333, 694.3e-9	
21.	.double	6.0221e2, 6.0221e23	
22.	.double	0, 0	@ End of list
23. nl:	.asciz	"\n"	@ Line ending characters.
24.	.end		

Listing 14.0: Perform a calculation using double precision floating point values

Go ahead and run the sample code. As in previous prototype coding samples, substitute some different test values and try some different operations like VADD and VDIV.

The output provided in Listing 14.1 is identical to that from the previous lab using single precision. In this example double precision was of no benefit.

```
30.0
-0.25
123456.703125
-69834144.0
0.000002314103767275810241699218 75
3990.960693359375
1.439999938011169433 59375
0.0
```

Listing 14.1: Output from loop of floating point multiplications

Introductions

The new instructions introduced in this lab are almost identical to those previously seen except they now apply to 64-bit floating point numbers. Note: The 64-bit exclusive-or instruction is only available with NEON, so the source code must be compiled with the "-mfpu=neon" option.

	List of Floating Point instructions introduced in Lab 14		
14.0.3.	vldm	R6!,{D0,D1}	@ Load next two 64-bit values.
14.0.4.	vmul.f64	D0,D1	@ Multiply the two floating point operands
14.0.5.	vcvt.f32.f64	S0,D0	@ Convert product back into single precision
14.2.3.	veor.64	D2,D2	@ Initialize running total to zero.
14.2.5.	vmla.f64	D2,D0,D1	@ Multiply 2 operands and add to total.

	List of assembler directives introduced in Lab 14
.double	Generate double precision floating point number

One of the differences in the machine code between the double precision and single precision instructions is the coprocessor number as shown in Figure 14.0

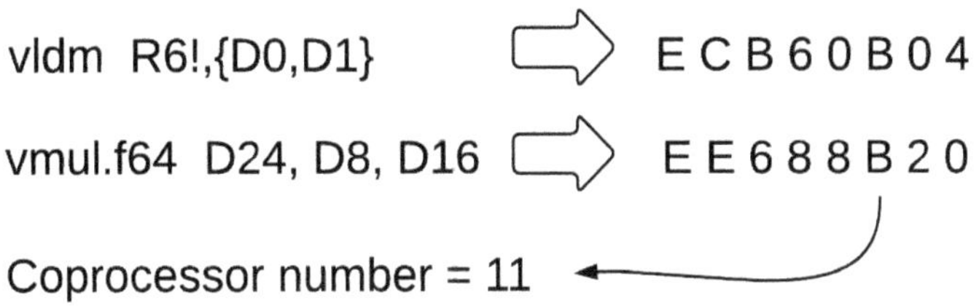

```
vldm  R6!,{D0,D1}            =>   E C B 6 0 B 0 4
vmul.f64  D24, D8, D16       =>   E E 6 8 8 B 2 0
Coprocessor number = 11  <-----------------'
```

Figure 14.0: Double precision uses coprocessor ID of 11

Principles

"I often say that when you can measure what you are speaking about and express it in numbers you know something about it; but when you cannot measure it, when you cannot express it in numbers, your knowledge is of a meagre and unsatisfactory kind: it may be the beginning of knowledge, but you have scarcely, in your thoughts, advanced to the stage of *science*, whatever the matter may be." Lord Kelvin, 1883

"Expressing a measurement in numbers" involves errors and uncertainty. Measurement involves comparing against a standard value. There is always an error or uncertainty in the measurement. Is the length 3.2 meters, or 3.21 meters, or 3.213 meters. etc? How close we need the measurement depends on the application. Errors related to computer processing of real number data can be categorized as the following:

1. Error in the original measurement
2. Error limitation of storing the number in a fixed number of bits
3. Error increasing with each arithmetic operation in a series of operations for a given calculation

We cannot do much about the first source of error once the measurement arrives at the computer, but we can minimize errors introduced by the latter two by increasing the number of bits for storing real numbers and reducing and reordering the arithmetic operations on the data.

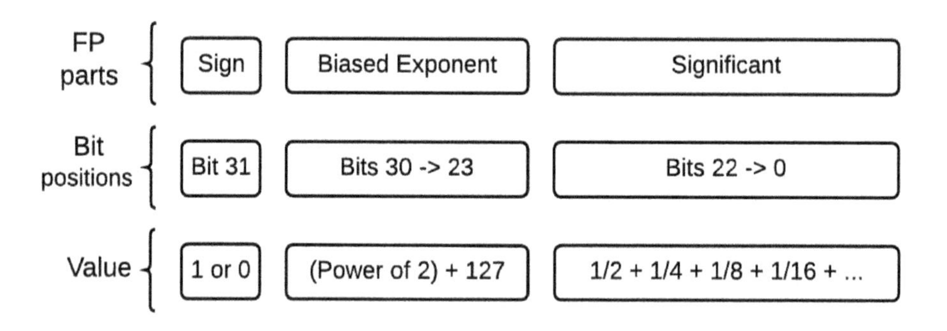

Figure 14.1: Single precision floating point fields in IEEE 754 format

Figures 14.1 and 14.2 show the IEEE 754 internal structure of single and double precision numbers, respectively. The sign bit definition is the same, but both the significant and exponent have more bits in the double precision format:

- The eleven bits allocated to the double precision exponent extends its range to 10^{308} and 10^{-308} compared to 10^{38} and 10^{-38} for single precision. Note: $\log(2)\times2^{10} = 308.25$ and $\log(2)\times2^{7}= 38.53$.
- The 53 bits of the double precision significant (including the "assumed" normalizing bit) can hold the number 9,007,199,254,740,992 which provides nearly 16 significant digits of precision in most cases. The single precision significant, having 24 bits, can hold the number 16,777,216 which is typically about 7 significant digits.

With double precision, we approximately double the precision, but increase the range by a factor of eight. Are precision and significant digits the same? No, but they are related. As long as we have the same exponent, they will be approximately the same.

With double precision format, we sacrifice some of the possible precision by giving up bits for the storage of the exponent. As will be considered in the next lab on performance, a 64-bit integer has approximately three more decimal significant digits of precision than what is possible in the 64-bit floating point format.

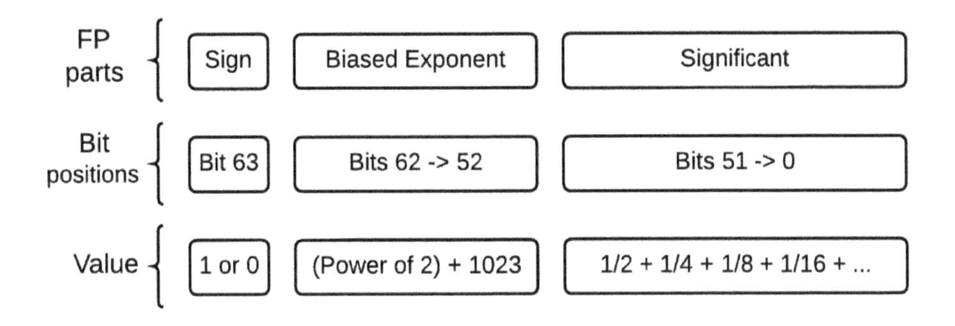

Figure 14.2: Double precision format in IEEE 754

Accuracy and Precision

Accuracy refers to how close the measurement is to some "true value" that is being investigated. Precision refers to the consistency of measurement: If the same quantity is measured 10 times, how much do the measurements differ from each other? For example, are all the measurements within a millimeter of each other?

Loss of accuracy is controlled by scientific or engineering design, but within a computer program, how can we minimize the loss of precision? Having enough bits available is one obvious answer, but also reducing and reordering the number of operations is also critical. Consider the following two examples for multiplication and subtraction:

- 101.1×102.4: In this example, the error is assumed to be plus/minus 0.1, so the error in the product spreads to $101.0 \times 102.3 = 10{,}332.3$ to $101.2 \times 102.5 = 10{,}373.0$, which is an error 39.7.
- $8.234567 - 8.234553 = 0.000014$: When we subtract two floating point numbers that are very close to each other, we can quickly lose several orders of magnitude of precision. In this example we start with two operands with seven significant digits and end with a result with only two.

There are techniques that may help. For example A(B–C) would be better executed as AB–AC, particularly if A is large and B nearly equals C. A second way to hold onto some precision is to use the VMLS instruction for "fused" multiply with subtract. Please see the example in the next section using "fused" multiply with addition to get some ideas.

One way to maintain precision is by reducing the number of calculations on the data. In this section, the example program will employ the floating point “fused” multiply with addition (VMLA) instruction, which is also available as an integer instruction in the ARM CPU. You might say that we saved a little time by loading two “instructions” at once, but we still performed two arithmetic operations and therefore still lost some precision with each.

In most implementations of the IEEE 754 standard, the floating point numbers are stored in either 32-bit or 64-bit registers, but the actual calculations are performed at a much higher resolution such as 80 bits. So with the “fused” instructions, the coprocessor does not have to lose any detectable precision between the multiply and the add in the VMLA and VMLS floating point instructions.

Dot Product

A very common calculation in physics, engineering, and graphical display applications is the “dot product” which can be calculated by the product of the magnitudes of two vectors and the cosine of the angle between them. It can also be calculated by the sum of the products of each of the respective vector components in a Cartesian coordinate system. In general, vectors can have more than three dimensions, and the sample program in Listing 14.2 calculates the dot product of two vectors in six dimensions.

Of course, you recognize this example as identical to the previous with two modifications:

1. Register D2 is initialized to zero on line 3. Note how I zeroed it: by taking the exclusive-or of the register with itself. This instruction is a special feature of the NEON coprocessor which can not only perform floating point operations, but has logical and integer capability as well.
2. The multiply has been changed to a “fused” multiply with addition on line 5. Basically, this program is calculating a running total of products, and the current total is being displayed as it progresses through the loop.

```
1.          .global       _start        @ Program starting address for linker
2. _start:  ldr           R6,=tstval    @ Point to list of test values.
3.          veor.64       D2,D2         @ Initialize running total to zero.
4. sample:  vldm          R6!,{D0,D1}   @ Load next two 64-bit values.
5.          vmla.f64      D2,D0,D1      @ Multiply 2 operands and add to total.
6.          vcvt.f32.f64  S0,D2         @ Convert running total to single precision
7.          vmov          R0,S0         @ Move the total to ARM for display.
8.          bl            v_flt         @ Subroutine displays floating point number.
```

9.	ldr	R1,=nl	@ Pointer to line ending characters.
10.	bl	v_ascz	@ Separate the floating point test values.
11.	cmp	R1,R6	@ Test if end of list of test values.
12.	bgt	sample	@ Stop test loop at end of list.
13.	mov	R0,#0	@ Status 0 indicates "normal completion"
14.	mov	R7,#1	@ Command code to terminate this program.
15.	svc	0	@ Issue Linux command to terminate program
16.	.data		
17. tstval:	.double	3.0, 10.0	@ Double precision floating point
18.	.double	0.25, -1.0	
19.	.double	100.0, 1234.567	
20.	.double	-9876.543, 7070.7070	
21.	.double	3.3333, 694.3e-9	
22.	.double	6.0221e2, 6.0221e23	
23.	.double	0, 0	@ End of list
24. nl:	.asciz	"\n"	@ Line ending characters.
25.	.end		

Listing 14.2: Calculate dot product using fused multiply with addition

Listing 14.3 shows each step of the dot product being constructed using a “fused” multiply with addition. The first five lines seem reasonable, but the final two lines of 0.0 result from the v_flt subroutine not being able to display a number outside its range. Actually, in the real world this example would not occur because the physical values would not vary by such extreme values.

```
30.0
29.75
123486.453125
-69710656.0
-69710656.0
0.0
0.0
```

Listing 14.3: Running total being displayed with each pass through loop

Maintenance

Review Questions

1. A normalized double precision value can be smaller than a normalized single precision value. Can a de-normalized single precision number be smaller than a normalized double precision number?
2. What are the sizes of the exponent and significant in the IEEE 754 16-bit floating point format (used for storage)? Check the Internet for this answer.
3. * Is getting that extra 1-bit of precision in the significant more important to the single precision, double precision, or half precision format numbers?

Programming Exercises

1. Write an ARM subroutine that converts from single to double precision floating point format.
2. Write an ARM subroutine that converts from double to single precision floating point format.

Lab 15
Performance Using Vectors

One way to improve performance is to perform multiple operations in parallel. The NEON and VFPv3 coprocessors provide SIMD (Single Instruction Multiple Data) capability where the same instruction, such as floating point multiplication, is performed on multiple pairs of numbers "at the same time." Do we really get better performance using vectors? Do these operations really take place "at the same time" or not?

Prototype

In the Prototype section, we will multiply three pairs of numbers simultaneously using the VFPv3 coprocessor as a warm up exercise. In the remainder of this lab, the NEON coprocessor will be used to not only perform this same floating point operation more efficiently, but also process integer and logical operations in parallell.

```
1.          .global     _start        @ Provide program starting address to linker
2._start:   ldr         R6,=tstval    @ Point to list of test values.
3.          vldr        S8,[R6]       @ Load pair of floating point numbers to test
4.          vldr        S16,[R6,#4]   @
5.          vmul.f32    S24,S8,S16    @ Multiply S8 and S16 with product into S24.
6.          vmov        R0,S24        @ Move the product back to ARM for display.
7.          bl          v_flt         @ Display floating point number.
8.          ldr         R1,=nl        @ Pointer to line ending character.
9.          bl          v_ascz        @ Separate the floating point result.
10.         mov         R0,#0         @ Status code 0 indicates "normal completion"
11.         mov         R7,#1         @ Service command code 1 terminates this
                                      program.
12.         svc         0             @ Issue Linux command to terminate program
13.tstval:  .float      -3.5, 10.5    @ Floating point test values

14.nl:      .asciz      "\n"          @ Line ending characters.
15.         .end
```

Listing 15.0: Multiply two numbers using VFPv3 "vector" registers.

Listing 15.0 repeats the floating point multiply example from Lab 13, except it is using a special combination of VFPv3 "vector" registers. Go ahead and run this example which will display the result of –36.75. Use the same edit/compile/link/execute sequence as you did in the previous lab, which is repeated in Listing 15.1.

Change the floating point input data to get a little variety, and also use other operations like addition and subtraction instead of the multiplication presented in the example. If you want to run it with the debugger, be sure to try out the memory dump command (x) with the floating point display format (f).

```
~$ nano model.s
~$ as -g -o model.o model.s -mfpu=vfp3
~$ ld -o model model.o view.a
~$ ./model
-36.75
```

Listing 15.1: Edit, compile, link, and execute for floating point

Listing 15.2 provides an example of three pairs of floating point operands being set up to be executed "simultaneously" "in parallel" by a single floating point instruction. Go ahead and run this example which will display the three results of -36.75, 10106.25, and 1050.0. Use the same edit/compile/link/execute sequence as you did in the previous example.

```
1.          .global   _start          @ Provide program starting address to linker
2. _start:  ldr       R6,=tstval      @ Point to list of test values.
3.          vldr      S8,[R6]         @ Load pair of floating point numbers to test
4.          vldr      S16,[R6,#4]     @
5.          vldr      S9,[R6,#8]      @* Load second pair of numbers.
6.          vldr      S17,[R6,#12]    @*
7.          vldr      S10,[R6,#16]    @* Load third pair of numbers.
8.          vldr      S18,[R6,#20]    @*
9.          vmrs      R0,FPSCR        @* Copy floating point FPSCR to update it.
10.         mov       R1,#2           @* Number of "simultaneous" operations - 1
11.         orr       R0,R1, lsl #16  @* Fill the "length" field in FPSCR with 3-1
12.         vmsr      FPSCR,R0        @* Restore the floating point PSCR.
13.         vmul.f32  S24,S8,S16      @ Multiply the three pairs of floating point
                                      operands
14.         vmov      R0,S24          @ Move product to ARM register for display.
15.         bl        v_flt           @ Display floating point number.
16.         ldr       R1,=nl          @ Pointer to line ending characters.
17.         bl        v_ascz          @ Separate test values with new lines.
18.         vmov      R0,S25          @* Move product to ARM register for display.
19.         bl        v_flt           @* Display floating point number.
20.         bl        v_ascz          @* Separate test values with new lines.
21.         vmov      R0,S26          @* Move product to ARM register for display.
22.         bl        v_flt           @* Display floating point number.
```

23.	bl	v_ascz	@* Separate test values with new lines.
24.			
25.	mov	R0,#0	@ Status code 0 indicates "normal completion"
26.	mov	R7,#1	@ Command code 1 will terminate program
27.	svc	0	@ Issue Linux command to terminate program
28.			
29. tstval:	.float	-3.5, 10.5	@ Floating point test values
30.	.float	101.0625, 100.0	@*
31.	.float	-5.25, -200.0	@*
32. nl:	.asciz	"\n"	@ Line ending characters.
33.			
34.	.end		

Listing 15.2: Multiplying three pairs of numbers using a single multiply instruction

The new lines appearing in Listing 15.2 are marked with an asterisk in the comments section. Note that although three multiplications have been performed, only the one multiplication instruction on line 13 is present. This example will give the same results, even if compiled to use the NEON coprocessor by assembling it with the "-mfpu=neon" option.

Introductions

New instructions are introduced for performing vector operations in the VFPv3 and NEON coprocessors.

List of VFP instructions introduced in Lab 15			
15.2.9.	vmrs	R0,FPSCR	@* Copy floating point FPSCR to update it.
15.2.12.	vmsr	FPSCR,R0	@* Restore the floating point PSCR.
15.3.6.	vand	D0,D1,D2	@ Perform the logical operation
15.5.22.	vadd.s8	D0,D0,D2	@ Perform 8 additions in parallel
15.11.22.	vaddw.s8	Q0,Q0,D2	@ Perform 8 additions in parallel
15.13.6.	vstm	R1,{D0-D2}	@ Save exact copy for display.
15.13.8.	vst1.8	{D0-D2},[R1]	@ Save as 1 "element" of 8 bits each
15.13.10.	vst3.8	{D0-D2},[R1]	@ Save as 3 "elements" of 8 bits each
15.13.12.	vst3.16	{D0-D2},[R1]	@ Save as 3 "elements" of 16 bits each

List of assembler directives introduced in Lab 15		
15.15.39.	.byte	2,5,127,-70,16,120,-120,130

Principles

When we program in assembly language, we lose portability. Likewise, when we target a particular floating point processor, we also lose portability. In the previous two labs, we found it to be pretty easy to use floating point. In Lab 15, we will dive deeper into its use by setting up a vector of multiple floating point operations to take place "at the same time."

Scalars and Vectors

A scalar value consists of a single number, and a vector value consists of a group of numbers. Examples of vectors are the position and velocity of an object in a three-dimensional coordinate system which would have X, Y, and Z components, for example. An example of a scalar is the mass of an object (i.e., one number).

Why all the bother? Are vectors of floating point numbers used that extensively that it's worth adding confusion to push for a performance gain? Consider the following:

1. In physics and engineering, quantities like position, velocity, and acceleration are all vectors that are measurable quantities (i.e., "real" numbers requiring the floating point or binary point representations).
2. Many scientific and engineering problems are solved using matrix transformations and inversions which require many vector-type multiplications and additions.
3. Graphics applications which display the 3-D world mapped onto a 2-D screen require many matrix multiplications which consist of a lot of vector floating point or binary point operations.
4. Digital signal processing and many analog to digital conversions require extensive floating point or binary point operations.

VFPv3 Banks, Len, and Stride

You probably noticed two unusual changes appearing in Listings 15.0 and 15.2:

1. The particular set of floating point registers used
2. The updating of a control register (FPSCR) within the coprocessor

The VFPv3 architecture divides its registers into four banks, where a vector operation must have each of its two operands coming from a separate bank and the result delivered to a third bank.

- Bank 0: Scalar bank consisting of registers S0 - S7 (D0 - D3)
- Bank 1: Vector bank consisting of registers S8 - S15 (D4 - D7)
- Bank 2: Vector bank consisting of registers S16 - S23 (D8 - D11)
- Bank 3: Vector bank consisting of registers S24 - S31 (D12 - D15)

The "VMUL.F32 S24,S8,S16" vector multiply on line 13 of Listing 15.2, has its operands in registers S8 (bank 1) and S16 (bank 2) with the result delivered to S24 (bank 3). The instruction thereby specifies the operation to be performed and the first set of registers. The Floating Point Processor Status and Control Register (FPSCR) specifies the registers for the remaining two multiplications of the vector operation. The len (bits 16-18) and stride (bits 20, 21) fields within the FPSCR specify how many pairs of registers will be multiplied and the offset to get the next group of registers, respectively. By default, both fields are set to zero indicating only one operation is to be performed. Line 11 of Listing 15.2 sets len to be 2, thereby indicating three sets of registers are to be processed. I left the stride field at zero indicating that the second and third sets of registers were adjacent to the first set specified in the instruction. Note that both the len and stride settings are one less than the actual value used.

In the event, you want to add or multiply a vector by a scalar constant, the second operand can be placed in the scalar bank. For example, "VMUL.F32 S24,S8,S0" will multiply each of the contents in registers S8, S9, and S10 by the same value in register S0, and deliver the results to registers S24, S25, and S26, respectively. Of course, this assumes that the len is still set to 2 (3 sets of data) and stride to 0 (subsequent registers offset by 1).

The single precision and double precision registers use the same physical bits within the coprocessor. Register D0 has its 64 bits overlapping with registers S0 and S1, for example. Each instruction specifies whether single (.F32) or double (.F64) precision is used, and this can be changed from one instruction to the next. Because floating point format is a package consisting of three parts, there is no meaningful overlap of content as there is with binary integers. There are no automatic castings (conversions) between single and double precision done by the hardware. If you use the lower 32 bits of the significant of a double precision number as a whole single precision number, you will get some very strange values.

NEON Coprocessor

The NEON coprocessor consists of a 2048-bit register file along with logical, integer, and floating point operations. How are these 2048 bits grouped? As registers, they are grouped as 16 128-bit Quad registers, 32 64-bit Double registers, and 32 32-bit Single registers. All three of these register types share the same physical bits. Figure 15.0 shows the correspondence of the first two Q registers with the D and S registers.

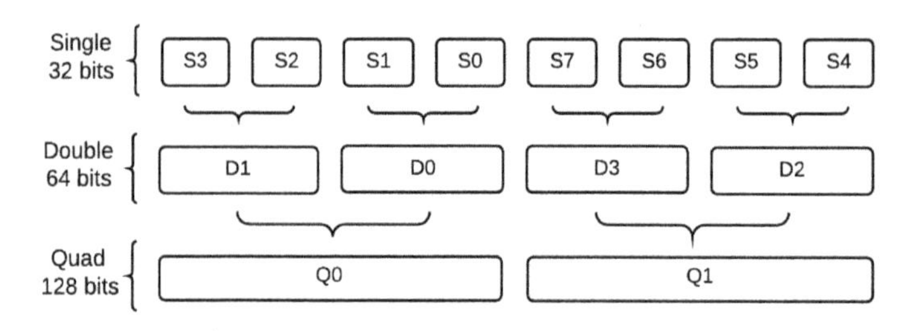

Figure 15.0: Overlap of NEON registers: Bits 31 of S3, 63 of D1, and 127 of Q0 are the same bit.

In Lab 2, we examined bit-by-bit logical operations, such as OR, AND, and exclusive OR, which worked on multiple bits in parallel. NEON supports bit-by-bit logical operations on either 64 or 128 bits at a time. The program in Listing 15.3 demonstrations 64 AND operations performed in parallel between two D registers. The 64-bit hexadecimal pattern 4847464544434241 is loaded into register D1 from the ASCII string "ABCDEFGH." Using a loop, D2 is loaded with three different patterns that are ANDed with D1 and the results displayed in hexadecimal.

```
1.            .global   _start          @ Set program starting address to linker.
2. _start:    ldr       R6,=tstval      @ Point to list of test values.
3.            vldm      R6!,{D1}        @ Load next 64-bit test value.
4.            mov       R2,#32          @ Subroutine v_hexto display 32 bits.
5. sample:    vldm      R6!,{D2}        @ Load second operand for operation.
6.            vand      D0,D1,D2        @ Perform the logical operation
7.            vmov      R0,S1           @ Move the high word into ARM for display.
8.            bl        v_hex           @ Call subroutine to display output.
9.            vmov      R0,S0           @ Move the low word into ARM for display.
10.           bl        v_hex           @ Display 2nd 32 bits.
11.           ldr       R1,=nl          @ Pointer to line ending character.
12.           bl        v_ascz          @ Separate the test values.
13.           cmp       R6,R1           @ Test for end of list of test values.
14.           blt       sample          @ Cotinue loop for next value in list.
15.           mov       R0,#0           @ Status code 0 for "normal completion"
16.           mov       R7,#1           @ Command code to terminate program.
17.           svc       0               @ Issue Linux command to terminate program
18.           .data
19. tstval:   .ascii    "ABCDEFGH"      @ Double words for logical tests
20.           .word     0xFFFFFFFF,0xFFFFFFFF
21.           .word     0xFFFFFFFF,0xFF00FF00
22.           .word     0xFF00FF00,0xFFFFFFFF
23. nl:       .asciz    "\n"            @ End of list and new line character.
24.           .end
```

Listing 15.3: NEON coprocessor performing 64 bit-by-bit logical AND operation

The first 3 lines of Listing 15.4 show the output from executing this program with the vector AND instruction. If the VAND instruction on line 6 is replaced by a VEOR or VORR instruction, the other two outputs will be displayed.

```
4847464544434241
4800460044434241
4847464544004200
------------------------eor below
B7B8B9BABBBCBDBE
B747B945BBBCBDBE
B7B8B9BABB43BD41
--------------------------orr below
FFFFFFFFFFFFFFFF
FF47FF45FFFFFFFF
FFFFFFFFFF43FF41
```

Listing 15.4: Output from 64-bit VAND, VEOR, and VORR NEON instructions

Lanes

NEON also provides integer arithmetic operations. How do the results from an arithmetic operation such as addition or multiplication compare to one of the logical operations? Logical operations are bit-by-bit, and the results stay in each bit "column," but arithmetic operations must expand to use more bits. Even an example such as $1_2+1_2=10_2$ shows addition can have a carry that requires another bit column. In order to provide multiple simultaneous parallel arithmetic operations, the NEON coprocessor forces arithmetic operations to stay in fixed-sized "lanes" and does not allow the results from one lane to carry (or overflow) into the next.

NEON arithmetic lane type and size combinations are specified by the instruction suffix:

- S8, S16, S32, S64: Signed integer: High order bit is the sign bit.
- U8, U16, U32, U64: Unsigned integer: High order bit is a data bit.
- I8, I16, I32, I64 : Integer, either signed or unsigned
- F32: Floating point, single precision

In a NEON arithmetic operation, the total number of bits involved is specified by whether a D (64 bits) or a Q (128 bits) register is used. The number of parallel lanes is therefore either 128 or 64 divided by one of the above lane sizes. For example, "VADD.U16 D0,D1" has four lanes (64/16). Although there can be 128 simultaneous 1-bit logical operations performed, the number of simultaneous arithmetic operations is reduced to the number of lanes.

The maximum number of lanes is sixteen (128-bit Q register divided by an 8-bit lane width). Of course, the minimum number of lanes is one, where a 64-bit D register is divided by a 64-bit lane width. As an example of simultaneous computation, the program in Listing 15.5 adds eight 8-bit integers. It's output demonstrates the results of staying in each lane without overflowing into the next.

```
 1.@          Macro "d_int8 leftbit,space" displays 8-bit integer "lane" in R1.
 2.@                   leftbit:        Bit position of left side of lane
 3.@                   space:          String to be output after integer
 4.@                   Note:           All register contents are preserved.
 5.
 6.           .macro   d_int8          leftbit,space
 7.           .set     left,31-\leftbit
 8.           push     {R0,R1,LR}      @ Preserve registers used by macro call.
 9.           mov      R0,R1,lsl #left @ "Shake off" unwanted bits on the left.
10.           asr      R0, #24         @ Right justify with sign extended.
11.           bl       v_dec           @ Display signed integer.
12.           ldr      R1,=\space      @ String to output after integer
13.           bl       v_ascz          @ Display text string.
14.           pop      {R0,R1,LR}      @ All register contents are restored.
15.           .endm
16.
17.           .global  _start          @ Set program starting address to linker.
18._start:    ldr      R6,=tstval      @ Point to list of test values.
19.           ldr      R2,=tab         @ Point to end of list of test values.
20.           veor     Q0,Q0           @ Zero out register Q0 (same as D0, D1).
21.sample:    vldm     R6!,{D2}        @ Load new 8 "second" operands.
22.           vadd.s8  D0,D0,D2        @ Perform 8 additions in parallel
23.           vmov     R1,S0           @ Move high word into ARM for display.
24.           d_int8   7,tab           @ Display 1st 8-bit integer
25.           d_int8   15,tab          @ Display 2nd 8-bit integer
26.           d_int8   23,tab          @ Display 3rd 8-bit integer
27.           d_int8   31,tab          @ Display 4th 8-bit integer
28.           vmov     R1,S1           @ Move the low word for display.
29.           d_int8   7,tab           @ Display fifth 8-bit integer
30.           d_int8   15,tab          @ Display sixth 8-bit integer
31.           d_int8   23,tab          @ Display seventh 8-bit integer
32.           d_int8   31,nl           @ Display eighth 8-bit integer
33.           cmp      R2,R6           @ Test for end of list of test values.
34.           bgt      sample          @ Cotinue loop for next value in list.
35.           mov      R0,#0           @ Status code 0 for "normal completion"
36.           mov      R7,#1           @ Code terminates program.
37.           svc      0               @ Linux cservice terminates program.
```

Listing 15.5a: Program to demonstrate staying in lanes

```
38.         .data
39. tstval:  .byte    2,5,127,-70,16,120,-120,130
40.         .byte    0,100,-5,-15,-16,115,-115,180
41. tab:     .asciz   "\t"     @ Tab to separate integers.
42. nl:      .asciz   "\n"     @ End of list and new line character.
43.         .end
```

Listing 15.5b: Sample data illustrating overflow within lanes

The program in Listing 15.5 uses eight 8-bit wide lanes for signed integer addition. The following highlights features of this program:

- Lines 1-15: Macro d_int8 will display an 8-bit "lane" in register R1 that has been copied from one of the NEON registers.
- Line 10: Algebraic shift handles sign correctly
- Line 18: Pointer to array of 64 bit test values
- Line 20: An exclusive OR initializes a Q register to zero.
- Line 21: Load next 64-bit test value from ARM memory.
- Line 22: Eight integer additions are performed in parallel.
- Lines 23-27: The first four sums are displayed.
- Lines 28-32: The second four sums are displayed.
- Lines 39-40: Test data demonstrating "staying in the lane"

This program produces eight running totals. Beginning with line 39, each text line contains a set of eight 8-bit integers to be added to the current eight running totals.

```
2    5     127   -70   16   120   -120   -126
2    105   122   -85   0    -21   21     54
```

Listing 15.6: Output from program with 8 simultaneous integer additions

The output from running the program is provided in Listing 15.6. The first line displayed shows the eight sums: 0 + 2 = 2, 0 + 5 = 5, 0 + 127 = 127, 0 + –70 = –70, 0 + 16 = 16, 0 + 120 = 120, 0 + –120 = –120, and 0 + 130 = –126. The sum in the eighth lane, 0 + 130 = –126, is really a misinterpretation of an unsigned value as a signed value. Decimal 130 in eight bits is 100000010_2. By two's complement, 100000010 is – 01111110 or -126_{10}.

The second display line shows the next eight sums: 2 + 0 = 2, 5 + 100 = 105, 127 + –5 = 122, –70 + –15 = –85, 16 + –16 = 0, 120 + 135 = –21, –120 + – 115 = 21, and –126 + 180 = 54. Here, we have more sign-related problems due to the sums overflowing into the sign bit. However, the main point is that the data in each individual lane stays in each individual lane.

The second group of eight simultaneous additions are illustrated in Figure 15.1. The sums in the first five lanes look correct, but the last three lanes have overflow problems (within each lane). In the sixth lane, 120 + 115 really should equal 235, which is 11101011_2 having set the sign bit. By two's complement, 11101011 is –00010101 or $–21_{10}$. The seventh and eighth lanes have similar sign-related problems in their input operands as well.

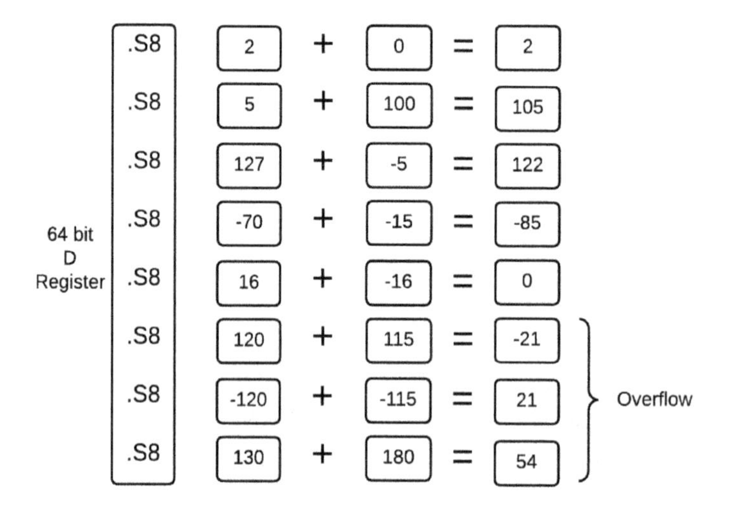

Figure 15.1: Eight lanes added in parallel with three resulting in overflows

S8, U8, or I8

What would be the results if we used VADD.U8 on line 22 and LSR on line 10 so that we would be working with 8-bit unsigned integers? The output is shown in Listing 15.7. Some of the numbers look different, but that's due to the LSR instruction not extending the sign bit. Actually, you'll discover that all of the data bits and the arithmetic are exactly the same in this example.

```
2    5     127   186   16    120   136   130
2    105   122   177   0     235   21    54
```

Listing 15.7: Display of 8 lanes of unsigned integer additions

All of the results will be the same whether you used an I8, S8, or U8 suffix, and this would also be true of 16-bit, 32-bit, and 64-bit lanes. If you run the debugger, you will also notice that exactly the same instruction is generated for VADD.U8, VADD.S8, and VADD.I8. So are these three suffixes synonymous, and is there nothing that can be done about overflow conditions giving ridiculous looking results? The answer to both of these questions is no. Figure 15.2 provides two modifications to NEON instructions to alleviate the overflow problem: Saturation (Q) and Wide (W) lanes.

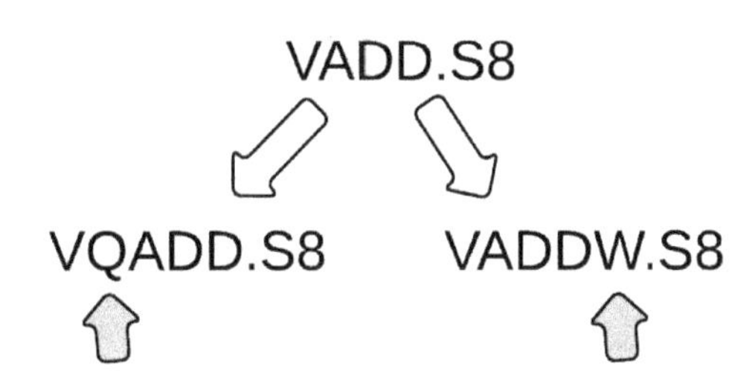

Figure 15.2: Saturation and wide modifications to instruction

Saturation Integer Arithmetic

By using VQADD instead of VADD, the NEON coprocessor still can't provide the correct answer in the case of an overflow, but it will keep the answer as close as possible. In other words, this "saturation" Q modification will prevent the sum of two negative numbers from being positive and the sum of two positive numbers from being negative. Figure 15.3 shows the results of Listing 15.5 line 22 being VADD.S8, VQADD.S8, and VQADD.U8, respectively. For signed S8 operations, the extreme limits are -128 and 127. For unsigned U8 operations, the limits are 0 and 255. What are the limits for the generalized integer code of I8? Since the limits aren't specified in I8, NEON cannot know the limits. Therefore, the I8, I16, I32, and I64 are not allowed to be used with the Q saturation modification.

```
2    5    127  186  16   120  136  130
2    105  122  177  0    235  255  255
```

Listing 15.8: Output from saturated U8 additions example

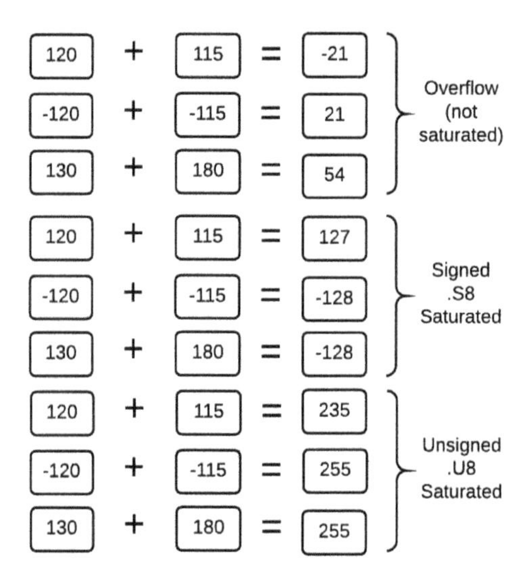

Figure 15.3: Comparison of saturated additions

Let's continue with the same test program and change the three data lines beginning with line 39 to the values in Listing 15.9. Recompile, link, and execute the program and watch the sums march to their saturation limits.

```
39. tstval:     .byte     -80,-40,-20,0,20,40,60,80
40.             .byte     -80,-40,-20,0,20,40,60,80
41.             .byte     -80,-40,-20,0,20,40,60,80
42.             .byte     -80,-40,-20,0,20,40,60,80
```

Listing 15.9: Another set of data to demonstrate saturation addition

Listing 15.10 shows the output where the lower and upper limits for signed 8-bit (S8) are -128 and 127, respectively.

```
-80    -40    -20    0    20    40    60    80
-128   -80    -40    0    40    80    120   127
-128   -120   -60    0    60    120   127   127
-128   -128   -80    0    80    127   127   127
```

Listing 15.10: Saturation output using data from Listing 15.9

Promotions to Wide Lanes

The prototype program in Lab 11 uses an ARM UMULL, unsigned long multiply instruction, to produce a 64-bit product from two 32-bit factors. Although addition of two large binary numbers can overflow by a single bit into the sign bit, it is not unusual for the product of two very large numbers to require double the number of bits of each of the factors. This "long" multiply is very common and is available on most CPU architectures. NEON, however, takes it a step further and provides this "long" modification for most of its instructions. As an example, the VADD instruction has the following modifications:

- VADDL: Long: The result is placed in a lane that is twice the size of operands.
- VADDN: Narrow: The result is placed in a lane that is half the size of operands.
- VADDW: Wide: The result is the same size as that of the first operand, but the second operand is half the size of the first.

The VADD on line 22 will now be modified for a wide result. The list of numbers being summed will remain with a size of 8 bits, but the eight running totals will now be maintained in 16-bit lanes.

Since the running totals now being displayed are 16 bits, macro d_int16 will replace d_int8.

```
1.@         Macro "d_int16 leftbit,suffix" displays 16-bit integer "lane" in R1.
2.@                    leftbit:        Bit position of left side of lane
3.@                    suffix:         String to be output sfter integer
4.@                    Note:           All register contents are preserved.
5.
6.          .macro     d_int16         leftbit,suffix
7.          .set       left,31-\leftbit
8.          push       {R0,R1,LR}      @ Preserve register used by macro call.
9.          mov        R0,R1,lsl #left @ "Shake off" unwanted bits on the left.
10.         asr        R0, #16         @ Right justify with sign extended.
11.         bl         v_dec           @ Display signed integer.
12.         ldr        R1,=\suffix     @ String to output after integer
13.         bl         v_ascz          @ Display text string.
14.         pop        {R0,R1,LR}      @ All register contetns are restored.
15.         .endm
16.
17.         .global    _start          @ Set program starting address to linker.
18._start:  ldr        R6,=tstval      @ Point to list of test values.
19.         ldr        R2,=tab         @ Point to end of list of test values.
20.         veor       Q0,Q0           @ Zero out register Q0 (same as D0, D1).
21.sample:  vldm       R6!,{D2}        @ Load new 8 "second" operands.
22.         vaddw.s8   Q0,Q0,D2        @ Perform 8 additions in parallel
23.         vmov       R1,S0           @ Move high word into ARM for display.
24.         d_int16    15,tab          @ Display 1st 16-bit integer
```

```
25.            d_int16     31,tab         @ Display 2nd 16-bit integer
26.            vmov        R1,S1          @ Move high word into ARM for display.
27.            d_int16     15,tab         @ Display 3rd 16-bit integer
28.            d_int16     31,tab         @ Display 4th 16-bit integer
29.            vmov        R1,S2          @ Move the low word for display.
30.            d_int16     15,tab         @ Display fifth 16-bit integer
31.            d_int16     31,tab         @ Display sixth 16-bit integer
32.            vmov        R1,S3          @ Move high word into ARM for display.
33.            d_int16     15,tab         @ Display seventh 16-bit integer
34.            d_int16     31,nl          @ Display eighth 16-bit integer
35.            cmp         R2,R6          @ Test for end of list of test values.
36.            bgt         sample         @ Cotinue loop for next value in list.
37.            mov         R0,#0          @ Status code 0 for "normal completion"
38.            mov         R7,#1          @ Code terminates program.
39.            svc         0              @ Linux service terminates program
40.            .data
41. tstval:    .byte       -80,-40,-20,0,20,40,60,80
42.            .byte       -80,-40,-20,0,20,40,60,80
43.            .byte       -80,-40,-20,0,20,40,60,80
44.            .byte       -80,-40,-20,0,20,40,60,80
45. tab:       .asciz      "\t"           @ Tab to separate integers.
46. nl:        .asciz      "\n"           @ End of list and new line character.
47.            .end
```

Listing 15.11: Promotion of result to “wide” lanes

```
-80    -40    -20    0    20    40    60    80
-160   -80    -40    0    40    80    120   160
-240   -120   -60    0    60    120   180   240
-320   -160   -80    0    80    160   240   320
```

Listing 15.12: Output from Listing 15.11 using promotion to wide lanes

The NEON design is both powerful and versatile, and compared to previous designs like VFPv2, VFPv3, and MMX, it is very straight forward and easy to use. In regards to VFPv3 not really running simultaneously, it wasn’t the intention. This is consistent with previous vector implementations. The objective was not so much to run at the same time, but to set up a pipelined operation so that a single floating point processor could run as fast as possible without interruptions.

Coding and Debugging

Using parallel computation, the NEON processor can substantially improve total throughput in processing data stored in tables and arrays. However, setting up the proper combinations of operands and loading them into the right sets of registers may be awkward and inefficient in itself. NEON has interleaved load and store instructions to meet this challenge of array data which is stored in various patterns in memory. They can transfer up to four quad words (512 bits) at a time between the NEON's registers and the ARM's memory.

```
1.              .global   _start          @ Starting address for linker
2._start:       ldr       R1,=A2X         @ Memory buffer for output from NEON.
3.              vldm      R1,{D0-D2}      @ Load "AB...WX" into 3 64-bit registers.
4.              ldr       R1,=result      @ Memory buffer for "interleaved" output.
5.
6.              vstm      R1,{D0-D2}      @ Save exact copy for display.
7.              bl        v_ascz          @ Display
8.              vst1.8    {D0-D2},[R1]    @ Save as 1 "element" of 8 bits each
9.              bl        v_ascz          @ Display
10.             vst3.8    {D0-D2},[R1]    @ Save as 3 "elements" of 8 bits each
11.             bl        v_ascz          @ Display
12.             vst3.16   {D0-D2},[R1]    @ Save as 3 "elements" of 16 bits each
13.             bl        v_ascz          @ Display
14.             vst3.32   {D0-D2},[R1]    @ Save as 3 "elements" of 32 bits each
15.             bl        v_ascz          @ Display
16.
17.             mov       R0,#0           @ Status 0 indicates "normal completion"
18.             mov       R7,#1           @ Service command terminates this program.
19.             svc       0               @ Issue Linux command to stop program.
20.             .data
21. A2X:        .ascii    "ABCDEFGHIJKLMNOPQRSTUVWX" @ 24 bytes
22. result:     .ds       12              @ Reserve room for "interleaved" store.
23.             .asciz    "\n"            @ Line feed and null to terminate string.
24.             .end
```

Listing 15.13: NEON interleaved storage example

The program in Listing 15.13 demonstrates some of the possible storage patterns available with the VST (interleaved vector store) instruction. A test pattern of 24 characters, A through X, is loaded into registers D0, D1, and D2 using a load multiple instruction like we've used in previous examples:

- D0: "HGFEDCBA"
- D1: "PONMLKJI"
- D2: "XWVUTSRQ"

Listing 15.14 displays the output from running this program. The first line simply stores the register data back into memory using the STM store multiple instruction on line 6. The remaining display lines result from four interleaved storage instructions on lines 8, 10, 12, and 14, which will be explained in the following paragraphs.

```
ABCDEFGHIJKLMNOPQRSTUVWX
ABCDEFGHIJKLMNOPQRSTUVWX
AIQBJRCKSDLTEMUFNVGOWHPX
ABIJQRCDKLSTEFMNUVGHOPWX
ABCDIJKLQRSTEFGHMNOPUVWX
```

Listing 15.14: Output from interleaved storage example

Output Line	Instruction
ABCDEFGHIJKLMNOPQRSTUVWX	VSTM R1,{D0-D2}
ABCDEFGHIJKLMNOPQRSTUVWX	VST1.8 {D0-D2},[R1]
AIQBJRCKSDLTEMUFNVGOWHPX	VST3.8 {D0-D2},[R1]
ABIJQRCDKLSTEFMNUVGHOPWX	VST3.16 {D0-D2},[R1]
ABCDIJKLQRSTEFGHMNOPUVWX	VST3.32 {D0-D2},[R1]

Table 15.0: Output and the instructions that generated each line

Interleaved storage instructions are analogous to eating dinner. I will eat all the carrots, beans, and fish on my plate, but in what order should I eat them? Do I eat all the carrots before moving on to eat all the beans, and then all the fish, or do I mix them together by taking a bite of carrots, then a bite of beans, then a bite of fish, and back to the carrots until everything is gone? The VST1.8 on line 8 is like eating all my carrots before starting on my beans. The "1" in VST1 indicates I want all the bits transferred from one register before moving onto the next. The "8" is the element size which is analogous to the size of a mouthful, and here indicates that VST1 will be transferring 8 bits at a time. As you might suspect, the VST1.8 storage pattern will look identical to that from VSTM.

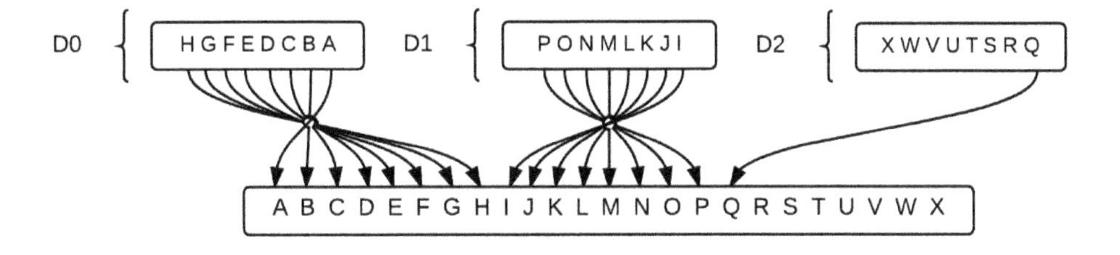

Figure 15.4: VST1.8 {D0-D2},[R1]

On line 10 of the program, the VST3.8 instruction says I will still be moving data in groups of 8-bits (the element size), but I will be taking them from a subset of three adjacent registers. This is analogous to eating a bite of carrots, then a bite of beans, and finally a bite of fish, before going back to get the next round of carrots, beans, and fish, etc.

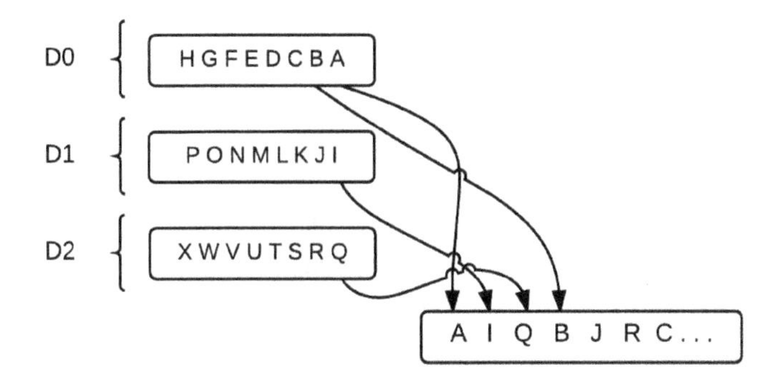

Figure 15.5: VST3.8 {D0-D2},[R1]

Figure 15.6 is the same as 15.5 except I'm taking bigger "mouthfuls": 16 bits instead of 8. The VST3.16 on line 12 says I will be moving data in as 16-bit elements, and I will be taking them from a subset of three adjacent registers.

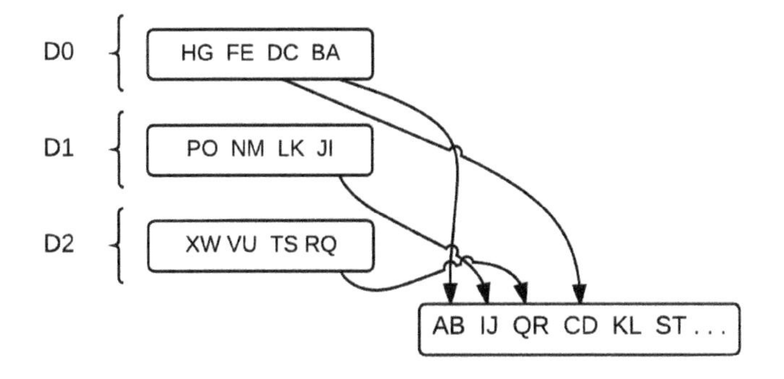

Figure 15.6: VST3.16 {D0-D2},[R1]

Figure 15.7 again uses 3 registers, but now 32 bits are moved at a time.

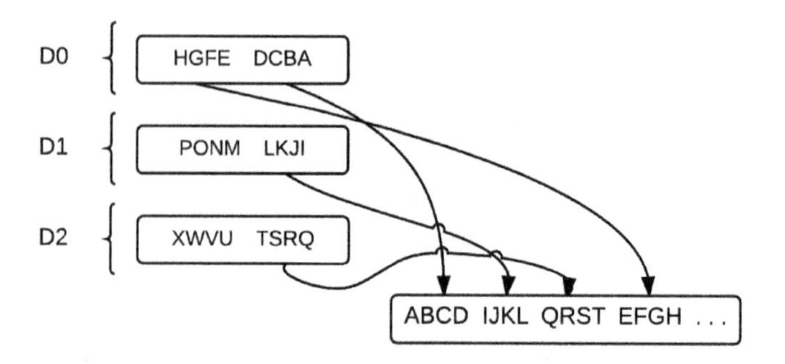

Figure 15.7: VST3.32 {D0-D2},[R1]

Can I use a VST2 instruction? Yes, but not with three registers in the list. My list would have to be either {D0,D1} or {D0-D3}. "Left overs" are not allowed. The number of registers in the list can be between one and four, but must be a multiple of the sublist size specified adjacent to the "VST."

Figure 15.8 summarizes the interleaved store and load instructions that are available with NEON.

- As few as one 64-bit D register and as many as 4 adjacent 128-bit Q registers are processed at one time (minimum of 64 bits and maximum of 512 bits).
- A memory array is composed of multiple elements of the same size which can be 8, 16, 32, or 64 bits in length.
- Elements are moved sequentially between ARM memory and one or more NEON registers.
- An ARM R register acts as an index and contains the byte address of the next element to load/store in the ARM's memory (little endian format).
- The ARM index register may be updated if the write-back bit is set ([R4]!).

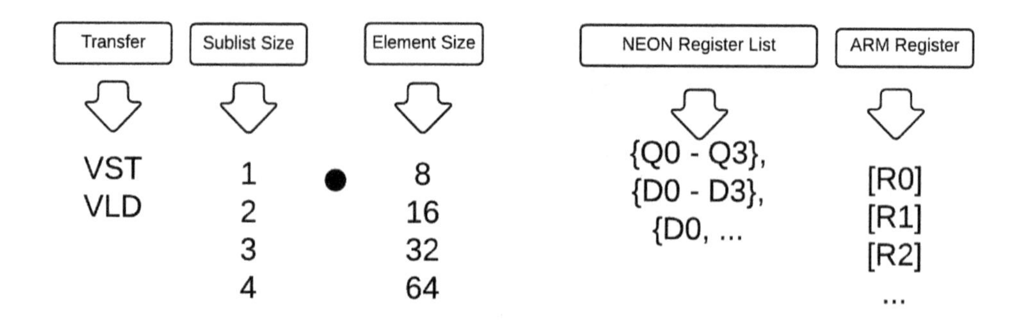

Figure 15.8: Interleaved load and store instructions

NEON Floating Point

We started this lab with the VFPv3 coprocessor floating point examples, and now we'll end it with a couple of examples using NEON's floating point. Listing 15.15 is the NEON equivalent to the VFPv3 example in Listing 15.2.

```
1.          .global     _start        @ Program starting address for linker
2.
3.@         Macro "flt_d SX" displays register SX contents as floating point
4.
5.          .macro      flt_d  SX     @ SX is S0, S1, ..., S31
6.          vmov        R0,\SX        @ Move register to ARM for display.
7.          bl          v_flt         @ Display floating point number.
8.          ldr         R1,=nl        @ Pointer to line ending characters.
9.          bl          v_ascz        @ Separate test values with new lines.
10.         .endm
11.
12._start:  ldr         R6,=tstval    @ Point to list of test values.
13.         vld2.32     {Q0,Q1},[R6]  @ Load "four" pairs of 32-bit numbers.
14.         vmul.f32    Q0,Q0,Q1      @ Multiply the "three" pairs of numbers.
15.         flt_d       S0            @ Display first product.
16.         flt_d       S1            @ Display second product.
17.         flt_d       S2            @ Display third product.
18.
19.         mov         R0,#0         @ Status 0 indicates "normal completion"
20.         mov         R7,#1         @ Command code 1 will terminate program
21.         svc         0             @ Issue Linux command to terminate program
22.
23.tstval:  .float      -3.5, 10.5    @ Floating point test values
24.         .float      101.0625, 100.0
25.         .float      -5.25, -200.0
26.nl:      .asciz      "\n"          @ Line ending character.
27.         .end
```

Listing 15.15: Multiply three pairs of floating point numbers

As you might expect, four floating point multiplies are taking place in parallel on line 14. The three, whose products are shown in Listing 15.16, are desired, while the fourth is simply excess "baggage" that is ignored.

```
-36.75
10106.25
1050.0
```

Listing 15.16:Three products output from program in Listing 15.15

The program in Listing 15.17 has been enhanced to multiply as many pairs of floating point numbers as can fit in memory. Line 12 loads the next four pairs of 32-bit floating point numbers into Q0 and Q1. Note that register R6 is updated and ready to load the next four pairs of 32-bit numbers. The bottom of the loop is on line 19 which checks to see if all pairs have been processed. If the list is not a multiple of four, then there will be as many as three strange products at the end of the list.

```
1.              .global     _start          @ Program starting address for linker
2.
3. @            Macro "flt_d SX" displays register SX contents as floating point
4.              .macro      flt_d SX        @ SX is S0, S1, ..., S31
5.              vmov        R0,\SX          @ Move register to ARM for display.
6.              bl          v_flt           @ Display floating point number.
7.              ldr         R1,=nl          @ Pointer to line ending characters.
8.              bl          v_ascz          @ Separate test values with new lines.
9.              .endm
10.
11. _start:     ldr         R6,=tstval      @ Point to list of test values.
12. sample:     vld2.32     {Q0,Q1},[R6]!   @ Load 4 pairs of 32-bit numbers.
13.             vmul.f32    Q0,Q0,Q1        @ Multiply the 4 pairs of numbers.
14.             flt_d       S0              @ Display first product of set.
15.             flt_d       S1              @ Display second product of set.
16.             flt_d       S2              @ Display third product of set.
17.             flt_d       S3              @ Display fourth product of set.
18.             cmp         R1,R6           @ Test if end of list of test values.
19.             bgt         sample          @ Cintinue loop untill all tests done.
20.
21.             mov         R0,#0           @ Status 0 indicates "normal completion"
22.             mov         R7,#1           @ Command code 1 will terminate program.
23.             svc         0               @ Issue Linux command to terminate program.
24.
25. tstval:     .float      3.0, 10.0       @ Floating point test values
26.             .float      0.25, -1.0
27.             .float      100.0, 1234.567
28.             .float      -9876.543, 7070.7070
29.             .float      3.3333, 694.3e-9
30.             .float      11.08, 360.195
31.             .float      12.0e20, 12.0e-22
32.             .float      6.0221e2, 6.0221e23
33. nl:         .asciz      "\n"            @ Line ending character.
34.             .end
```

Listing 15.17: Multiplying a variable length list of factors using NEON

```
30.0
-0.25
123456.703125
-69834144.0
0.00000231410376727581024169921875
3990.960693359375
1.43999993801116943359375
0.0
```

Listing 15.18: Output showing list of products from NEON example

NEON has dozens of vector operations. Bit-shifting within lanes is one example. Finding the lane with the maximum or minimum value is another. Please check the Internet or a NEON technical reference for a full listing.

Maintenance

Review Questions

1. In NEON, there are 16 128-bit Q registers ($16 \times 128 = 2048$ bits), 32 64-bit D registers ($32 \times 64 = 2048$ bits), and 32 32-bit S registers ($32 \times 32 = 1024$ bits). Why are there not 64 S registers to use the full 2048 bits available?
2. In a later release of the NEON coprocessor, there are 32 128-bit Q registers. How do you think these extra 2048 bits within the coprocessor could be accessed as D and S registers?
3. * If the lane size for integer addition within NEON could be one bit wide, it would be exactly the same as which logical operation?

Programming Exercises

1. Modify Listing 15.17 to produce a multidimensional dot product using NEON's fused multiply with addition (VMLA). Note: I did not provide an example of the VMLA, but I bet you could guess the format. If not, a quick search of the Internet will provide one.
2. Modify Listing 15.5 using the NEON VMAX instruction to find the highest character code of 8 ASCII characters (one per lane).

Lab 16
Text and Logical Input

It's time to reinforce previously introduced topics by examining them from a different angle. Lab 16 will develop input subroutines c_ascz and c_bool which will complement display subroutines v_ascz and v_bool, respectively.

Prototype

Subroutine c_ascz reads a line of ASCII text characters from the standard input device (usually the keyboard).

```
 1. @         Subroutine c_ascz will read a string of characters terminated by a null.
 2. @                  R1: Points to first byte of buffer to receive
 3. @                  R2: Maximum number of bytes in above buffer
 4. @                  LR: Contains the return address
 5.
 6. @         Returned register contents:
 7. @                  R0: Number of characters read.
 8. @                           A null will mark end of string
 9. @                  All other register contents will be preserved
10.
11.           .global  c_ascz     @ Pass entry point name to linker.
12.
13. c_ascz:   push     {R2,R7}    @ Save contents of registers to be used.
14.           mov      R0,#0      @ Stdin (standard input, i.e., the keyboard)
15.           mov      R7,#3      @ Linux service command code to read a string.
16.           sub      R2,#1      @ Leave room for the null for end of string.
17.           svc      0          @ Issue command to read string from stdin.
18.           mov      R7,#0      @ Null to mark end of string
19.           strb     R7,[R1,R0] @ Put a null at end of input line.
20.           pop      {R2,R7}    @ Restore saved register contents
21.           bx       LR         @ Return to the calling program
22.           .end
```

Listing 16.0: Subroutine c_ascz reads text line from keyboard.

The main program in Listing 16.1 will test subroutine c_ascz. There is a dramatic, yet subtle, difference between the two arguments that are passed to the subroutine. Up until this point, we have been using "pass by value" to provide the arguments to subroutines (such as the maximum buffer size in register R2). The memory buffer to be filled is using the second approach for passing arguments: "pass by reference."

- Pass by value: The value is passed in a register or the stack to the subroutine, and the subroutine has no access to the source variable.
- Pass by reference: The memory address of a variable is passed, and the subroutine can actually update the variable in the calling routine's data area.

"Pass by value" is generally preferred because the subroutine cannot directly alter the data in the calling program. However, "pass by reference" is many times more efficient and allows any structure of any length to be passed to a subroutine.

```
1.              .global   _start          @ Provide program starting address to linker
2.              .equ      BUFSIZE,50      @ Size of input buffer can be changed
3.
4. @            Loop to get next line from user.
5.
6. _start:      ldr       R1,=prompt      @ Message asking for input of text line
7.              bl        v_ascz          @ Display text line on monitor
8.              ldr       R1,=inpbuf      @ Set pointer to input buffer.
9.              mov       R2,#BUFSIZE     @ Maximum number of bytes to receive.
10.             bl        c_ascz          @ Call subroutine to get keyboard input.
11.
12. @           Echo line back to user.
13.
14.             bl        v_ascz          @ Redisplay text line just entered.
15.             cmp       R0,#1           @ Test for 1 byte (/nl) to end loop.
16.             bgt       _start          @ Go prompt user for another line.
17.
18.             mov       R0,#0           @ Exit code 0 indicates "normal completion"
19.             mov       R7,#1           @ Command code to terminate program
20.             svc       0               @ Issue Linux command to terminate program
21.
22.             .data
23. inpbuf:     .ds       BUFSIZE         @ Reserve storage for input data characters
24. prompt:     .asciz    "Please enter some text. Enter just ENTER to stop.\n"
25.             .end
```

Listing 16.1: Main program in file model.s to test text line input

Go ahead and run this demonstration program. You will need the view.a library in addition to the model.o and c_ascz object files when you link.

```
Please enter some text. Enter just ENTER to stop.
Hello World
Hello World
Please enter some text. Enter just ENTER to stop.
Floating point
Floating point
Please enter some text. Enter just ENTER to stop.
```

Listing 16.2:Output from test run of the c_ascz subroutine

Let's make the above test a little more interesting by outputting a hexadecimal dump for each character in the input line.

```
1.          .global   _start          @ Provide program starting address to linker
2.          .equ      BUFSIZE,50      @ Size of input buffer can be changed
3.
4. @        Loop to get next line from user.
5.
6. _start:  ldr       R1,=prompt      @ Message asking for input of text line
7.          bl        v_ascz          @ Display text line on monitor
8.          ldr       R1,=inpbuf      @ Set pointer to input buffer.
9.          mov       R2,#BUFSIZE     @ Maximum number of bytes to receive.
10.         bl        c_ascz          @ Call subroutine to get keyboard input.
11.         cmp       R0,#1           @ Test for 1 byte (\nl) to end loop.
12.         ble       exit            @ Exit if nothing but ENTER entered.
13.
14. @       Display the characters of input line in hexadecimal.
15.
16.         mov       R4,R1           @ R4 will point to next character to display
17.         mov       R5,R0           @ R5 will "count down" the number of bytes.
18.         ldr       R1,=spacer      @ A blank will separate each hex number
19.         mov       R2,#2           @ Display 2 hex digits for each byte.
20. nxtchr: ldrb      R0,[R4],#1      @ Get next byte from input buffer
21.         bl        v_hex           @ Display contents of R0 as 2 hex digits.
22.         bl        v_ascz          @ Put a blank after the number
23.         subs      R5,#1           @ Decrement number of bytes to display.
24.         bgt       nxtchr          @ Go display next character in input buffer.
25.         ldr       R1,=newln       @ Pointer to end of line ASCII code.
26.         bl        v_ascz          @ Mark end of output with a new line code.
27.         b         _start          @ Go prompt user for another line.
28.
29. exit:   mov       R0,#0           @ Exit code 0 indicates "normal completion"
30.         mov       R7,#1           @ Command code to terminate program
31.         svc       0               @ Issue Linux command to terminate program
32.
33.         .data
34. inpbuf: .ds       BUFSIZE         @ Reserve storage for input data characters
```

```
35. prompt:   .asciz    "Please enter some text. Enter just ENTER to stop.\n"
36. spacer:   .asciz    " "                @ Separator for display (comma, blank)
37. newln:    .asciz    "\n"               @ End of line (It will be 0x0A)
38.           .end
```

Listing 16.3: Main program to display each character of input line in hexadecimal

Go ahead and run the new test. Notice that the last character in the input text is the Enter key itself (hex 0A, a.k.a., "line feed" or "new line").

```
Please enter some text or only ENTER to stop.
Hello World
48 65 6C 6C 6F 20 57 6F 72 6C 64 0A
Please enter some text or only ENTER to stop.
Floating Point
46 6C 6F 61 74 69 6E 67 20 50 6F 69 6E 74 0A
Please enter some text or only ENTER to stop.
```

Listing 16.4: Sample conversion of text input to hexadecimal

Introductions

	List of assembler directives introduced in Lab 16		
16.1.2.	.equ	BUFSIZE,50	@ Size of input buffer can be changed
16.1.23.	.ds	BUFSIZE	@ Reserve storage for input data characters

	List of Linux service calls introduced in Lab 16
3	Read array of bytes from device

Principles

What is truth? I thought we already addressed that question in Lab 4. We did, but now we have to look at how to input Boolean values rather than display them. Does an ASCII string like "True," "T," or "1" indicate True? What if the word "True" is misspelled? When we wrote display subroutines, we knew what to do:

1. We knew where to find the data: It was in register R0.
2. We knew there were no errors in the data. There was some discussion about what might or should be considered True or False. But, once it was defined, there was no mystery.
3. We knew what display format was required: It was determined by which subroutine was called.
4. We knew how many characters to display.

The following is characteristic of input subroutines which convert a string of characters to a number or Boolean value:

1. Do we know the format? Is it simply a string of ASCII characters or is it a number in binary, hex, or decimal format?
2. Where does the data begin and end? Are there leading blanks? Is it a fixed format with a definite width or number of columns? If it has a free format, is it terminated by a null, line ending key, blank, tab, or semicolon?

Basically, we have to look at existing standards and "best practices." We then have to make a decision, and *document* what we expect.

Coding and Debugging

Let's build a c_bool program that will have a rather broad value of Truth: If the next word on the input text line begins with "T" or "t" or begins with the digit "1," c_bool will return a value of True, otherwise it will return False. Listing 16.5 contains the c_bool subroutine which will not read the text line directly, but locate it by the "pass by reference" argument in register R1.

```
1.@           Subroutine c_bool will look for True in a text line
2.@                     R1: Points to buffer containing True
3.@                     R2: Maximum number of bytes in above buffer
4.@                     LR: Contains the return address
5.
6.@           Returned register contents:
7.@                     R0: =1 if first non-blank character is T, t, or 1.
8.@                               =0 otherwise
9.@                     All other register contents will be preserved
10.
11.           .global   c_bool    @ Pass entry point name to linker.
12.
13.c_bool:    push      {R1-R4}   @ Save contents of registers used.
```

```
14.          mov      R0,#0        @ Assume False (until T, t, or 1)
15.
16. nxtdig:  subs     R2,#1        @ Check for end of buffer.
17.          blt      exit         @ Assume False if nothing found.
18.          ldrb     R3,[R1],#1   @ Load next character from buffer.
19.          cmp      R3,#' '      @ Test if it's a blank
20.          cmpne    R3,#9        @ Test for horizontal tab (ASCII 09)
21.          beq      nxtdig       @ Search for non-blank
22.
23. @        Consider anything beginning with a "T" or 1 to indicate True
24.
25.          cmp      R3,#'T'      @ Assume "True" "Truth" ...
26.          cmpne    R3,#'t'      @ "true"
27.          cmpne    R3,#'1'      @ Let "1" also indicate true
28.          moveq    R0,#1        @ Return value of True
29.
30. exit:    pop      {R1-R4}      @ Reload saved register contents.
31.          bx       LR           @ Return to the calling program.
32.          .end
33.
```

Listing 16.5: Subroutine c_bool will "attempt" to determine True and False

Neither c_ascz nor c_bool seem appropriate for the view.a library which contains all of the previous display subroutines. Listing 16.6 shows the creation of library controller.a and the insertion of the two new subroutines.

```
~$ as -o c_bool.o c_bool.s
~$ as -o c_ascz.o c_ascz.s
~$ ar rvs controller.a c_ascz.o c_bool.o
```

Listing 16.6: Build new library with input subroutines

The main program in Listing 16.7 tests subroutine c_bool with a loop that prompts for input from the keyboard. It will then print "True" or "False" depending on what it finds in the text. A blank line will terminate the program.

```
1.           .global  _start            @ Provide program starting address to linker
2.           .equ     BUFSIZE,50        @ Size of input buffer can be changed
3.
4. @         Loop to get next line from user.
5. _start:   ldr      R1,=prompt        @ Message asking for input of integer
6.           bl       v_ascz            @ Display text line on monitor
7.           ldr      R1,=inpbuf        @ Set pointer to input buffer.
8.           mov      R2,#BUFSIZE       @ Maximum number of bytes to receive.
9.           bl       c_ascz            @ Call subroutine to get keyboard input.
```

```
10.
11.@          Convert text line to Boolean value and display True or False.
12.           bl      c_bool          @ R0 gets 0: False, 1: True
13.           bl      v_bool          @ Display True or False.
14.           ldr     R1,=newln       @ Pointer to end of line ASCII code.
15.           bl      v_ascz          @ Mark end of output with a new line code.
16.           cmp     R0,#1           @ Test for one byte (\n) to end loop.
17.           bgt     _start          @ Go prompt user for another line.
18.           mov     R0,#0           @ Exit code 0 indicates "normal completion"
19.           mov     R7,#1           @ Command code to terminate program
20.           svc     0               @ Issue Linux command to terminate program
21.
22.           .data
23.inpbuf:    .ds     BUFSIZE         @ Reserve storage for input data characters
24.prompt:    .asciz  "Please enter logical value or ENTER to stop.\n"
25.newln:     .asciz  "\n"            @ End of line (It will be 0x0A)
26.           .end
```

Listing 16.7: Main loop program to test True and False text input

Listing 16.8 shows the linking of the test program using two libraries. The order that the libraries appear on the ld command line is not important in this example since both are independent of each other.

```
~$ as -o model.o model.s
~$ ld -o model model.o view.a controller.a
```

Listing 16.8: Linking with two libraries

Figure 16.9 shows the sample test lines being examined as True or False.

```
Please enter logical value or ENTER to stop.
Hello World
False
Please enter logical value or ENTER to stop.
true git
True
Please enter logical value or ENTER to stop.
12345
True
Please enter logical value or ENTER to stop.

False
```

Listing 16.9:Execution of program to test subroutine c_bool

Maintenance

Review Questions

1. Name two characteristics of using "pass by value."
2. Name two advantages of using "pass by reference."
3. * What is a principal danger in using "pass by reference"?

Programming Exercises

1. Modify c_ascz subroutine so that the argument providing the maximum size of buffer is "pass by reference."
2. Modify c_bool so that the return value is -1 if the first word on the line begins with an "F" or a number beginning with a zero. Only send back a zero if neither True nor False is detected.

Lab 17
Integer Input

Lab 17 provides additional programming experience by reversing the "output display" examples from Lab 5 (binary), Lab 6 (hexadecimal) and Lab 7 (decimal). Here we will generate integers from an ASCII string representing numbers in decimal, binary, hexadecimal, and even octal formats. New material includes a comparison of 16-bit Thumb instruction format to the 32-bit ARM format we have been using.

Prototype

In this Prototype section, we will loop through decimal whole numbers being entered from the keyboard and display them in decimal, hexadecimal, and binary formats. We will then expand this program in the Principles section to include numbers also entered in hexadecimal, binary, and octal formats. Basically, this program is one that will convert integers among any of the bases commonly used in computer programing.

```
1.          .global   _start          @ Provide program starting address to linker
2.          .equ      BUFSIZE,50      @ Size of input buffer can be changed
3.
4. @        Loop to get next number, convert it, and display it in multiple formats.
5.
6. _start:  ldr       R1,=prompt      @ Message asking for input of integer
7.          bl        v_ascz          @ Display text line on monitor
8.          ldr       R1,=inpbuf      @ Set pointer to input buffer.
9.          mov       R2,#BUFSIZE     @ Maximum number of bytes to receive.
10.         bl        c_ascz          @ Call subroutine to get keyboard input.
11.         bl        c_int           @ Convert string at R1 to number in R0.
12.
13. @       Display integer in decimal, hexadecimal, and binary.
14.
15.         ldr       R1,=msgtxt      @ Document what is being displayed
16.         bl        v_ascz          @ Display text line on monitor
17.         ldr       R1,=spacer      @ Pointer to comma and blank
18.         bl        v_dec           @ Subroutine to view decimal value.
19.         bl        v_ascz          @ Put spacer between numbers being
                                      displayed.
```

```
20.            mov      R2,#0         @ Tell subroutine v_hex "no leading zeroes."
21.            bl       v_hex         @ Subroutine to view hex value of R0.
22.            bl       v_ascz        @ Put spacer after the hexadecimal format.
23.            mov      R2,#16        @ Tell subroutine v_bin to display only 16
                                      bits.
24.            bl       v_bin         @ Subroutine to view binary value of R0.
25.            ldr      R1,=newln     @ Set pointer to "new line" character string.
26.            bl       v_ascz        @ Put a line feed after binary display.
27.            cmp      R0,#0         @ Test for zero value to end loop.
28.            bne      _start        @ Go prompt user for another number.
29.
30.            mov      R0,#0         @ Exit code 0 indicates "normal completion"
31.            mov      R7,#1         @ Service command code to terminate
                                      program
32.            svc      0             @ Issue Linux command to terminate program
33.
34.            .data
35. inpbuf:    .ds      BUFSIZE       @ Reserve storage for input data characters
36. prompt:    .asciz   "Please enter an integer followed by the Enter key. Zero to
                        stop.\n"
37. msgtxt:    .asciz   "Decimal, hexadecimal, and binary: " @ Display message
38. spacer:    .asciz   ", "          @ Separator for display (comma, blank)
39. newln:     .asciz   "\n"          @ End of line (It will be 0x0A)
40.            .end
```

Listing 17.0: Main program in model.s to display integers in various formats

Listing 17.1 contains a first version of subroutine c_int, which converts an ASCII character string containing an integer expressed in base 10 to a 32-bit binary number. In the next sections, this subroutine will be modified to run in 16-bit Thumb mode as well as being enhanced to include input in binary, octal, and hexadecimal.

```
1. @           Subroutine c_int will convert input buffer from decimal digits to binary
2. @                    R1: Points to string of decimal digits terminated with a null byte.
3. @                    LR: Contains the return address
4. @                    R0: Returned integer value (converted from ASCII input string)
5. @                    All register contents except R0 will be preserved.
6.
7.             .global  c_int         @ Subroutine entry point.
8.
9. c_int:      push     {R1-R4}       @ Preserve working register contents.
10.            mov      R3,#0         @ Integer will be built in register R3..
11.            mov      R4,#10        @ Base 10 used to "shift" over each digit.
12.
13. nxtdig:    ldrb     R0,[R1],#1    @ Load next character from input buffer.
14.            subs     R0,#'0'       @ Subtract the ASCII character bias.
15.            blt      notdig        @ Check if end of digits has been reached.
```

```
16.            cmp     R0,#9           @ Check upper limit of digits range.
17.            bgt     notdig          @ Go exit if no more digits found.
18.            mla     R3,R4,R3,R0     @ Include this digit: [R3] = [R3]*[R4] + [R0]
19.            b       nxtdig          @ Continue loop with next digit.
20.
21.notdig:     mov     R0,R3           @ Return binary number in R0.
22.            pop     {R1-R4}         @ Reload saved register contents.
23.            bx      LR              @ Return to the calling program.
24.            .end
```

Listing 17.1: Subroutine c_int converts decimal integer in ASCII to binary

Go ahead and run the sample program. As shown in Listing 17.2, you'll need the view.a library from the previous labs for the v_dec and v_ascz subroutines.

```
~$ nano model.s
~$ as -o model.o model.s
~$ nano c_int.s
~$ as -o c_int.o c_int.s
~$ ld -o model model.o c_int.o c_ascz.o view.a
~$ ./model
```

Listing 17.2: Sequence of edit, compile, link, and execute

Listing 17.3 shows the program output for keyed in values of 123, 3780, and 3033. If you would like the hexadecimal numbers preceded by "0x" and the binary numbers by "0b" or other options, that can be easily done.

```
Please enter an integer followed by the Enter key. Zero to stop.
123
Decimal, hexadecimal, and binary: 123, 7B, 0000000001111011
Please enter an integer followed by the Enter key. Zero to stop.
3780
Decimal, hexadecimal, and binary: 3780, EC4, 0000111011000100
Please enter an integer followed by the Enter key. Zero to stop.
3033
Decimal, hexadecimal, and binary: 3033, BD9, 0000101111011001
Please enter an integer followed by the Enter key. Zero to stop.
0
Decimal, hexadecimal, and binary: 0, 0, 0000000000000000
```

Listing 17.3: Sample test of subroutine c_int

Introductions

The floating point version of the Multiply and Accumulate (MLA) instruction has been previously examined, and now it will be showcased in both ARM and Thumb mode integer applications. Assembler directives .thumb and .arm dictate whether 16-bit Thumb or 32-bit ARM instructions are produced. Please note: At execution time, the ARM CPU is switched between Thumb and ARM states using the Branch Exchange (BX) instruction.

	List of ARM instructions introduced in Lab 17
17.1.18.	mla R3,R4,R3,R0 @ Include this digit: [R3] = [R4]*[R3] + [R0]

	List of assembler directives introduced in Lab 17
.thumb	Start generating Thumb machine code
.arm	Return to generating 32-bit ARM machine code

Principles

We displayed an integer in decimal format in Lab 7. Here in Lab 17, we are basically reversing that technique by creating a 32-bit integer from an ASCII string of decimal digits. We will have a loop that reads the digits from left to right, multiplies the current value by base 10, and adds each new digit as we go. A decimal number is really a short notation for a polynomial of powers of 10. For example: 3274 is $3\times10^3 + 2\times10^2 + 7\times10^1 + 4\times10^0$. There will be four passes through the loop, one for each digit of the "3274" example :

1. Start with 3
2. Multiply 3 by 10 and add 2 = 32
3. Multiply 32 by 10 and add 7 = 327
4. Multiply 327 by 10 and add 4 = 3274

MLA R3,R4,R3,R0

Using the Multiply and Accumulate (MLA) instruction seems natural for this subroutine. As we saw in floating point examples, this instruction is very handy for calculating running totals of products. Here we have another example of a fairly complicated looking instruction within a RISC architecture.

Integers Input in Different Bases

A computer program's source code routinely has numbers expressed in binary, octal, and hexadecimal. Unless otherwise stated, a number is assumed to be expressed in decimal. Octal format numbers have traditionally been identified by a leading zero. There are multiple ways of designating binary and hexadecimal depending on the application development language as indicated in the following examples:

- **Binary:** 0b10101, 10101b, %10101
- **Hexadecimal:** 0x1A, 1Ah, $1A

The original c_int will now be extended to accept binary, hexadecimal, and octal using the leading zero formats.The general flow of the new c_int follows:

1. Ignore any leading blanks or tabs.
2. Determine the base and load it into R4.
    a. Default is base 10 (load R4 with 10)
    b. If 0x or 0X, then load R4 with 16
    c. If 0b or 0B, then load R4 with 2
    d. Otherwise, a leading zero will load R4 with 8
3. Loop to load each digit
    a. Subtract off the character bias: "6" - "0" = 6, for example
    b. Check proper range: 3 not allowed in binary, for example
    c. Use the MLA instruction: multiply the previous running total by the base value and then add the new digit value

```
1. @        Subroutine c_int will convert input buffer from decimal, hexadecimal (0x...),
2. @                octal (0...), and binary (0b...) digits to integer.
3. @                Leading blanks and tabs will be ignored.
4. @                R1: Points to string of digits terminated with a non-digit.
5. @                LR: Contains the return address
6. @                R0: Returned integer value (converted from ASCII input string)
7. @                All register contents except R0 will be preserved.
8.
9.          .global   c_int             @ Externalize subroutine entry point for linker.
10.
```

```
11. c_int:    push     {R1-R4}        @ Preserve working register contents.
12.
13. @         Skip over any leading blanks or horizontal tabs.
14.
15. skpblk:   ldrb     R0,[R1],#1     @ Load next character from input buffer.
16.           cmp      R0,#9          @ Check ASCII code for horizontal tab.
17.           beq      skpblk         @ Continue search for non-blank, non-tab.
18.           cmp      R0,#' '        @ Check ASCII code for blank.
19.           beq      skpblk         @ Continue search for non-blank, non-tab.
20.           mov      R3,#0          @ Integer will be built in register R3.
21.
22. @         Check whether binary, octal, decimal, or hexadecimal.
23.
24.           mov      R4,#10         @ Base 10 used to "shift" over each digit.
25.           cmp      R0,#'0'        @ Test for a leading zero (such as 0x or 0b).
26.           bne      chkdig         @ If no leading zero, use the default base 10.
27.           ldrb     R0,[R1],#1     @ Load the character after the zero.
28.           mov      R4,#16         @ Assume it's going to be hexadecimal.
29.           cmp      R0,#'X'        @ 0X... says it's hexadecimal.
30.           cmpne    R0,#'x'        @ 0x... says it's hexadecimal.
31.           beq      nxtdig         @ Go process base 16 digits.
32.           mov      R4,#8          @ Assume it's going to be octal (no b or h).
33.           cmp      R0,#'B'        @ 0B... says it's binary.
34.           cmpne    R0,#'b'        @ 0b... says it's binary.
35.           bne      chkdig         @ Go process base 8 digits.
36.           mov      R4,#2          @ B and b says it's binary.
37.
38. @         Load next character and test if it's 0..9 or A..F or a..f
39.
40. nxtdig:   ldrb     R0,[R1],#1     @ Load next character from input buffer.
41. chkdig:   cmp      R0,#'9'        @ Check whether a posible digit or letter
42.           ble      chk09          @ Go finish check on 0..9 range
43.
44. @         For hexadecimal, map A..F and a..f to 10..15.
45.
46.           orr      R0,#0x20       @ Map upper case letters to lower case.
47.           cmp      R0,#'f'        @ Check upper limit of "a through f".
48.           bgt      exit           @ Go exit if end of number reached.
49.           subs     R0,#'a'        @ Check lower limit of "a through f".
50.           blt      exit           @ Go exit if end of number reached.
51.           add      R0,#10+'0'     @ 'A' .. 'F' is 10 .. 15
52.
53. @         Map character 0..9 to number 0..9.
54.
55. chk09:    subs     R0,#'0'        @ Subtract off the character bias.
56.           blt      exit           @ Go exit if end of number reached.
57.
58. @         Include this digit in number being constructed.
59.
```

```
60.         cmp     R0,R4           @ Test upper limit of base: 2, 8, 10, or 16.
61.         bge     exit            @ Go exit if digit is out of range.
62.         mla     R3,R4,R3,R0     @ Include this digit: [R3] = [R3]*[R4] + [R0]
63.         b       nxtdig          @ Continue loop with next digit.
64.
65. exit:   mov     R0,R3           @ Return binary number in R0.
66.         pop     {R1-R4}         @ Reload saved register contents.
67.         bx      LR              @ Return to the calling program.
68.         .end
```

Listing 17.4: Allow binary, hex, octal, and decimal integer formats

Go ahead and reassemble the new version of c_int. Link it with the same main test program as before. Listing 17.5 shows the output from running the test program with the new c_int supporting the multiple input formats. Try a few different input values of your own.

```
Please enter an integer followed by the Enter key. Zero to stop.
123
Decimal, hexadecimal, and binary: 123, 7B, 0000000001111011
Please enter an integer followed by the Enter key. Zero to stop.
0x34
Decimal, hexadecimal, and binary: 52, 34, 0000000000110100
Please enter an integer followed by the Enter key. Zero to stop.
0xabc
Decimal, hexadecimal, and binary: 2748, ABC, 0000101010111100
Please enter an integer followed by the Enter key. Zero to stop.
055
Decimal, hexadecimal, and binary: 45, 2D, 0000000000101101
0b11001010
Decimal, hexadecimal, and binary: 202, CA, 0000000011001010
```

Listing 17.5: Test of subroutine c_int with multiple input formats

Thumb Code

Thumb is an alternate "instruction set" for the ARM CPU. It uses the same registers, but has a different instruction format and can be interspersed between regular ARM instructions. Because Thumb instructions are mostly 16-bits wide, they take up less room in memory and will load faster in hardware configurations having a 16-bit data bus or smaller. Today's configurations, having 32-bit data buses and large memories, could actually run slower in Thumb-coded programs due to the sacrifice of features present in the 32-bit ARM instruction format. There is also the overhead of switching between ARM and Thumb modes. More details of Thumb code are provided in Appendix N.

Many of the characteristics of Thumb code can be seen in Figure 17.0:

- Most Thumb instructions are 16-bits wide.
- Only branch instructions have conditional execution.
- Most 16-bit instructions use only one register, sometimes two.
- Because the register field is only three bits, only registers R0 through R7 can be used. However, there are special Thumb instructions for the SP, LR, and PC registers.
- There is no built-in rotate of the second operand for every arithmetic instruction. However, there are individual instructions that shift register contents.
- The maximum value of immediate data is only 255 (8 bits).

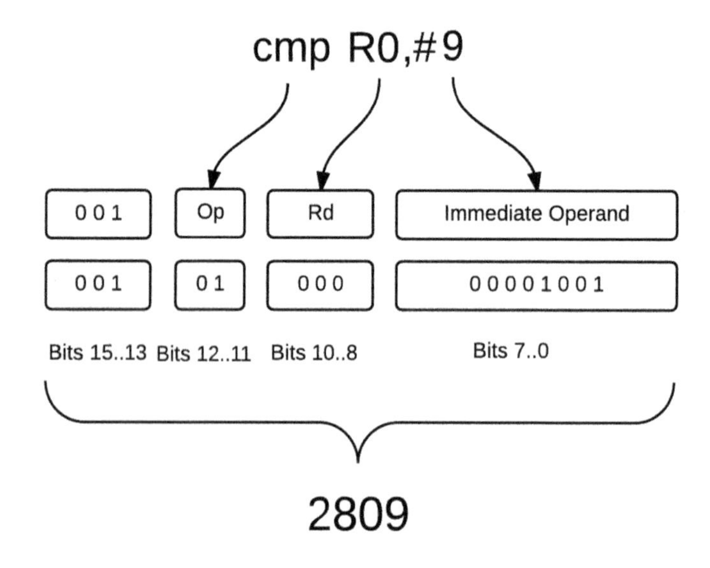

Figure 17.0: "Compare" instruction in Thumb machine code format

Coding and Debugging

This section will be devoted to demonstrating Thumb code. What's an easy way to convert a section of ARM code to Thumb code? Let's let the GNU assembler tell us what we have to change. In Figure 17.6, I've modified the original c_int subroutine source code with a .thumb assembler directive on line 8 and an .arm directive on line 24. Actually I do not need the .arm directive since I have no further code being assembled after that point.

```
 1.@            Subroutine c_int will convert input buffer from decimal digits to binary
 2.@                        R1: Points to string of decimal digits terminated with a null
                            byte.
 3.@                        LR: Contains the return address
 4.@                        R0: Returned integer value (converted from ASCII input string)
 5.@                        All register contents except R0 will be preserved.
 6.
 7.             .global     c_int           @ Subroutine entry point.
 8.             .thumb
 9.c_int:       push        {R1-R4}         @ Preserve working register contents.
10.             mov         R3,#0           @ Integer will be built in register R3.
11.             mov         R4,#10          @ Base 10 used to "shift" over each digit.
12.
13.nxtdig:      ldrb        R0,[R1],#1      @ Load next character from input buffer.
14.             subs        R0,#'0'         @ Subtract the ASCII character bias.
15.             blt         notdig          @ Check if end of digits has been reached.
16.             cmp         R0,#9           @ Check upper limit of digits range.
17.             bgt         notdig          @ Go exit if no more digits found.
18.             mla         R3,R4,R3,R0     @ Include this digit: [R3] = [R3]*[R4] + [R0]
19.             b           nxtdig          @ Continue loop with next digit.
20.
21.notdig:      mov         R0,R3           @ Return binary number in R0.
22.             pop         {R1-R4}         @ Reload saved register contents.
23.             bx          LR              @ Return to the calling program.
24.             .arm
25.             .end
```

Listing 17.6: Subroutine c_int converts decimal integer in ASCII to binary

The assembler was able to generate Thumb machine code for all but two instructions.

1. The LDRB instruction on line 13 is post-indexed mode.
2. The SUBS instruction on line 14 needs to be just SUB.

```
c_int.s: Assembler messages:
c_int.s:13: Error: Thumb does not support this addressing mode -- `ldrb R0,[R1],#1'
c_int.s:14: Error: instruction not supported in Thumb16 mode -- `subs R0,#48'
```

Listing 17.7: Error messages from GNU assembler for c_int

We notice from Figure 17.1 that load/store byte/word instructions are always in pre-indexed mode, but the offset is only a 5-bit field (maximum offset of 31). There is no write-back possible for the offset. Bits 13 - 15 indicate the instruction type.

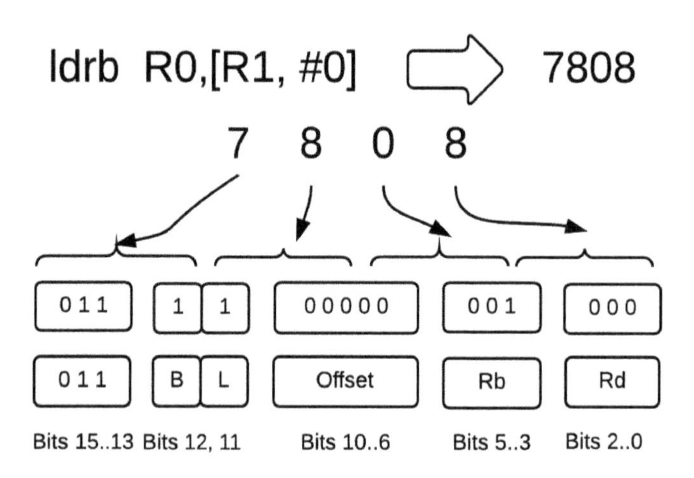

Figure 17.1: Thumb instruction to load byte in pre-indexed mode

If we examine the format in Figure 17.2 for arithmetic instructions, we see an immediate value for the second operand is possible with a range 0 through 255 (8-bits). These instructions always generate the condition codes. Note: The GNU debugger will show the "S" suffix, but the assembler requires it to be omitted (like the CMP instruction).

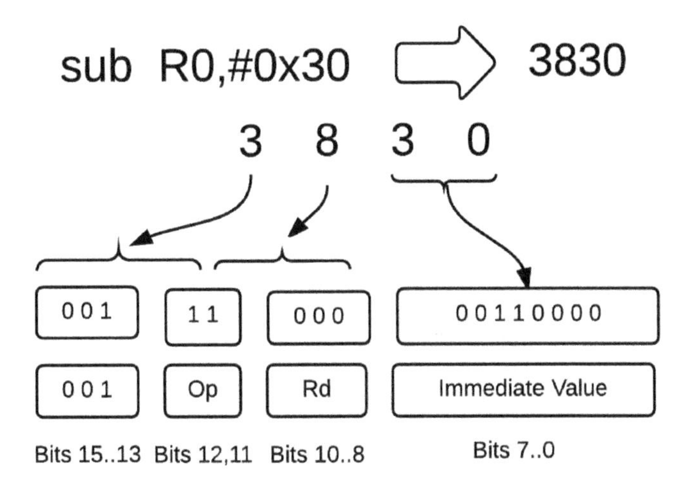

Figure 17.2: "S" suffix is assumed for Thumb arithmetic instructions.

These two problems found by the assembler are easily remedied with the substitutions shown in table 17.0. In the first case, we needed an additional ADD to update the index register. In the second case, only the "S" had to be removed because all arithmetic and logical instructions generate the condition codes.

LDR	R0,[R1],#1	LDR ADD	R0,[R1] R1,#1
SUBS	R0,#0x30	SUB	R0,#0x30

Table 17.0: Substitute Thumb code for ARM code

Switching the CPSR to Thumb State

Now that we have the subroutine c_int converted into Thumb machine code, how do we execute it? Serious errors will occur if the CPU executes Thumb code instructions thinking they are ARM format. Simply setting the CPSR T-bit using an MSR instruction is not appropriate.

ARM instructions must be aligned on a 4-byte boundary and Thumb instructions must be aligned on a 2-byte boundary. Therefore, bit zero is always zero for instruction addresses. Since bit 0 of the "branch to" address is not needed, the BX (Branch exchange) instruction uses it to set a new value in the T-bit of the CPSR enabling the BX to both branch and switch between ARM and Thumb states.

Listing 17.8 shows The new c_int modified with corrections, and includes a BX instruction on lines 10 and 11 to switch into Thumb state. The subroutine automatically returns to ARM mode with the BX on line 27 which it uses to return to the calling program.

```
1. @          Subroutine c_int will convert input buffer from decimal digits to binary
2. @                     R1: Points to string of decimal digits terminated with a null
                         byte.
3. @                     LR: Contains the return address
4. @                     R0: Returned decimal integer value (converted from ASCII
                         input string)
5. @                     All register contents except R0 will be preserved.
6.
7.            .global    c_int          @ Subroutine entry point.
8.
9. c_int:     push       {R1-R4}        @ Preserve working register contents.
10.           add        R3,PC,#1       @*Address to begin Thumb state + 1
11.           bx         R3             @*Switch CPU into Thumb state
12.           .thumb
13.           mov        R3,#0          @ Integer will be built in register R3.
14.           mov        R4,#10         @ Base 10 used to "shift" over each digit.
15.
16. nxtdig:   ldrb       R0,[R1]        @*Load next character from input buffer.
17.           add        R1,#1          @*Set pointer to next byte
18.           sub        R0,#'0'        @*Subtract the ASCII character bias.
19.           blt        notdig         @ Check if end of digits has been reached.
20.           cmp        R0,#9          @ Check upper limit of digits range.
21.           bgt        notdig         @ Go exit if no more digits found.
22.           mla        R3,R4,R3,R0    @ Include this digit: [R3] = [R3]*[R4] + [R0]
23.           b          nxtdig         @ Continue loop with next digit.
24.
25. notdig:   mov        R0,R3          @ Return binary number in R0.
26.           pop        {R1-R4}        @ Reload saved register contents.
27.           bx         LR             @ Return to the calling program.
28.           .end
```

Listing 17.8: Decimal c_int coded in Thumb code

How do the instructions on lines 10 and 11 set the CPSR T-bit? The Thumb mode instructions begin on line 13 with the MOV instruction following the .thumb directive. I could have put a label on that line, such as THMCOD, and then used a LDR R3,=THMCOD+1 instruction instead of the ADD R3,PC,#1? That would have worked, but so does the code I used. Since every ARM instruction is four bytes long, why didn't I add 9 on line 10 rather than 1? That first Thumb instruction is two instructions (8 bytes) ahead plus the "1" to set the T-bit. The ARM processor is pipelined, so it is prefetching instructions before the current instruction has completed. The PC has already been incremented to subsequent instructions before the current instruction has finished.

Go ahead and assemble listing 17.8 and relink it like you did in the previous two examples. When executed, it will operate just like the example in the Prototype section. If you're curious about the Thumb code generated, I have provided a gdb dump of subroutine c_int in both disassembler format and half-word hex dumps.

```
(gdb) x/14wi 0x80f0
=>  0x80f0 <c_int+12>:     movs   r3, #0
    0x80f2 <c_int+14>:     movs   r4, #10
    0x80f4 <nxtdig>:       ldrb   r0, [r1, #0]
    0x80f6 <nxtdig+2>:     adds   r1, #1
    0x80f8 <nxtdig+4>:     subs   r0, #48 ; 0x30
    0x80fa <nxtdig+6>:     blt.n  0x8108 <notdig>
    0x80fc <nxtdig+8>:     cmp    r0, #9
    0x80fe <nxtdig+10>:    bgt.n  0x8108 <notdig>
    0x8100 <nxtdig+12>:    mla    r3, r4, r3, r0
    0x8104 <nxtdig+16>:    b.n    0x80f4 <nxtdig>
    0x8106 <notdig>:       adds   r0, r3, #0
    0x8108 <notdig+2>:     pop    {r1, r2, r3, r4}
    0x810a <notdig+4>:     bx     lr
(gdb) x/14hx 0x80f0
    0x80f0 <c_int+12>:     0x2300 0x240a 0x7808 0x3101 0x3830 0xdb04 0x2809 0xdc02
    0x8100 <nxtdig+12>:    0xfb04 0x0303 0xe7f5 0x1c18 0xbc1e 0x4770
```

Listing 17.9: Hex and disassembled display of Thumb code

Table 17.1 reformats the gdb output from Listing 17.9 and confirms the following:

1. Almost all of the Thumb instructions are half-words (16 bits).
2. The Thumb 32-bit MLA and ARM 32-bit MLA instructions are different.
3. The ALU instructions, including the MOV, always set the condition codes.
4. Pre-indexed mode is possible in a 16-bit LDR/STR instruction.

Thumb Machine Code	Thumb Assembly Language
2300	movs r3, #0
240A	movs r4, #10
7808	ldrb r0, [r1, #0]
3101	adds r1, #1
3830	subs r0, #48 ; 0x30
DB04	blt.n 0x8108 <notdig>
2809	cmp r0, #9
DC02	bgt.n 0x8108 <notdig>
FB04 0303	mla r3, r4, r3, r0
E7F5	b.n 0x80f4 <nxtdig>
1C18	adds r0, r3, #0
BC1E	pop {r1, r2, r3, r4}
4770	bx lr

Table 17.1: Thumb instructions and machine code generated

```
c_int.s: Assembler messages:
c_int.s:15: Error: Thumb does not support this addressing mode -- `ldrb R0,[R1],#1'
c_int.s:27: Error: Thumb does not support this addressing mode -- `ldrb R0,[R1],#1'
c_int.s:30: Error: Thumb does not support conditional execution
c_int.s:34: Error: Thumb does not support conditional execution
c_int.s:40: Error: Thumb does not support this addressing mode -- `ldrb R0,[R1],#1'
c_int.s:46: Error: unshifted register required -- `orr R0,#0x20'
c_int.s:49: Error: instruction not supported in Thumb16 mode -- `subs R0,#97'
c_int.s:55: Error: instruction not supported in Thumb16 mode -- `subs R0,#48'
```

Listing 17.10: Error messages from GNU assembler for c_int (full hex, octal, binary input version)

In the event you want to modify the final version of c_int from Listing 17.4 which allows input in different bases, the assembler output in Listing 17.10 shows you the instructions that must be modified.

Maintenance

Review Questions

1. Why was Thumb code more important 20 years ago than today?
2. Why is it still important to be familiar with thumb code even though it is not needed for performance anymore?
3. Thumb also has a 32-bit MLA instruction. Why must it be a different machine code than the ARM version of the same instruction?
4. * We switch between Thumb mode and ARM mode with the BX instruction. Although we could set the T-bit in the CPSR directly, why would this lead to problems if we did?
5. * What simple change can be made to the c_int subroutine so that it works regardless of whether it is called with the T-bit already set or not?
6. Octal, hexadecimal, and binary are all indicated by a number beginning with the digit "0." Why do you think octal appears to be the default?

Programming Exercises

1. Subroutine c_int accepts integers expressed in different bases. If a number begins with a zero digit, then the next character indicates the base: X for hexadecimal, B for binary, or a digit for octal. Modify the subroutine to also accept a leading % to indicate binary and $ to represent hexadecimal.
2. Modify subroutine c_int to distinguish hexadecimal and binary numbers by a suffix of H and B, respectively (i.e., 134H, 101b).

Lab 18
Floating Point Input

In Lab 12, I introduced how a real number could be represented in IEEE 754 floating point format. At that time, we did not construct the floating point value, but displayed it using ARM integer based instructions. Now that we have some experience using the floating point processor in the Raspberry Pi, we will convert real numbers represented as a string of ASCII characters, such as 3.1416 and 6.02e23 to the floating point format.

Prototype

This Prototype section programming exercise converts a real number represented as a decimal fraction in an ASCII string to IEEE 754 floating point format. The program will have the following parts:

- A main program in file model.s and Listing 18.0. It has a main loop that will prompt the user to input a real number from the keyboard, then call subroutine c_flt to convert it to IEEE 754 format, and finally display the conversion back in ASCII.
- The c_flt subroutine in Listing 18.1 that will convert an ASCII text string to IEEE 754 format.
- Previously developed subroutines, such as v_flt, residing in the view.a and controller.a libraries.

A real number, such as 6.02×10^{23}, is typically formatted as 6.02E+23 for character-mode input and output in computer applications. Lower case "e" may also precede the base ten exponent. The plus sign may be omitted for positive exponents, and is replaced by a minus sign for negative exponents. The exponent will not be supported in this first version of c_flt, but will be added later in this lab.

The main program appearing in Listing 18.0 has a loop that will repeatedly call subroutine c_ascz to get another input text line, subroutine c_flt to convert it into floating point format, and subroutine v_flt to display it. An input value of 0.0 or a blank line will terminate the loop and exit the program. This same main program will also be used in the Coding and Debugging section to test the enhanced version of v_flt that includes base ten exponents.

```
1.          .global   _start         @ Provide program starting address to linker
2.          .equ      BUFSIZE,50     @ Size of input buffer can be changed
3.
4. @        Write message prompting for input from keyboard.
5.
6. _start:  ldr       R1,=prompt     @ Set R1 pointing to prompt message
7.          bl        v_ascz         @ Display message prompting user for input
                                     or exit
8.
9. @        Read a line from the keyboard
10.
11.         ldr       R1,=msgtxt     @ Set R1 pointing to input buffer
12.         mov       R2,#BUFSIZE    @ Maximum number of bytes to enter.
13.         bl        c_ascz         @ Fill ASCII input buffer
14.
15. @       Convert text to floating point and echo the value back to user.
16.
17.         bl        c_flt          @ String to convert to floating point.
18.         bl        v_flt          @ Display floating point value.
19.         cmp       R0,#0          @ Test if zero to force exit from loop.
20.         bne       _start         @ Continue the loop and get next input.
21.
22. @       Exit this program with no errors flagged.
23.
24.         mov       R0,#0          @ Status code 0 says "normal completion"
25.         mov       R7,#1          @ Command code 1 terminates program.
26.         svc       0              @ Issue Linux command to terminate program
27.
28.         .data
29. prompt: .asciz    "\nEnter next number or 0 to quit.\n"
30. msgtxt: .ds       BUFSIZE        @ Reserve storage for input data characters
31.         .end
```

Listing 18.0: “Main” program to test subroutine c_flt

Listing 18.1 provides the first version of subroutine c_flt. The first 40 lines shown in Listing 18.1a will construct two values in registers: R3 gets the input number, but without the decimal point and R2 gets the number of decimal digits (for 3.14, R3 would be 314 and R2 would be 2).

```
 1.@          Subroutine c_flt will convert input buffer to floating point format
 2.@                  R1: Points to string containing "real number" of form 123.456
 3.@                  LR: Contains the return address
 4.@                  R0: Returned with floating point value (converted from ASCII
                      input string)
 5.@                  R1: Returned pointing to first character after number.
 6.@                  All register contents except R0 and R1 will be preserved.
 7.
 8.           .global c_flt           @ Enable linker to see subroutine entry point.
 9.c_flt:     push    {R2-R6}         @ Save contents of registers R2 through R6.
10.
11.@          Register usage in this subroutine
12.@                  R0: Contents of next ASCII "digit" from input buffer
13.@                  R1: Points to next byte in input buffer
14.@                  R2: Number of digits right of decimal point
15.@                  R3: Accumulated value of significant being constructed
16.@                  R4: =10: Decimal base value needed to "shift" digits
17.@                  R5: "Flag" (=1) indicating decimal point present
18.
19.           mov     R2,#0           @ Count of number of digits right of point
20.           mov     R3,#0           @ Initialize R3 for value of real number.
21.           mov     R4,#10          @ Decimal, base 10, used to "shift" the
                                      significant
22.           mov     R5,#0           @ 0 => no decimal point, 1 => decimal point
                                      encountered
23.
24.@          Loop to read each of the significant digits of the floating point number
25.
26.nxtdig:    ldrb    R0,[R1],#1      @ Load next character from input buffer.
27.           subs    R0,#'0'         @ Subtract the ASCII character bias.
28.           blt     notdig          @ Check if end of string of digits is reached.
29.           cmp     R0,#9           @ Check upper limit of digits range.
30.           bgt     notdig          @ Go exit if end of string of digits found.
31.           mla     R3,R4,R3,R0     @ Shift accumulated value and add. [R3] =
                                      [R4]*[R3] + [R0]
32.           add     R2,R5           @ Increment number of  decimal places.
33.           b       nxtdig          @ Continue loop with next digit from input
                                      buffer.
34.
35.@          If non-digit is the decimal point, set flag to start counting digits
36.
37.notdig:    add     R0,#'0'         @ Restore the character bias.
38.           cmp     R0,#'.'         @ Test for decimal point
39.           moveq   R5,#1           @ Value of R5 to be added to count in R2
40.           beq     nxtdig          @ Continue converting base, but count decimal
                                      places, too.
```

Listing 18.1a: Build significant as an integer

41.			
42. @	Combine significant and decimal shift count using the floating point processor.		
43.			
44.	vmov	S3,R3	@ Copy integer value of significant.
45.	vcvt.f32.s32	S0,S3	@ Convert significant from to floating point.
46.	subs	R2,#1	@ Decrement number of decimal places.
47.	blt	cpy2R0	@ If none, then go copy "whole" floating point number.
48.	vldr	S1,point1	@ Load 0.1 into floating point register.
49.	beq	combbe	@ For 1 decimal place, go multiply significant times 0.1
50.	vmov	S2,S1	@ Extra copy of 0.10 needed for multiplication loop.
51. pow10:	vmul.f32	S1,S2	@ Change final value by a factor of 10
52.	subs	R2,#1	@ Decrement number of decimal places to "shift"
53.	bgt	pow10	@ Continue loop to "shift decimal point"
54. combbe:	vmul.f32	S0,S1	@ Combine significant and exponent.
55. cpy2R0:	vmov	R0,S0	@ Move finished floating pointer number for return.
56.	sub	R1,#1	@ Set R1 pointing to character that stopped scan.
57.			
58.	pop	{R2-R6}	@ Reload saved register contents.
59.	bx	LR	@ Return to the calling program.
60.			
61. point1:	.float	0.1	@ Floating point value to "shift" number to the right 1 place.
62.	.end		

Listing 18.1b: Combine significant and decimal shift count to finish floating point format

Listing 18.1b shows the significant being “shifted” to become the correct value by multiplying by 0.1_{10} for each digit appearing to the right of the decimal point.

Go ahead and run the sample program. For the link command, you’ll need the view.a and controller.a libraries from the previous labs for the v_flt, v_ascz, and c_ascz subroutines.

Listing 18.2 shows the output from several tests. Of course, try a few additional test values of your own. Do you think the extra digits (debris) to the right of the correct output value comes from errors in the conversion into floating point format or from converting the floating point back into character display format? Does the debris possibly come from both, or does it depend on the particular value being tested?

```
Enter next number or 0 to quit.
12.34
12.340001106262207031 25

Enter next number or 0 to quit.
12.5
12.5

Enter next number or 0 to quit.
12.1
12.1000003814697265625

Enter next number or 0 to quit.
12.25
12.25000095367431640625

Enter next number or 0 to quit.
12.75
12.75000095367431640625
```

Listing 18.2: Output from sample test run of subroutine c_flt with subroutine v_flt

Introductions

	List of VFP instructions introduced in Lab 18		
18.1.45.	vcvt.f32.s32	S0,S3	@ Convert from integer to floating point format.
18.1.48.	vldr	S1,point1	@ Load from memory into floating point register.
18.3.77.	vldrgt	S1,point1	@*Copy decimal 0.1 into floating point register.
18.3.78.	vldrlt	S1,dec10	@*Copy decimal 10.0 into floating point register.

Principles

Figure 18.0 shows four steps commonly used to generate a floating point number "package" from a number expressed in scientific notation:

1. Calculate the base 2 exponent: Convert the exponent from base 10 to base 2, but this will leave a coefficient that must be accommodated in the next step.
2. Calculate the significant: Multiply the base-10 coefficient with the base-2 coefficient from above.
3. Normalize the significant, which usually involves a slight modification to the base 2 exponent.
4. Set the sign bit to finish the package.

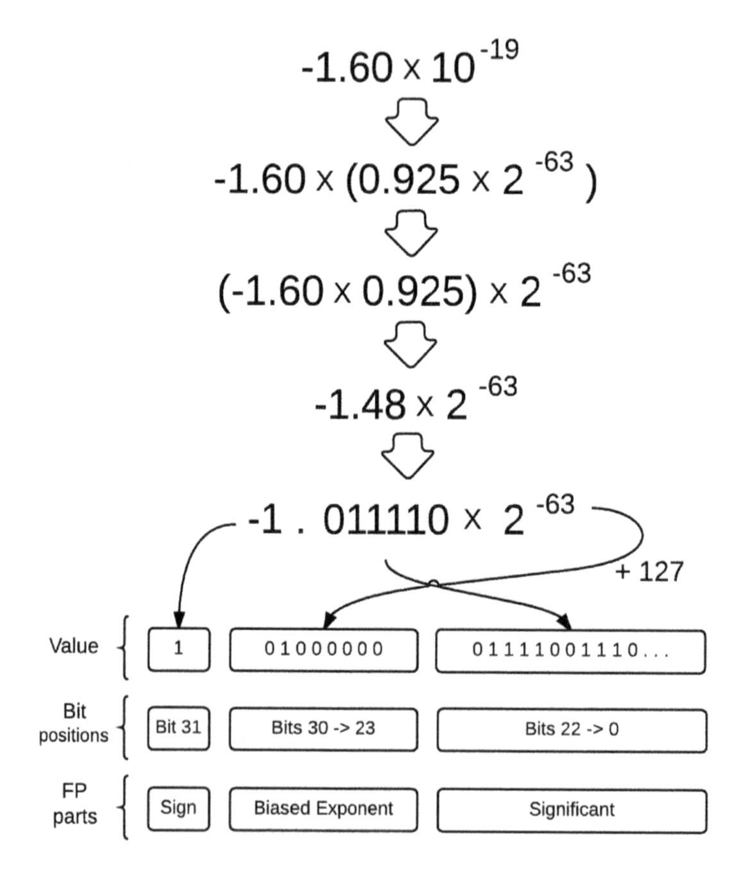

Figure 18.0: Single precision floating point fields in IEEE 754 format

The first step in the process is computationally the most demanding and thereby time consuming. Basically, we have to calculate logarithms, which involves several multiplications. In the Coding and Debugging section, I will use a simpler technique, but it is prone to loosing precision with larger exponents.

```
Enter next number or 0 to quit.
12345678
12345678.0
Enter next number or 0 to quit.
123456789
123456792.0
Enter next number or 0 to quit.
1.2345678
1.23456799983978271484375
Enter next number or 0 to quit.
12.345678
12.34567928314208984375
```

Listing 18.2: Example of running subroutines c_flt and v_flt together

Coding and Debugging

How should we input scientific notation in assembly language coding? There are a variety of trade-offs among execution speed, accuracy, and ease of coding. The program from the Prototype section only has to be modified to accommodate the occurrence of an exponent, so the processing of the significant will be the same.

```
 1.@        Subroutine c_flt will convert input buffer to floating point format
 2.@                   R1: Points to string containing "real number" of form 123.456
 3.@                   LR: Contains the return address
 4.@                   R0: Returned with floating point value (converted from ASCII
                       input string)
 5.@                   R1: Returned pointing to first character after number.
 6.@                   All register contents except R0 and R1 will be preserved.
 7.
 8.         .global    c_flt            @ Enable linker to see subroutine entry point.
 9.c_flt:   push       {R2-R6}          @ Save contents of registers R2 through R6.
10.
11.@        Register usage in this subroutine
12.@                   R0: Contents of next ASCII "digit" from input buffer
13.@                   R1: Points to next byte in input buffer
14.@                   R2: Number of digits right of decimal point
```

```
15. @                   R3: Accumulated value of significant being constructed
16. @                   R4: =10: Decimal base value needed to "shift" digits
17. @                   R5: "Flag" (=1) indicating decimal point present
18. @*                  R6: Value of exponent (power of ten) such as -34 in 6.6E-34.
19.
20.          mov        R2,#0          @ Count of number of digits right of decimal
                                       point
21.          mov        R3,#0          @ Initialize R3 for value of real number being
                                       converted.
22.          mov        R4,#10         @ Decimal, base 10, used to "shift" value in
                                       significant
23.          mov        R5,#0          @ 0 => no decimal point, 1 => decimal point
                                       encountered
24.
25. @        Loop to read each of the significant digits of the floating point number
26.
27. nxtdig:  ldrb       R0,[R1],#1     @ Load next character from input buffer.
28.          subs       R0,#'0'        @ Subtract the ASCII character bias.
29.          blt        notdig         @ Check if end of string of digits has been
                                       reached.
30.          cmp        R0,#9          @ Check upper limit of digits range.
31.          bgt        notdig         @ Go exit if end of string of digits found.
32.          mla        R3,R4,R3,R0    @ Shift accumulated value and add. [R3] =
                                       [R4]*[R3] + [R0]
33.          add        R2,R5          @ Increment number of places right of decimal
                                       point.
34.          b          nxtdig         @ Continue loop with next digit from input
                                       buffer.
```

Listing 18.3a: Build significant as an integer (same code as 18.1a)

As shown in Listing 18.3a, the construction of the significant is the same as that in the Prototype section. The additional coding appearing in Listing 18.3b shows the handling of non-digit characters such as the decimal point, the “E” for exponent, and the plus and minus signs of the exponent.

```
35.
36. @        If non-digit is the decimal point, set flag to start counting digits
37.
38. notdig:  add        R0,#'0'     @ Restore the character bias.
39.          cmp        R0,#'.'     @ Test for decimal point
40.          moveq      R5,#1       @ Value of R5 to be added to count in R2
41.          beq        nxtdig      @ Continue converting base, but count decimal
                                    places, too.
42.
43. @*       Non-digit of 'E' or 'e' indicates the beginning of a base ten exponent.
44.
45.          cmp        R0,#'E'     @* Test for the exponent being present.
```

46.	cmpne	R0,#'e'	@*Test for the exponent being present.
47.	bne	sigexp	@*End of number; Go combine signifant and exponent.
48.	ldrb	R0,[R1]	@*Load next character after the E or e
49.	cmp	R0,#'+'	@*Test for the Positive exponent being present.
50.	addeq	R1,#1	@*Load character after the plus sign.
51.	cmp	R0,#'-'	@*Test for a negative exponent being present.
52.	moveq	R5,#-1	@*Set flag that exponent is negative.
53.	mov	R6,#0	@*Initial value of exponent, 10^0 = 1
54.			

Listing 18.3b: Find non-digit while reading significant

The value of the exponent will be built in register R6 and will then be combined with the number of decimal places right of the decimal point.

55. @	Read the value of the exponent (such formats as 6.02E23, 6.02E+23, 6.6E-34)		
56.			
57. nxtexp:	ldrb	R0,[R1],#1	@*Load next character from input buffer.
58.	subs	R0,#'0'	@*Subtract the ASCII character bias.
59.	blt	finexp	@*Check if end of string of digits has been reached.
60.	cmp	R0,#9	@*Check upper limit of digits range.
61.	bgt	finexp	@*Go exit if end of string of digits found.
62.	mla	R6,R4,R6,R0	@*Shift accumulated value and add. [R6] = [R4]*[R6] + [R0]
63.	b	nxtexp	@*Continue loop with next digit from input buffer.
64.			
65. @*	Finish the exponent by combining number of decimal places with exponent value.		
66.			
67. finexp:	cmp	R5,#0	@*Set Z-flag if exponent is negative
68.	addlt	R2,R6	@*Combine negative exponent with number of decimal places.
69.	subge	R2,R6	@*Combine positive exponent with number of decimal places.
70.			

Listing 18.3c: Exponent determines number of decimal digits to “shift”

In figure 18.3d, the exponent will be included in the final floating point number by a somewhat strong arm, yet straightforward, approach. If the exponent is negative, a loop of multiplying by one tenth will be used, and if the exponent is positive a loop to multiply by ten will provide the correct result. The obvious problem with this technique is that if the absolute value of the exponent is large, each successive multiplication will introduce errors. In the Maintenance section, alternatives will be considered.

```
71. @        Combine significant and decimal shift count using the floating point
             processor.
72.
73. sigexp:  vmov          S3,R3        @ Move integer significant into floating
                                        point register.
74.          vcvt.f32.s32  S0,S3        @ Convert significant from integer to
                                        floating point.
75.          cmp           R2,#0        @*Test if decimal "shift" will be left or right.
76.          beq           cpy2R0       @*If none, then go copy "whole" number.
77.          vldrgt        S1,point1    @*Copy decimal 0.1 into floating point
                                        register.
78.          vldrlt        S1,dec10     @*Copy decimal 10.0 into floating point
                                        register.
79.          rsblt         R2,R2,#0     @*Take absolute value of exponent.
80.          subs          R2,#1        @ Decrement number of places right of
                                        decimal point.
81.          beq           combbe       @ For 1 decimal place, go multiply significant
                                        one time.
82.          vmov          S2,S1        @ Extra copy of 0.10 or 10.0 for multiplication
                                        loop.
83. pow10:   vmul.f32      S1,S2        @ Change final value by a factor of 10
84.          subs          R2,#1        @ Decrement number of decimal places to
                                        "shift"
85.          bgt           pow10        @ Continue loop to "shift decimal point"
86. combbe:  vmul.f32      S0,S1        @ Finish by combining significant and
                                        exponent.
87. cpy2R0:  vmov          R0,S0        @ Move finished floating pointer number for
                                        return.
88.          sub           R1,#1        @ Set R1 pointing to character that stopped
                                        scan.
89.
90.          pop           {R2-R6}      @ Reload saved register contents.
91.          bx            LR           @ Return to the calling program.
92.
93. point1:  .float        0.1          @ Floating point value to "shift" number to
                                        the right 1 place.
94. dec10:   .float        10.0         @*Floating point value to "shift" number to
                                        the left 1 place.
95.          .end
96.
```

Listing 18.3d: Combine significant and decimal shift count to finish floating point format

Go ahead and run this new version of c_flt. Some examples of testing it are shown in Listing 18.4.

```
Enter next number or 0 to quit.
1e5
100000.0
Enter next number or 0 to quit.
123e5
12300000.0
Enter next number or 0 to quit.
123e-5
0.00123000005260109901428222656 25
Enter next number or 0 to quit.
123.456e+2
12345.6005859375
Enter next number or 0 to quit.
123.456e-2
1.2345601320266723632 8125
```

Listing 18.4: Example of running c_flt using exponents

Maintenance

Review Questions

1. *The loop which multiplies the significant by either 10.0 or 0.1 to accommodate the base ten exponent results in some loss of precision in the conversion of ASCII character format to floating point. What two relatively simple modifications to that technique can greatly improve the resulting precision?
2. *Why will multiplying by 0.1 always result in a loss of precision in a binary computer?
3. How could the CLZ (count leading zeroes) instruction be used to approximate a base two logarithm?

Programming Exercises

1. Make a floating point power-of-ten table (10.0, 100.0, 1000., ...) similar to what was done for subroutine v_dec. Instead of the loop in c_flt that multiplies 10.0 times itself to calculate the scientific notation exponent, just index into the table instead. Do you think this will improve the accuracy for numbers with large exponents? Will it improve performance?
2. The above programming exercise will lead to a table of 77 values (10^{-38} to 10^{38}) and if done for double precision will require a table of hundreds of words. Modify c_flt to use a combination of the two techniques. For example, let the table be in multiples of 10^5, such as 10^5, 10^{10}, 10^{15}, ...
3. Modify c_flt to produce a double precision floating point format number.

Lab 19
Model View Controller

There are many design patterns, with Model View Controller (MVC) being the most well known. The motives for using design patterns are to improve software design, reduce programming errors, and provide for much more effective long term maintenance.

Prototype

Here's one final example program to showcase one of my favorite machine code instructions: Count Leading Zeroes (CLZ). In this Prototype section, it will be used as an ARM instruction, which together with a shift instruction, will left justify a bit pattern.

```
1.          .global  _start      @ Set program starting address to linker.
2. _start:  ldr      R6,=tstval  @ Point to list of test values.
3.          ldr      R3,=nl      @ Point to end of list of test values.
4. sample:  ldr      R0,[R6],#4  @ Load next integer (bit pattern).
5.          clz      R1,R0       @ Get number of zero bits left of first one.
6.          lsl      R0,R1       @ Left justify original bit pattern.
7.          mov      R2,#8       @ Display the full 32-bit word.
8.          bl       v_hex       @ Display left justified pattern
9.          ldr      R1,=nl      @ display one value per line.
10.         bl       v_ascz
11.         cmp      R3,R6       @ Test for end of list of test values.
12.         bgt      sample      @ Cotinue loop for next value in list.
13.         mov      R0,#0       @ Status code 0 for "normal completion"
14.         mov      R7,#1       @ Code terminates program.
15.         svc      0           @ Linux service terminates program.
16.         .data
17. tstval: .word    1,0b10,0b101,0b1111
18.         .word    1<<30,-1,1108,6600
19. nl:     .asciz   "\n"        @ End of list and new line character.
20.         .end
21.
```

Listing 19.0: Left justify an array of bit patterns

Compile the short program in Listing 19.0 and link it along with the view.a library. Listing 19.1 shows the program output with resulting register contents that are left-justified in a 32-bit word. Add a few additional test numbers and patterns of your own as well as display the result in binary if you like.

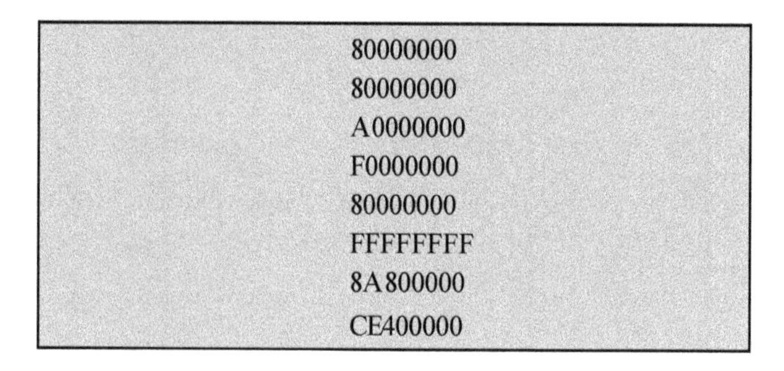
```
80000000
80000000
A0000000
F0000000
80000000
FFFFFFFF
8A800000
CE400000
```

Listing 19.1: Output of program to left justify eight 32-bit words

Introductions

	List of ARM instructions introduced in Lab 19		
19.0.5.	clz	R1,R0	@ Get number of zero bits left of first one.

	List of NEON instructions introduced in Lab 19		
19.2.14.	vclz.U32	Q1,Q0	@ Count the leading zeroes
19.2.15.	vshl.U32	Q0,Q1	@ Left justify each 32-bit word.

Principles

Along with the ultimate freedom available in assembly language programming comes the possibility of ultimate chaos in software development. But just because one can write horrible code, doesn't mean it has to be done. There have been numerous techniques attempting to restrict the writing of bad code: not putting a GOTO statement in the C language, information hiding in object oriented languages, and "no pointers" in Java, to name a few. Although design patterns were formally introduced in the 1970s with higher level languages, they are even more important in assembly language programming because there are few restrictions in assembly language to rein in sloppy programming.

There's nothing wrong with having fun with computers: Write short programs that do entertaining things, and then move on to the next fun activity. Who cares about comments? Who cares about the program structure? Who cares about whether the program can be expanded or corrected one or two years from now? I don't. It's disposable code. Production code, however, has a job to do, objectives to meet, responsibilities to achieve, and must be maintained over a period of months, years, and sometimes even decades. Production code is typically very large and complicated. Production code will probably require several significant enhancements during its lifetime. Production code will probably have several programmers working on it over a period of many years.

A design pattern is not the program code itself, but rather a way to organize the functioning or structure of a program. Problems can be solved many different ways. A design pattern provides a general way of solving a problem that has been successful in the past. Over the years, there have been multiple definitions and interpretations as to what each design pattern means and how it should be implemented, and the Model View Controller introduced with the Smalltalk object oriented language at Xerox's PARC during the 1970s is no exception. Figure 19.0 is one perspective where the user interacts directly with the Controller and the View.

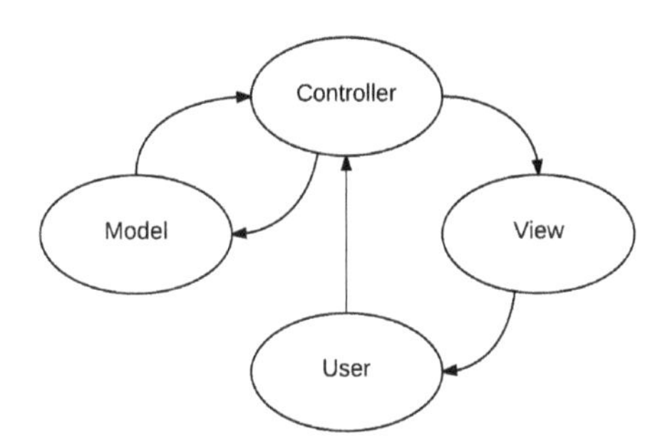

Figure 19.0: User interacting with Model View Controller architectural design pattern

I recall the Late Rear Admiral Grace Murray Hopper telling us there were two important considerations when dividing a program into components:

1. Make the subroutine interfaces as simple and clean as possible and
2. Once defined, never change the interface. If you need to enhance the interface, make a new one, but don't change what is already in production.

This leads us to the Adapter pattern: a case where a subroutine or group of subroutines is given a second "face." It's analogous to the Thumb instruction set being a different approach to use the same registers, memory, and instructions as the ARM CPU 32-bit format.

Another popular pattern today is the Layered pattern where an application consists of a user interface layer, a business logic layer, and a database layer. Don't go nuts! We do not need 100 layers. Don't build in complexity for the sake of complexity. The patterns are not the goal, good clean maintainable code is.

Listing 19.2 is a rewrite of Listing 19.0, but using the NEON coprocessor. The count of leading zeroes will be performed in parallel in four independent 32-bit lanes. Likewise, the resulting four counts will simultaneously shift the contents of four 32-bit lanes to left jutify the pattern in each. Compile, link, and execute it, and you will get the same answers as before.

```
1. @        Macro d_bin displays value in NEON 32-bit register.
2.          .macro    d_bin reg
3.          vmov      R0,\reg       @ Pointer to string to output before integer
4.          mov       R2,#8         @ Display all 32 bits.
5.          bl        v_hex         @ Display left justified binary pattern.
6.          ldr       R1,=nl        @ Display one value per line.
7.          bl        v_ascz        @ Call one of the display subroutines in
                                    view.o.
8.          .endm
9.
10.         .global   _start        @ Set program starting address to linker.
11. _start: ldr       R6,=tstval    @ Point to list of test values.
12.         ldr       R3,=nl        @ Point to end of list of test values.
13. sample: vldm      R6!,{Q0}      @ Load next 4 32-bit words.
14.         vclz.U32  Q1,Q0         @ Count the leading zeroes
15.         vshl.U32  Q0,Q1         @ Left justify each 32-bit word.
16.         d_bin     S0            @ Display the four left justified patterns.
17.         d_bin     S1            @
18.         d_bin     S2            @
19.         d_bin     S3            @
20.         cmp       R3,R6         @ Test for end of list of test values.
21.         bgt       sample        @ Cotinue loop for next value in list.
22.         mov       R0,#0         @ Status code 0 for "normal completion"
23.         mov       R7,#1         @ Code terminates program.
24.         svc       0             @ Linux cservice terminates program.
25.         .data
26. tstval: .word     1,0b10,0b101,0b1111
27.         .word     1<<30,-1,1108,6600
28. nl:     .asciz    "\n"          @ End of list and new line character.
29.         .end
30.
```

Listing 19.2: Left justify an array of bit patterns using NEON parallel procerssing

Maintenance

Maintenance really is the essential element of a product. There are two issues in software and hardware maintenance: bug fixes and enhancements. I've emphasized design patterns and documentation in order to provide an approach to successful product development that will result in successful maintenance. It is better to have not brought a product to market than develop one that cannot be maintained. Who needs the nightmare and expense of recalls (i.e., bugs)? Who needs the frustration of seeing a new market that is almost within grasp, but cannot be achieved (i.e., enhancements)?

Review Questions

1. Search the Internet and list five design patterns besides MVC, adapter, and layers.
2. * Name four operations available in NEON that are not available in VFPv3.

Programming Exercises

1. Modify program 19.2 to convert an integer to floating point. Both the significant and exponent are already present, but not quite in the final format needed.
2. Modify program 19.2 to perform the same task of left justifying a bit pattern, but use the instruction to convert from integer to floating point, followed by a fixed size (9 bits) shift to clear out the sign and exponent. Of course, it won't work on a few examples such as negative one unless you use double precision.

Epilogue

What about the bridge? Assembly language is certainly a low-level bridge between CPU hardware and software, but I also wrote this book to be a bridge of understanding between hardware and software engineers. What is the objective of this book, and what should have been learned by those reading it and working through its examples? The objective is to assist electronics engineers who want to bridge the gap toward understanding software development and software engineers who want to bridge the gap toward understanding CPU hardware operation. It's also for non-engineers who simply want a bridge to a deeper understanding of computer architecture and assembly language programming. After crossing the bridge, where do you want to go from here?

Aristotle is credited with the saying, "The more you know, the more you know you don't know." I look at learning assembly language not so much as a target, but as a springboard. While closing a few loose ends in this epilogue, I will open up a few new topics you may want to pursue.

The Hardware

I did not cover the following three embedded systems hardware topics in this book:

- CPU user mode and supervisor mode
- I/O devices
- Interrupts

The Software

The software components corresponding to the above three hardware topics are the following:

- Which operating system to use
- Device drivers
- Interrupt handlers

The Bridge

What are the bridges between the above three topics?

CPU User Mode and Supervisor Mode

The ARM processor design enables multiprogramming where multiple users can be sharing the same CPU and memory at virtually the same time. One responsibility of an operating system, such as Linux, is to protect one user from another while they are in the same memory space and taking turns using the CPU. This capability requires the operating system to run in a privileged, or supervisory, state where it can have access to all of memory, while user programs are restricted to their own data areas. The ARM processor has four modes in which it can operate. In this book, we've operated entirely in the user mode, where we are restricted from using a few powerful instructions. The other three modes are in supervisory mode.

When Unix, the metaphorical "grandfather" of Linux through Minix, first appeared, I was doing real-time programming related to gathering high-energy particle physics data. Unix was much too slow for our application, so other operating systems and approaches were needed. Although today's processors are much faster, the overhead of a general purpose operating system might still be too great for your embedded system, depending on what you're doing.

Possible problems with using Linux as your embedded systems operating system:

- It might be too slow due to its complexity.
- Accessing I/O ports requires device drivers to do it properly, which is a lot more difficult than just reading or writing an I/O port.

There are a variety of other operating system available for the Raspberry Pi. The NOOBS distribution that you used to initially load Linux already includes some. So, am I saying don't use Linux for imbedded systems? No. I am saying keep your options open. The purpose of this book was to acquaint the reader with professional grade assembly language programming. From there, further choices will have to be made depending on the application.

I/O Devices

In a real-time embedded systems program, there are basically two approaches that can be used to determine when a device requires attention from the software:

- Polling
- Interrupts

Examples of when a device needs attention:

- A sensor has changed state: a door has opened, a limit-switch has been hit, a temperature has changed, a light has gone out, etc.
- A task has been completed: a byte has been written and the buffer is ready for

more data, a motor has reached its limit, a time-delay has expired, etc.
- A message has been received: a message from a Local Area Network, a message coming through the Internet, etc.

In the polling approach, the software must loop though all possible devices, reading the status of each and deciding what to do with any status changes. This approach is very controlled and relatively easy to implement. The problem is it wastes a lot of time checking devices that do not need attention while devices that do need immediate attention have to wait their turn.

The interrupt approach is like a "news flash" coming in the middle of watching a ball game. It's also like being interrupted at work when someone comes to your office to ask a question. Polling would be like going from office to office asking if anyone needs assistance.

There are basically two types of interrupts on most CPUs:

- Software: The SVC instruction we've been using to call Linux kernel services in every program in this book. The SVC (service call) instruction is also known as SWI (software interrupt).
- Hardware: A change in state of an I/O device essentially "calls" a device driver (subroutine) using an "instruction" that behaves almost identical to the SVC (a.k.a, SWI) instruction.

Why Assembly Language?

In summary, I believe experience with assembly language is important today, but somewhat different than fifty years ago. If you know how to write good assembly code, you can write even better higher level language code. By "good" I mean placing an emphasis on the following:

- Structure: Use patterns when possible and have a low diet on spaghetti code.
- Documentation: Indicate not so much *what* is done but *why* it is being done.
- Efficiency: Don't count on Moore's law to bail out poorly designed and programmed applications.

Here's a few directions to try for improving your working knowledge of assembly language:

- Finish the ARM disassembler
- Write ARM and NEON emulators
- Write graphics matrix transformations using binary point
- Attend programming classes and join user groups

Appendix A
Raspberry Pi Setup

Most readers of this book will have their Raspberry Pi computers already working. For those of you who do not, I've included this brief appendix. If you're totally new to computers and the Raspberry Pi, then you will want to follow more detailed setup instructions from the Internet or a book aimed more at beginners.

This appendix also includes Linux configuration parameters related to assembly language programming and download instruction for obtaining the source code for the assembly language programs developed in this book.

GNU Assembler

This book is a tutorial for learning assembly language using the Raspberry Pi computer. The information presented and examples given are specifically for the Raspberry Pi 2 and 3, but should also work fine for previous and future versions with the possible exception for the examples using the floating point and vector coprocessors.

The main constituent of the Raspberry Pi is the ARM CPU, so this book uses the GNU assembler from the Free Software Foundation that is included with the Raspian Linux kernel distribution for the Raspberry Pi. This software is not part of the Raspberry Pi itself, but is included with almost all packages sold that include the Raspberry Pi circuit board, power supply, case, and SD memory card. If the Raspian Linux distribution software is not included with the NOOBS (New Out Of the Box Software) present on the SD card, it may be downloaded over the Internet for free from various Internet sites.

The Hash Symbol (#)

The GNU "as" assembler program uses the hash symbol (#) to identify immediate constants. If the British currency symbol appears instead of the hash, it is probably due to the keyboard not being configured correctly. For users in the United States, this configuration is normally adjusted by entering the following from the Linux command line and selecting the US Keyboard option as follows.

```
sudo raspi-config

Language English US Keyboard(9) US
```

Source Code

This book contains over 50 program listings as examples of ARM, VFPv3, and NEON coding. I have made them available on the Internet so they can be easily downloaded using the git utility and GitHub website. GitHub "is a code hosting platform for version control and collaboration." It is composed of multiple public and private "repositories" holding text, image, and video files. Enter the following command from the Linux command prompt to create directory *RPi_Asm_Bridge* and load all the program listings for this book into it.

```
git clone https://github.com/robertdunne/RPi_Asm_Bridge.git
```

Of course, you must already have your Raspberry Pi configured for Internet access. In the event the git utility is not already present on your system, it can be downloaded and installed using the following command line:

```
sudo apt install git
```

Once you have the program listings, please display details related to their use from the README file by entering the following command or use any text editor of your choosing.

```
cat README.md
```

Warning: The assembler source code that appears in this book and is available for download is for learning to program in assembly language. Some of these subroutines are incomplete and even contain problems that need to be corrected in exercises at the end of the labs. No guarantee of their commercial utility is expressed or implied.

Alternatives to the I/O formatting subroutines presented in this book are contained in the CLIB library, but there are conditions and restrictions for their use in commercial applications as well.

BeagleBone Black

Although this book was designed for use with the Raspberry Pi, other Linux-based systems having an ARM CPU, NEON coprocessor, and GNU utilities will also work fine. For the BeagleBone Black, I recommend running Linux command mode using the PuTTY program. The IP address of the BeagleBone Black attached to a PC through the USB cable is (192.168.7.2).

Appendix B
Binary Numbers

To be precise, it's not the numbers that are binary, but the written representation of numbers. For example, we currently count eight planets in the solar system. This has been "written down" as 8, VIII, 10_8, 1000_2, as well as a variety of other representations throughout history.

What's Binary?

Binary means *two* like a binary star system consisting of a pair of stars. In the case of binary "numbers," the *two* refers to the base, also known as the radix, which indicates how many different symbols (or digits) can be used. In our every day decimal (base 10) system, there are ten symbols available {0, 1, 2, 3, 4, 5, 6, 7, 8, 9} so we can represent a number in a form like 3274, 1620, and 36. While in binary (base 2), we have only two symbols available {0, 1} so we are restricted to representing numbers in a form like 1100, 10101, 1, and 111. Other popular bases that have been used in the computer industry are octal (base 8) having eight symbols {0, 1, 2, 3, 4, 5, 6, 7} and hexadecimal (base 16) having sixteen symbols {0, 1, 2, 3, 4, 5, 6, 7, 8, 9, A, B, C, D, E, F}.

Why Binary?

The simple answer is that the logical building blocks (i.e., electronics in today's systems) are simpler and more efficient in binary than they are in our everyday decimal. The electronic logic circuits have two states: High and Low (voltage levels) which can model a variety of binary states like True and False, Yes and No, and of course One and Zero. This system follows the logic attributed to Aristotle thousands of years ago.

In the following table, we compare the written representations of counting from 0 to 12 in five different bases. Notice how the rightmost column (one's place) is incremented through all of the possible symbols available in the base before the next column to its left is incremented.

base 10 10 symbols {0123456789}	base 2 2 symbols {01}	base 3 3 symbols {012}	base 4 4 symbols {0123}	base 5 5 symbols {01234}
0	0	0	0	0
1	1	1	1	1
2	10	2	2	2
3	11	10	3	3
4	100	11	10	4
5	101	12	11	10
6	110	20	12	11
7	111	21	13	12
8	1000	22	20	13
9	1001	100	21	14
10	1010	101	22	20
11	1011	102	23	21
12	1100	110	30	22

Table B.0: Counting from 0 to 12 in bases 10, 2, 3, 4, and 5

Column	3	2	1	0
Base 10	$10^3=1000$	$10^2=100$	$10^1=10$	$10^0=1$
Base 2	$2^3=8$	$2^2=4$	$2^1=2$	$2^0=1$
Base 3	$3^3=27$	$3^2=9$	$3^1=3$	$3^0=1$
Base 4	$4^3=64$	$4^2=16$	$4^1=4$	$4^0=1$
Base 5	$5^3=125$	$5^2=25$	$5^1=5$	$5^0=1$

Table B.1: Value of each column in bases 10, 2, 3, 4, and 5

The Problems with Binary

The problems with binary are not with computers, but with us humans:

1. We are comfortable with base ten and have used it daily for most of our lives.
2. Binary numbers are awkward for us due to the large number of columns required. Who would prefer replacing the decimal representation of 7094, 1620, 1108, 6600, 3033, and 7800 with their binary equivalents 1101110110110, 11001010100, 10001010100, 1100111001000, 101111011001, and 1111001111000?
3. Conversion between binary and decimal is difficult to do "in our heads." The difficulty stems from the fact that 10 is not an integer power of two.

Superscripts and Subscripts

In math books, the base (or radix) used to represent a number is given as a subscript. For example: a number written in decimal would be like 257_{10} and in binary it would be like 10000001_2. If no subscript is provided, we assume it is decimal unless it is stated in the text that the numbers are expressed in a different base such as binary. When working with computer programs, whether assembler or higher level, subscripts are not commonly available so binary is generally entered as 0b10101, 10101b, or %10101 depending on the computer system or application being used.

Superscripts indicate a number raised to a power. For example, 4^3 means 4×4×4 equaling 64 and 2^8 is 2×2×2×2×2×2×2×2 equaling 256. Also recall that 2^0, 10^0, 16^0, and any non-zero number raised to the zeroth power equals one.

A decimal number is really a short notation for a polynomial of powers of 10. For example: 137_{10} is $1\times10^2 + 3\times10^1 + 7\times10^0$ which is 100 + 30 + 7. Likewise, a binary number is really a short notation for a polynomial of powers of 2. For example: 110101_2 is $1\times2^5 + 1\times2^4 + 0\times2^3 + 1\times2^2 + 0\times2^1 + 1\times2^0$. By the way, this polynomial structure is the main reason we label and count bits within a byte or word from right to left starting with zero.

Bit Position	3	2	1	0
Power of 2	2^3=8	2^2=4	2^1=2	2^0=1
Binary example	1	0	1	1
$1011_2 = 1\times2^3 + 0\times2^2 + 1\times2^1 + 1\times2^0 = 8+0+2+1 = 11_{10}$				

Table B.2: Bit position example: $1011_2 = 2^3 + 0 + 2^1 + 2^0 = 8 + 0 + 2 + 1 = 11_{10}$

Conversion to Any Base

A popular way to convert a number to a particular base is successive division. The remainders from each division will provide the digits (i.e., symbols) beginning with rightmost digit. For example, converting the number 3274 to decimal follows:

1. 3274 / 10 = 327 Remainder 4
2. 327 / 10 = 32 Remainder 7
3. 32 / 10 = 3 Remainder 2
4. 3 / 10 = 0 Remainder 3

So the "number" 3274 is represented in decimal as the sequence of remainders "3" "2" "7" and "4." By the way: This technique of successively dividing a number by the desired base works regardless of how the “computer” internally stores numbers. It could be binary, decimal, or any conceivable internal structure that would permit division.

Converting the same number 3274 to binary follows:

1. 3274 / 2 = 1637 Remainder 0
2. 1637 / 2 = 818 Remainder 1
3. 818 / 2 = 409 Remainder 0
4. 409 / 2 = 204 Remainder 1
5. 204 / 2 = 102 Remainder 0
6. 102 / 2 = 51 Remainder 0
7. 51 / 2 = 25 Remainder 1
8. 25 / 2 = 12 Remainder 1
9. 12 / 2 = 6 Remainder 0
10. 6 / 2 = 3 Remainder 0
11. 3 / 2 = 1 Remainder 1
12. 1 / 2 = 0 Remainder 1

So the “number” 3274 is represented in binary as the sequence of remainders “1” “1” “0” “0” “1” “1” “0” “0” “1” “0” “1” and “0.” As an exercise, try converting 3274 to base five by successively dividing by five until the quotient is zero (3274/5 = 654 remainder 4, ...). The answer will be 101044_5.

Multiplying and Dividing by Shifting

If we want to multiply by ten “in our heads” in our everyday decimal system, we just append a zero. For example to multiply 709 by 10, we append “0” to “709” and get “7090.” Likewise, when we multiply by 100 (i.e., 10^2), we append two zeroes, and for 1000, we append 3 zeroes, etc. For dividing by powers of ten, we do the reverse: we remove zeroes on the right. What if there are not enough zeros present on the right? Then we move the decimal point. For example to divide 1108 by 100, we move the decimal point to the left two places giving us 11.08.

When we shift a number to the left in base two, we are multiplying by a power of two, and when we shift to the right, we are dividing by a power of two. This means that conversion into and from binary format is done very efficiently using shifting rather than division. Converting the same number 3274 (110011001010_2) to binary by shifting is below. Note: The notation “>> 1” means shift 1 bit position to the right, and the “Carry out” refers to the rightmost bit that is lost when the value is shifted.

1. 110011001010 >> 1 = 11001100101 with Carry out 0
2. 11001100101 >> 1 = 1100110010 Carry out 1
3. 1100110010 >> 1 = 110011001 Carry out 0
4. 110011001 >> 1 = 11001100 Carry out 1

5. 11001100 >> 1 = 1100110 Carry out 0
6. 1100110 >> 1 = 110011 Carry out 0
7. 110011 >> 1 = 11001 Carry out 1
8. 11001 >> 1 = 1100 Carry out 1
9. 1100 >> 1 = 110 Carry out 0
10. 110 >> 1 = 11 Carry out 0
11. 11 >> 1 = 1 Carry out 1
12. 1 >> 1 = 0 Carry out 1

Converting Digits Into a Number

To convert "written digits" into a number, run the above process in reverse: Do successive multiplications. For example in base 10: the sequence of digits "1" "6" "2" "2" could be used to "build" the number 1622 as follows:

1. Start with 0
2. $0 \times 10 + 1 = 1$
3. $1 \times 10 + 6 = 16$
4. $16 \times 10 + 2 = 162$
5. $162 \times 10 + 2 = 1622$

In binary, it is simply a matter of shifting to the left one bit position to "multiply" by two. In the following example, the number expressed as a sequence of digits "110011001010" is built by a series of logical left shifts notated by "<< 1" combined with a logical OR notated by "+":

1. Start with 0
2. 0 << 1 + 1 = 1
3. 1 << 1 + 1 = 11
4. 11 << 1 + 0 = 110
5. 110 << 1 + 0 = 1100
6. 1100 << 1 + 1 = 11001
7. 11001 << 1 + 1 = 110011
8. 110011 << 1 + 0 = 1100110
9. 1100110 << 1 + 0 = 11001100
10. 11001100 << 1 + 1 = 110011001
11. 110011001 << 1 + 0 = 1100110010
12. 1100110010 << 1 + 1 = 11001100101
13. 11001100101 << 1 + 0 = 110011001010

Negative Binary Numbers

When we include negative numbers, we effectively double how many numbers we have to be able to represent in binary. For every positive number, we have a corresponding negative number. This requires an additional bit, a "sign" bit, that has to be associated with every binary number in registers and storage.

Rather than append an additional bit to each numeric storage type, computer manufacturers have chosen to steal a bit from the positive range. Instead of an 8-bit byte supporting numbers in the range of 0 through 255, it supports -128 through +127 for "signed" bytes. Likewise, signed half-words have a range of -32,768 to +32,767 rather than 0 through 65,535 for the unsigned format. The range is actually the same, but it has been shifted by 50%.

There have been four formats popular for representing signed numbers in binary computers:

- **Bias:** Add ½ the total range to all numbers
- **Sign and magnitude:** High order (leftmost) bit is the sign: 1 for negative
- **One's complement:** Complement (i.e., toggle) all bits for negative.
- **Two's complement:** Add 1 to one's complement value

The question is, which one is popular in today's computers? Being even more specific, which are present in the Raspberry Pi? Three are used: two's complement represents signed integers in the ARM CPU while both sign/magnitude and bias are used in the floating point format. Table B.0 gives 8-bit binary examples where positive and negative 26_{10} are represented four ways. I've also included zero, including the rather unexpected negative zero case.

Decimal	+26	– 26	+0	– 0
Sign & Magnitude	00011010	10011010	00000000	10000000
One's Complement	00011010	11100101	00000000	11111111
Two's Complement	00011010	11100110	00000000	00000000
Biased	10011010	01100110	10000000	10000000

Table B.0: Comparison of +26, –26, +0, and –0 in four signed byte formats

Nine's complement

How can we subtract using an "adding machine"? This question was not new with electronic computers, but goes back to the days when accountants and human "computers" used mechanical adding machines. It involves converting the algebraic expression "A – B" to "A + (–B)" which transforms the question into how should we represent –B?

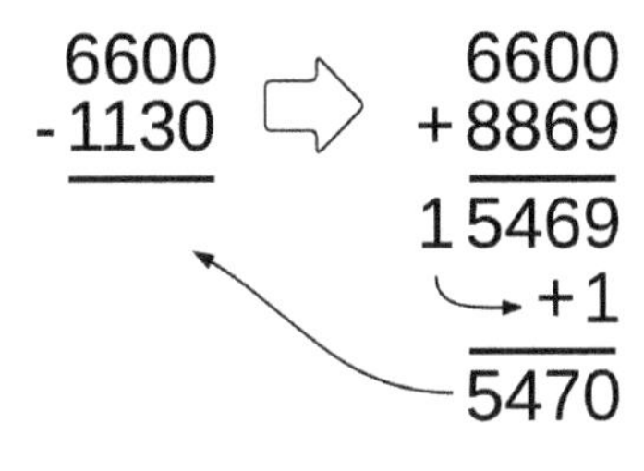

Figure B.0: Nine's complement example of subtraction by addition

Accountants, working in base ten, could represent a negative number by subtracting each of its digits from nine (one less than the base). On the left, we see an example where the negative of 1130 is 8869 in nine's complement (each 8 comes from 9 – 1, the 6 comes from 9 – 3, and the 9 comes from 9 – 0).

Obviously, since we're adding, rather than subtracting, the result is larger than we want, but if you do the algebra, you'll notice that the correct answer can be achieved. Notice how the first sum in Figure B.0 had a "carry out" that did not fit in the number of columns we were using. If you add this carry in a second step as shown, the correct answer appears. If there is no carry, do not add it, and there will be a large number, but it is really a negative number.

One's complement

One's complement is the same as nine's complement except the base is now two: every digit is subtracted from 1, instead of 9. Actually, this technique works in any base. Do the algebra if you like to prove it. By coincidence (or really because there's only two symbols in base 2), one's complement is achieved by simply inverting each bit as is shown in Figure B.1.

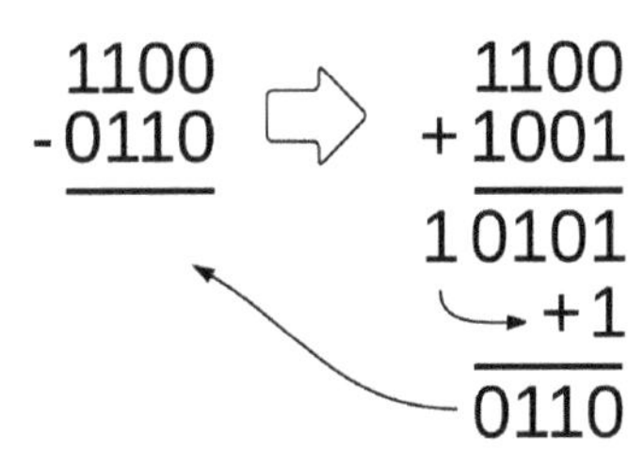

Figure B.1: One's complement example of subtraction by addition

Just like in nine's complement a subtraction is converted into an addition. Here the negative of "0110" is calculated to be "1001" where the value in each column is calculated by subtracting it from one less than the base. Notice that it's still a two-step process where the carry out is added back to obtain the correct answer.

If you follow the above naming convention, you would think that two's complement involves numbers expressed in base 3. Actually, the expression "two's complement" refers to the technique that eliminates the second step during a subtraction.

Two's complement

In two's complement, the negative of a number is generated by adding one to the one's complement and ignoring any caries. For example, the negative of 0110 is 1001 + 1 = 1010.

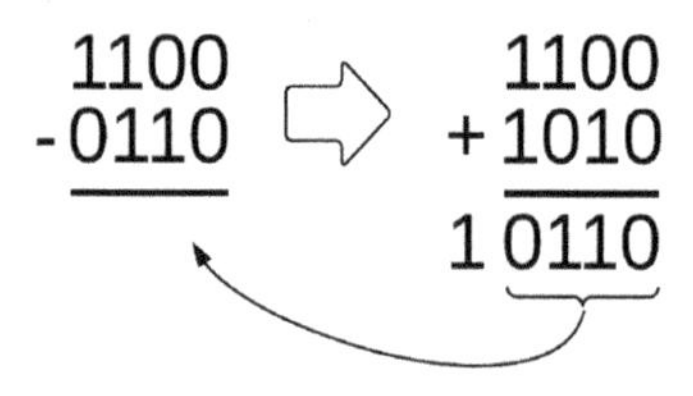

Figure B.2: Two's complement example of subtraction by addition

Rather than adding the "carry out" as a second step, a 1-bit is added preemptively when the negative is generated. Then during the subtraction, the carry is just ignored, making two's complement subtractions twice as fast as one's complement subtractions.

Floating point

Floating point is a package containing two signed binary numbers: a biased exponent, and a sign and magnitude significant.

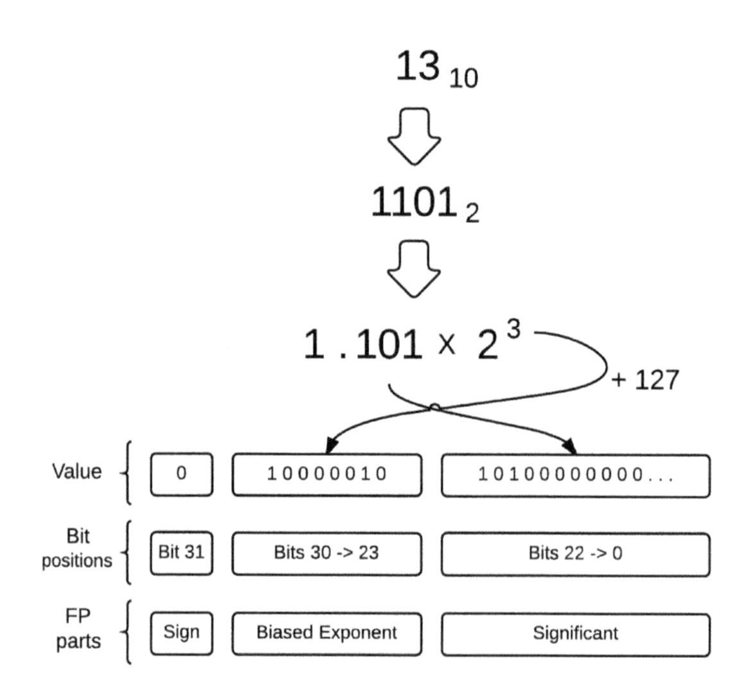

Figure B.3: Single precision floating point fields in IEEE 754 format

Appendix C
Hexadecimal Numbers

To be precise, it's not the numbers that are hexadecimal, but the written representation of numbers. Hexadecimal is a compact form of binary representation where we have sixteen symbols {0,1,2,3,4,5,6,7,8,9,A,B,C,D,E,F} to represent numbers. If you're not familiar with binary representation, please read Appendix B before studying hexadecimal. If it wasn't for binary, there would be negligible need for hexadecimal in the computer industry.

A decimal number is really a short notation for a polynomial of powers of 10. For example: 137_{10} is $1\times10^2 + 3\times10^1 + 7\times10^0$ which is 100 + 30 + 7. Likewise, a binary number is really a short notation for a polynomial of powers of 2. For example: 110101_2 is $1\times2^5 + 1\times2^4 + 0\times2^3 + 1\times2^2 + 0\times2^1 + 1\times2^0$. A hexadecimal number is really a short notation for a polynomial of powers of 16. For example: $5A732C_{16}$ is $5\times16^5 + 10\times16^4 + 7\times16^3 + 3\times16^2 + 2\times16^1 + 12\times16^0$ where A and C are digits representing values of 10 and 12, respectively.

Why Use Hexadecimal?

The simple answer is hexadecimal is compact, and it is very easy for us humans to convert between binary and hexadecimal. Consider the following:

1. Internally, almost all our computer systems are based in binary (see Lab 5 and Appendix B for an explanation).
2. Inputting and displaying numbers in the computer's natural binary notation is very efficient for the computer, but clumsy and inefficient for us humans. Who is comfortable reading and entering numbers like 100001101010010 or 1101101101101, and even much longer ones up to 64 bits in length?
3. Decimal is a rather compact form of representing numbers, and we are very comfortable with it because we use it in our daily lives. We can convert between decimal and binary by using successive divisions by ten. However, that is slow and cumbersome to do "in our heads." A division by sixteen is simply a four bit shift, but a division by ten cannot be achieved by shifting bits.
4. Do we humans actually need to use binary? As people working with computers at a detailed architectural level, we have to see the actual bits. We have to look at status words, IP addresses, instruction formats, and memory dumps.

Table C.0 shows counting from 0 to 17 in decimal, binary, hexadecimal, and octal. Notice how one hexadecimal digit fits exactly in four bits. Figure C.0 shows a binary number being "mapped" to hexadecimal digits, four bits at a time. starting from the right side.

base 10 10 symbols {0123456789}	base 2 2 symbols {01}	base 16 16 symbols {0123456789ABCDEF}	base 8 8 symbols {01234567}
0	0	0	0
1	1	1	1
2	10	2	2
3	11	3	3
4	100	4	4
5	101	5	5
6	110	6	6
7	111	7	7
8	1000	8	10
9	1001	9	11
10	1010	A	12
11	1011	B	13
12	1100	C	14
13	1101	D	15
14	1110	E	16
15	1111	F	17
16	10000	10	20
17	10001	11	21

Table C.0: Counting from 0 to 17 in four different bases

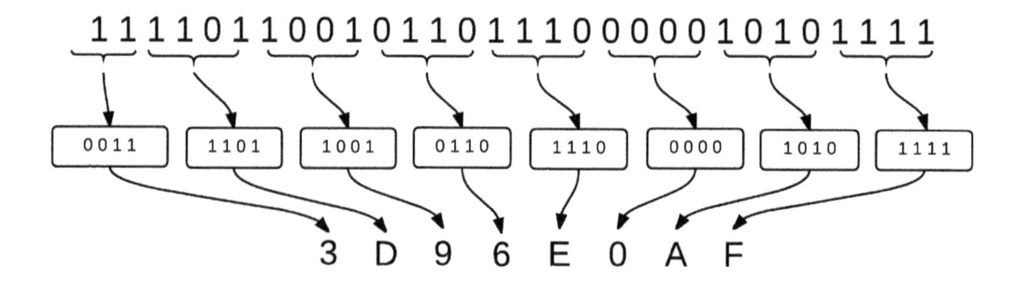

Figure C.0: Convert binary to hex, 4 bits at a time starting from the right (low order) side

Appendix D
ASCII

Why ASCII? Why not Baudot, BCD, Display Code, Fieldata, Unicode, XS3, or any other character code?

What is a Character Code

Binary computers store and manipulate bits (binary digits). Numbers are represented by "groups of bits" as either integers or real numbers. That's fine for science and engineering applications, but what's stored in "groups of bits" for business applications, such as correspondence, reports, and mailing lists? How is this text data consisting of letters, digits, and punctuation represented by "groups of bits"? A character code is a set that assigns each text character to a unique number.

This was not so much of a problem 3000 years ago. Several of the ancient languages including Assyrian, Hebrew, and Greek were "computer ready," but our modern written languages, such as English, are not. In these ancient languages, every symbol used to compose words was also used to compose numbers. The symbols alpha and beta in Greek were assigned both sounds to form words as well as numeric values to write numbers. In English, letters and digits are separate (i.e., the letter "R" does not have a numeric value). This means there was no "standard" for storing text data as a series of numbers.

In the 1960s, several companies were manufacturing mainframe computer systems. They were competing for sales and were interested in locking customers into their unique designs rather than making computer data files and applications portable from one system to another. There were basically two problems with character codes in the 1960s:

- Each character was stored in a byte, but the number of bits composing a byte varied from system to system.
- Each character was assigned a unique numeric code, but each computer system had a different set of character code assignments..

Several mainframe computer systems had 6-bit bytes, which supported a set of 64 different characters. BCD, Display Code, Fieldata, and XS3 are examples of 6-bit codes. Each of these sets contained 26 upper case letters, 10 digits, and a few punctuation marks and control characters. In order to include lower case letters, IBM switched from a 6-bit code to an 8-bit EBCDIC code in the mid 1960s. The size of the byte determines how many different characters can be represented as listed below:

- 6 bits: 64 characters
- 7 bits: 128 characters
- 8 bits: 256 characters
- 16 bits: 65,536 characters

The second compatibility problem was that the unique assignments were inconsistent among the different character code sets and computer systems. It took a presidential decree to alleviate some of the inconsistencies. On March 11, 1968, President Johnson signed ASCII (American Standard Code for Information Interchange) into existence.

Character Code	Letter A	Digit 5	blank
IBM BCD	11	05	30
CDC Display Code	01	20	2D
Univac Fieldata	06	25	05
XS3	14	08	33
EBCDIC	C1	F5	40
ASCII	41	33	20
Unicode	41	33	20

Table D.0: Example of three characters expressed in various character codes (in hexadecimal)

The 7-bit ASCII code from 1968 was fine for the English language, but it could not even support all the characters used in French, Spanish, and other Latin languages. In 1985, character set ISO 8859 was defined as an 8-bit code with 256 character codes defined, where the first 128 are identical to 7-bit ASCII. The remaining 128 character codes were assigned to accent characters for the Latin languages and a variety of special symbols like copyright and trademark.

0	NUL	10	DLE	20	
1	SOH	11	DC1	21	!
2	STX	12	DC2	22	"
3	ETX	13	DC3	23	#
4	EOT	14	DC4	24	$
5	ENQ	15	NAK	25	%
6	ACK	16	SYN	26	&
7	BEL	17	ETB	27	'
8	BS	18	CAN	28	(
9	HT	19	EM	29	)
0A	LF	1A	SUB	2A	*
0B	VT	1B	ESC	2B	+
0C	FF	1C	FS	2C	,
0D	CR	1D	GS	2D	-
0E	SO	1E	RS	2E	.
0F	SI	1F	US	2F	/

Table D.1a: ASCII and ISO codes in hexadecimal

30	0	40	@	50	P
31	1	41	A	51	Q
32	2	42	B	52	R
33	3	43	C	53	S
34	4	44	D	54	T
35	5	45	E	55	U
36	6	46	F	56	V
37	7	47	G	57	W
38	8	48	H	58	X
39	9	49	I	59	Y
3A	:	4A	J	5A	Z
3B	;	4B	K	5B	[
3C	<	4C	L	5C	\
3D	=	4D	M	5D	]
3E	>	4E	N	5E	^
3F	?	4F	O	5F	_

Table D.1b: ASCII and ISO codes in hexadecimal

60	`	70	p		
61	a	71	q		
62	b	72	r		
63	c	73	s		
64	d	74	t		
65	e	75	u		
66	f	76	v		
67	g	77	w		
68	h	78	x		
69	i	79	y		
6A	j	7A	z		
6B	k	7B	{		
6C	l	7C	|		
6D	m	7D	}		
6E	n	7E	~		
6F	o	7F			

Table D.1c: ASCII and ISO codes in hexadecimal

80	€	90		A0	
81		91	‘	A1	¡
82	‚	92	’	A2	¢
83	ƒ	93	“	A3	£
84	„	94	”	A4	¤
85	…	95	•	A5	¥
86	†	96	–	A6	¦
87	‡	97	—	A7	§
88	ˆ	98	˜	A8	¨
89	‰	99	™	A9	©
8A	Š	9A	š	AA	ª
8B	‹	9B	›	AB	«
8C	Œ	9C	œ	AC	¬
8D		9D		AD	
8E	Ž	9E	ž	AE	®
8F		9F	Ÿ	AF	¯

Table D.1d: ISO 8895 codes in hexadecimal

Hex Code	Character	Hex Code	Character	Hex Code	Character
B0	°	C0	À	D0	Ð
B1	±	C1	Á	D1	Ñ
B2	²	C2	Â	D2	Ò
B3	³	C3	Ã	D3	Ó
B4	´	C4	Ä	D4	Ô
B5	µ	C5	Å	D5	Õ
B6	¶	C6	Æ	D6	Ö
B7	·	C7	Ç	D7	×
B8	¸	C8	È	D8	Ø
B9	¹	C9	É	D9	Ù
BA	º	CA	Ê	DA	Ú
BB	»	CB	Ë	DB	Û
BC	¼	CC	Ì	DC	Ü
BD	½	CD	Í	DD	Ý
BE	¾	CE	Î	DE	Þ
BF	¿	CF	Ï	DF	ß

Table D.1e: ISO 8895 codes in hexadecimal

Hex Code	Character	Hex Code	Character		
E0	à	F0	ð		
E1	á	F1	ñ		
E2	â	F2	ò		
E3	ã	F3	ó		
E4	ä	F4	ô		
E5	å	F5	õ		
E6	æ	F6	ö		
E7	ç	F7	÷		
E8	è	F8	ø		
E9	é	F9	ù		
EA	ê	FA	ú		
EB	ë	FB	û		
EC	ì	FC	ü		
ED	í	FD	ý		
EE	î	FE	þ		
EF	ï	FF	ÿ		

Table D.1f: ISO 8895 codes in hexadecimal

What about those written languages like Hebrew and Greek that were "computer ready" thousands of years ago. Were they still computer ready in 1968 when ASCII was defined? They were ready by themselves, but in order to include them alongside ASCII and ISO 8895, a new character set has since been defined: Unicode. Casually speaking, Unicode is considered to be a 16-bit code supporting 65,536 different character code symbols, enough to encompass all the written symbols composing thousands of different languages. The first 128 characters of Unicode are the same as the ASCII character set.

Appendix E
Text Editors

Editing is the first step in the vicious cycle of edit-compile-link-execute used to test and debug software programs. A "word processor" program cannot be used to generate or update the assembler source code because it inserts many hidden commands for changing fonts, formatting pages, and including images that would not be meaningful to the assembler. Appendix E provides a brief introduction to three text editors included in the Raspian Linux distribution that are compatible with editing assembly language source code files.

Program Source Code

Assembly language source code files have a simple structure consisting of the following ASCII characters:

1. **"Printable" Characters:** A, B, C, ... 1, 2, 3, ..., #, !, [,], ... See Appendix D for a full list. If the hash symbol (#, "pound sign") is missing, see Appendix A for suggestions for Linux system configuration.
2. **Line Feed "control character":** Hexadecimal code 0x0A terminates each line.
3. **Horizontal Tab "control character":** Hexadecimal code 0x09 separates each column. One or more blanks would work as well, but tabs automatically form columns.
4. **No line numbers:** Although line numbers appear on listings and error diagnostics, they're not physically in the file. The text editors can count lines and display that count if the user chooses.

```
        .global _start  @ Indicate _start is global for linker
_start: mov     R0,#78  @ Move a decimal 78 value into register R0
        mov     R7,#1   @ Move a decimal 1 integer value into register R7
        svc     0       @ Perform Service Call to Linux
        .end
```

Listing E.0: Sample source code for assembly language program

Leafpad Editor

Most of my students prefer to use the Leafpad editor and like its features due to its similarity to PC-based text editors. Students usually keep the following two windows constantly open and switch back and forth between them:

1. Leafpad full screen editor for constructing the source code
2. LXTerminal (Lightweight X Terminal emulator) for command lines

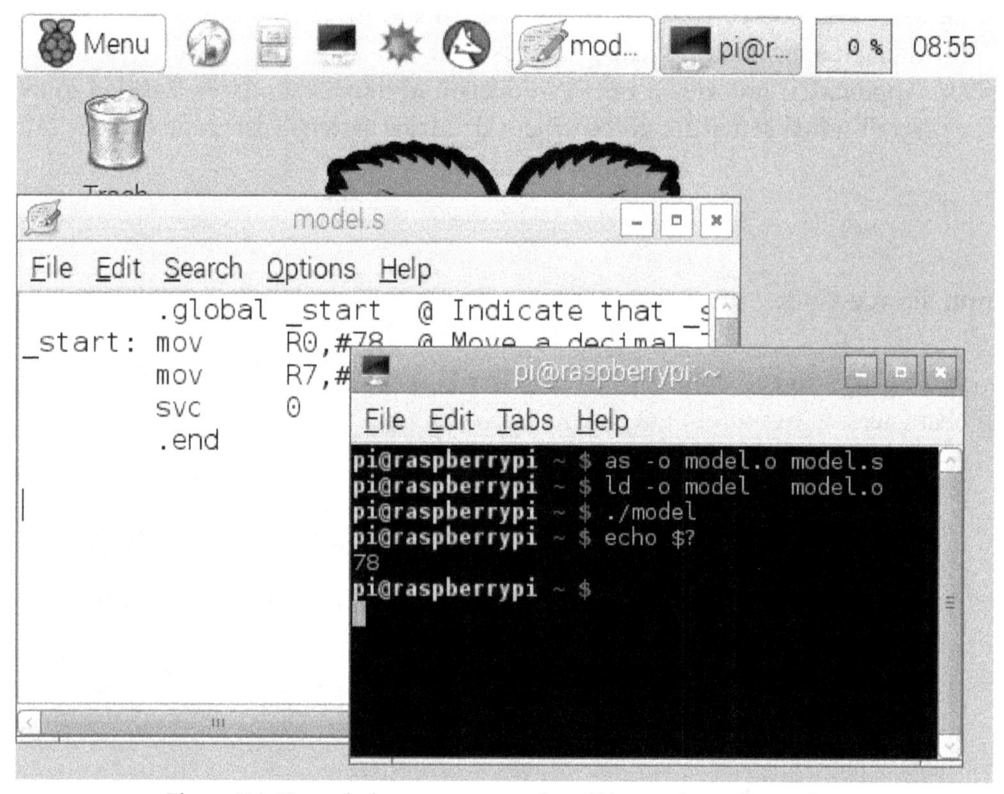

Figure E.0: Two windows open: one for editing and one for testing

The Leafpad editor is a "simple text editor" that can be opened from the accessories pull-down menu as show in Figure E.1. It can also be started by double clicking on the assembly language source code file displayed in the GUI file explorer.

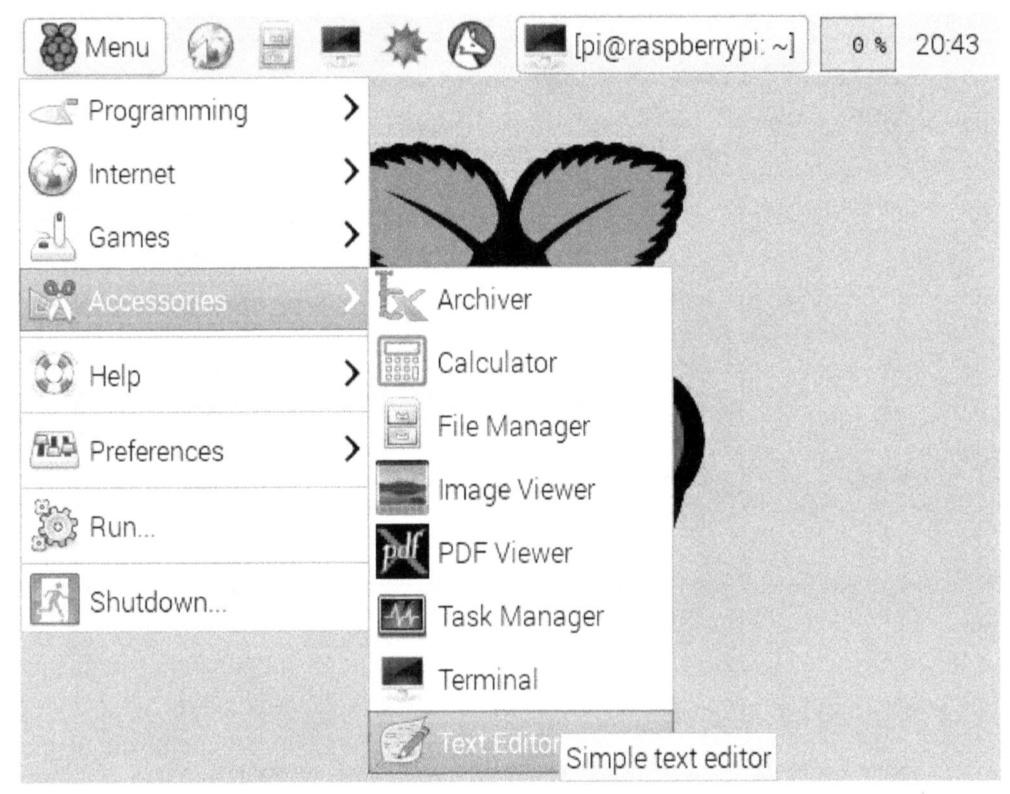

Figure E.1: LXTerminal and Leafpad in pull-down menu

Once open, the user can move around the screen with either the mouse or arrow keys. Editing is basically done by inserting new text at the current cursor position or replacing ("keying over") text at the current cursor position. The keyboard "Insert" key enables toggling between insert and replace mode.

As shown in Figure E.2, commonly used search and edit functions are accessed through pull-down menus as well as control keys.

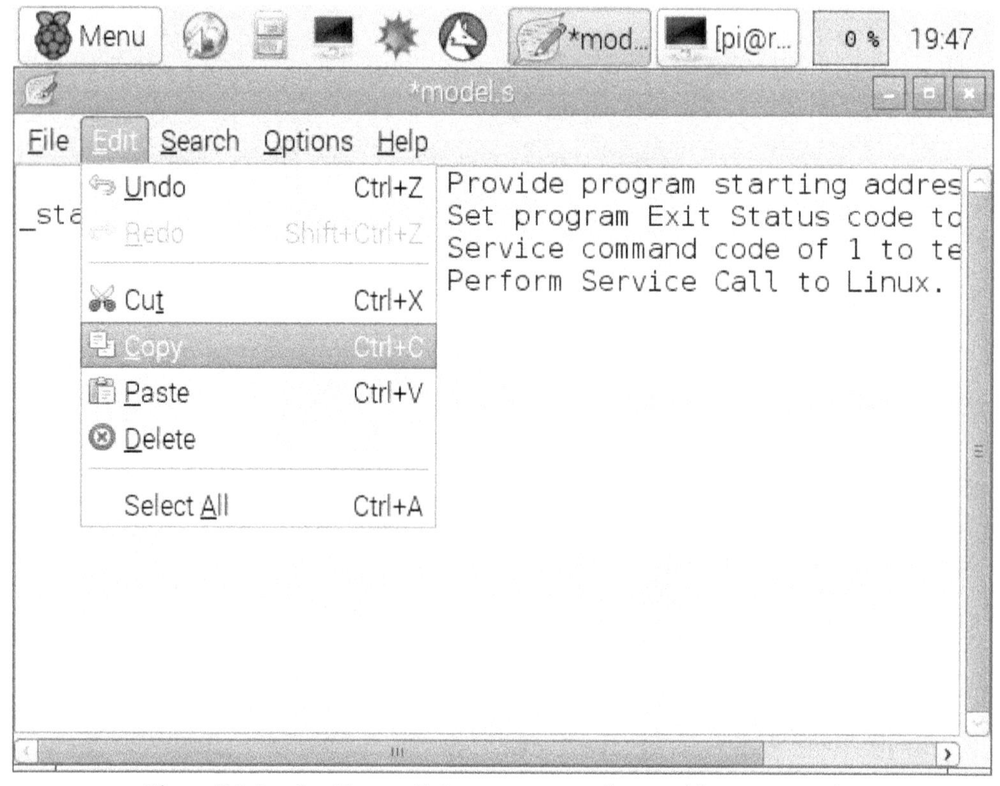

Figure E.2: Leafpad has pull-down menus and control-key commands.

Exiting the editor is done with either the control-q key or by "x-ing" out of the window. Most students don't exit, but use the control-s to save the file updates and then switch to the command window to assemble, link, and debug. After testing their updates, the students then return to the open editor window and continue modifying their programs.

Nano Editor

Many of my students prefer to work entirely in command line mode using an editor such as nano, vi, vim, or emacs. Of these, nano is by far the easiest one to begin using. It is started by entering “nano model.s” on the command line to edit a file named model.s.

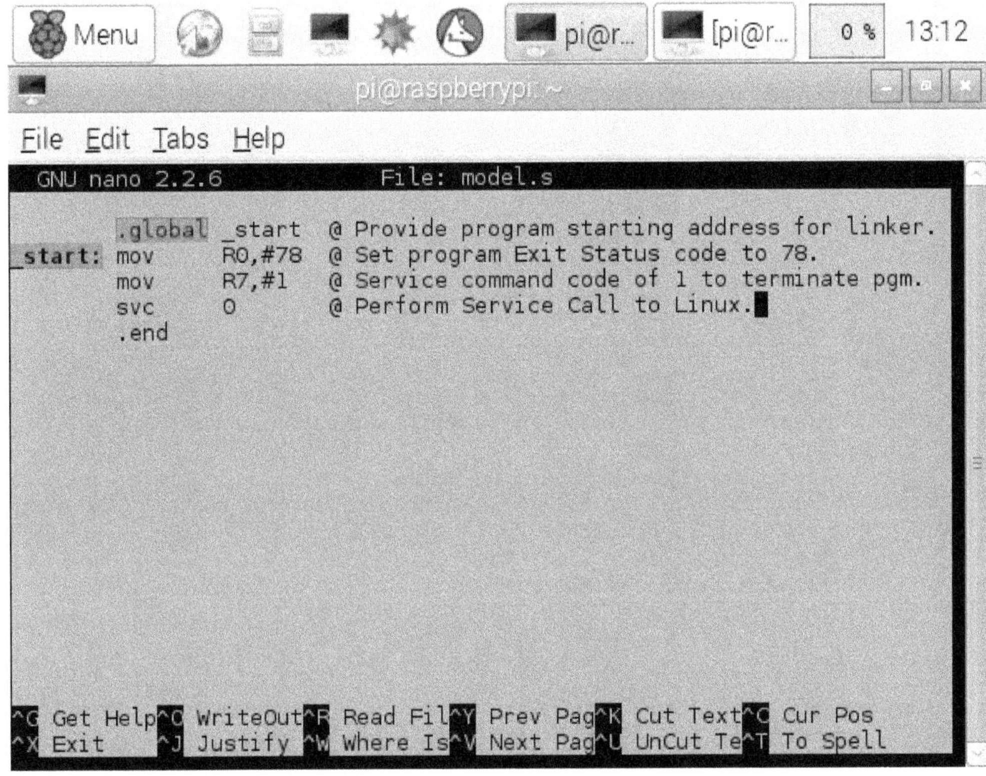

Figure E.3: Nano editor with first program displayed.

Several commands are available and each is invoked by holding down the control-key along with one of the letters listed on the bottom of the editing window as displayed in Figure E.3 and Table E.0.

^G Get Help	^O WriteOut	^R Read File	^Y Prev Page	^K Cut Text	^C Cur Pos
^X Exit	^J Justify	^W Where Is	^V Next Page	^U UnCut Text	^T To Spell

Table E.0: Nano control commands

The control-G "Get Help" command provides background on the nano editor and describes its editing commands as shown in Figure E.4. Some of the commands, such as control-J to justify text, are not very useful for an assembly language program. The copy-paste capability using control-K and control-U is present, but limited to one line of text at a time.

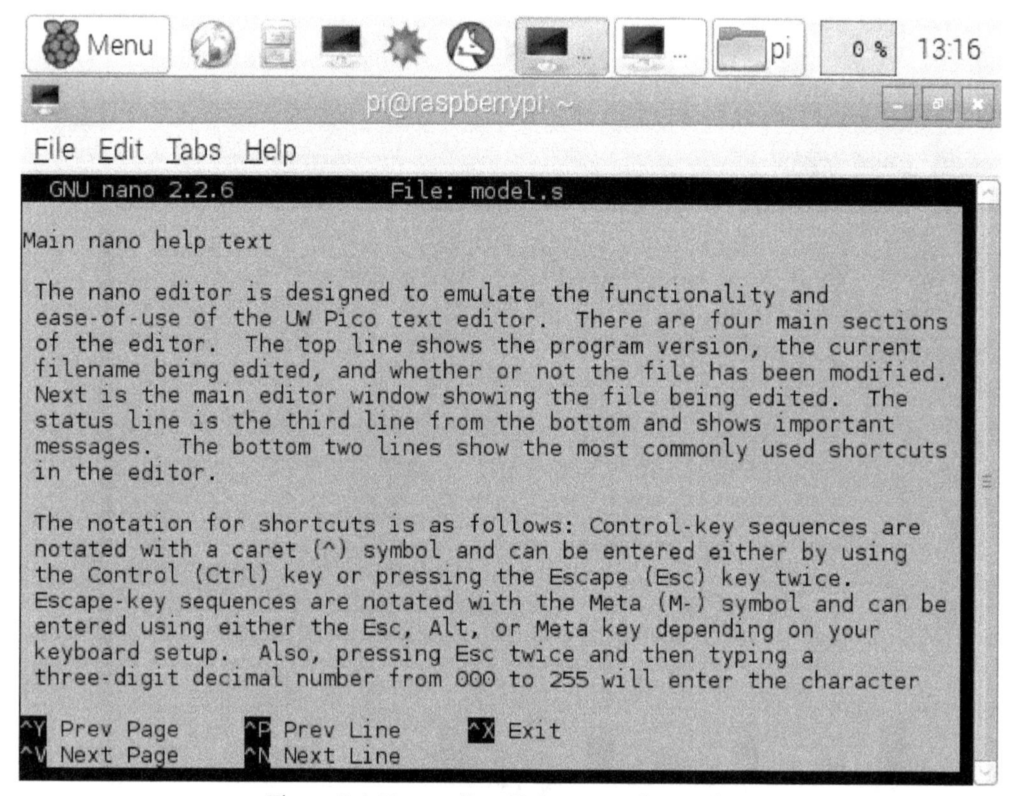

Figure E.4: Nano editor Help screen (control G)

Exiting from the nano editor is done using a control-X. If any updating was performed, the following prompt will appear. After answering "Y," a second prompt will appear to see if the file name to be written should be changed.

- Save modified buffer (ANSWERING "No" WILL DESTROY CHANGES) ?
- File Name to Write [DOS Format]: model.s

vi Editor

The vi editor is typically used by students who already have some experience with Linux or Unix. This will be an extremely brief introduction to vi. Once you get a little experience with it, please check the many Internet sites for many more features you might find useful.

The vi editor is started by entering "vi model.s" on the command line to edit a file named model.s. This editor has two modes: insert and command. Insert mode is similar to the full screen mode of nano and Leafpad where editing takes place by keying in new text at the current cursor location. The backspace key will delete the character to its left, and the delete key will delete the character at or to the right of the cursor. The keyboard "Insert" key will toggle between inserting and replacing ("key over") characters at the current cursor position.

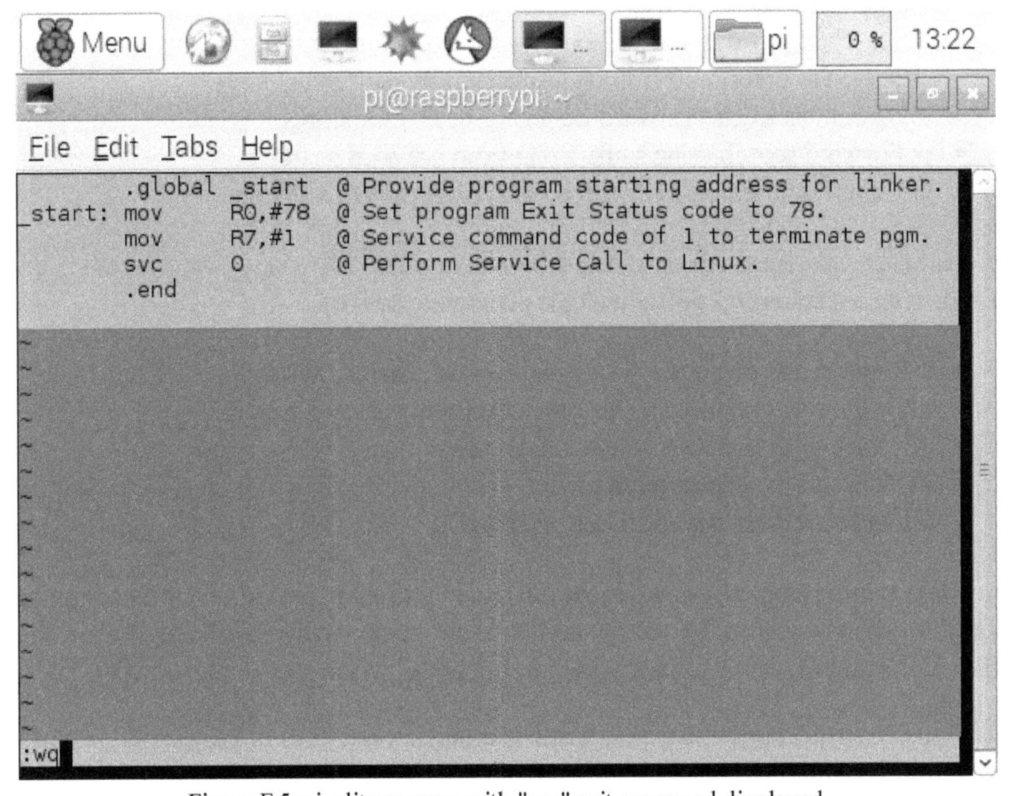

Figure E.5: vi editor screen with "wq" exit command displayed.

Command mode is entered by keying in the ESC key while in insert mode. One group of commands is initiated by keying in a colon character which brings the cursor down to a blank command line on the bottom of the screen. As shown in Figure E.5, the vi editor is typically exited while in command mode by entering one of the following responses to the "colon" command:

- :wq: Save updates and exit.
- :q!: Exit and ignore any changes.

Insert mode is entered when vi starts, and insert mode can be reentered from command mode by entering one of the following commands:

- i: Insert before current cursor, I: Insert at beginning of current line.
- a: Inset After current cursor, A: Append at end of current line
- o: Open a new line after current line, O: Open a new line before current line.
- dd: Delete current line and open new one in its place.
- R: Enter replace mode (i.e., "key over") at current cursor.

While in command mode, the cursor can be moved using the arrow, page-up, and page-down keys. It can also be moved using one of the following:

- G: Go to the end of the file. If preceded by a number, such as "7G" the cursor will go to beginning of line 7.
- /: Search forward in the file for the string (cursor temporarily drops to bottom of screen to enter search string terminated by the "enter" key).
- ?: Search backward (toward beginning of file) for string.
- n: Forward search using same string from previous search.
- N: Backward search using same string from previous search.

Single characters and lines can be deleted by entering the following keys. Preceding the commands with a number will delete multiple characters or lines.

- r: Replace one character (i.e., type over) at current cursor.
- x: Delete one character to the right of current cursor position.
- X: Delete one character to the left of cursor.
- D Delete to the end of the line.
- dd Delete current line (3dd deletes 3 lines).

Lines of text can be copied (yanked) into a buffer and later pasted. By preceding the following commands with a quote (") and a letter name, multiple unique buffers are available. For example, "a3yy copies 3 lines to buffer "a" which can be pasted with "ap.

- yy: Yank the current line (i.e., copy text into a temporary buffer).
- p: Put (or paste) saved buffer contents after the line of current cursor.
- P: Put saved buffer contents before the current line.

The search and replace command is entered following a colon. Its general format is %s/str1/str2/gc where occurrences of str1 will be replaced by str2. If the % is omitted, only the current line is searched, and if the g is omitted, only the first occurrence will be replaced. If the c is included, then all substitutions will be prompted for a confirmation.

Appendix F
ARM Instructions

Appendix F provides a list of ARM instructions used in this book and where they first appear in a program listing. The first column in Table F.0 contains a three number field indicating the lab number, listing number, and text line number where the instruction can be found. For example, 4.17.13 indicates the instruction is on line 13 of Listing 17 in Lab 4. This book's index also references locations of instructions by page number.

Not all ARM instructions are represented in this book. There are many very good technical reference manuals on the ARM processor available on the Internet that provide tables of all possible instructions.

1.0.2.	mov	R0,#78	@ Move "immediate" a decimal 78 value into register R0
1.0.4.	svc	0	@ Perform Service Call to Linux
2.1.4.	add	R0,R6	@ Add R6 contents to R0 ([R0] = [R0] + [R6]) or 17+2=19
2.2.4.	and	R0,R6	@ [R0] = [R0] & [R6] or 1011 & 1100 = 1000 (decimal 8)
2.3.3.	lsl	R0,#2	@ Shift R0 left 2 bits (i.e., multiply by 4)
2.4.4.	lsl	R0,R6	@ Shift R0 left by the value in R6 (i.e., multiply by 4
3.0.2.	ldr	R1,=msgtxt	@ Set R1 pointing to message to be displayed
3.0.3.	mov	R2,#10	@ Number of bytes in message
3.1.4.	bl	v_asc	@ Call subroutine (Branch and Link) to view text string
3.1.22.	bx	LR	@ Return (Branch eXchange) to the calling program
4.0.13.	cmp	R0,#0	@ Set condition flags for True or False.
4.0.14.	ldrne	R1,=T_msg	@ Load pointer to "True" if z-flag clear.
4.17.13.	push	{R0-R7}	@ Save contents of registers R0 through R7.
4.17.22.	pop	{R0-R7}	@ Reload contents of registers R0 through R7.
5.0.19.	and	R1,R4,R3,lsr R6	@ Logical AND with 4 registers and a shift
5.0.26.	subs	R6,#1	@ Subtract and set condition codes
5.0.27.	bge	nxtbit	@ Branch if greater or equal
5.8.18.	movhi	R6,#0	@ If bad range, default to displaying only 1 bit.
7.0.27.	rsb	R3,R3,#0	@ Reverse Subtract [R3] = 0 - {R3].
7.0.34.	ldr	R5,[R6],#4	@ Load next 32-bit word from table
			@ Load next character from string (and

			increment R2 by 1)
8.10.1.	stmeqdb	R8,{R0-R7}	@ Store R0-R7 only if Z-flag is set
8.10.2.	ldmgtdb	R8,{R0-R7}	@ Load R0-R7 only if !Z and (N = V)
8.12.17.	tst	R3,#1<<20	@ Test if this is a load or store instruction.
8.15.45.	bleq	v_ascz	@ Output "]" to close post-indexed format.
11.1.19.	umull	R3,R1,R4,R3	@ Unsigned 64-bit product: {R3},[R1] = [R4] * [R3].
12.3.19.	orr	R3,#0x80000000	@ Set the "assumed" high order bit.
17.1.18.	mla	R3,R4,R3,R0	@ Shift accumulated value and add. [R3] = [R4]*[R3] + [R0]
19.0.5.	clz	R1,R0	@ Get number of zero bits left of first one.

Listing F.0: Program location for first appearance of ARM instruction

Appendix G
Vector & Floating Point Instructions

Appendix G provides a list of coprocessor instructions used in this book for vector and floating point operations. The first column in Table G.0 contains a three number field indicating the lab number, listing number, and text line number where the instruction first appears. For example, 13.0.3 indicates the instruction is on line 3 of Listing 0 in Lab 13.

Not all floating point and vector instructions are represented in this book. There are many very good technical reference manuals on the VFPv2, VFPv3, and NEON processors available on the Internet that provide tables of all possible instructions.

13.0.3.	vldr	S0,[R6]	@ Load next 32-bit floating point test value.
13.0.5.	vmul.f32	S0,S1	@ Multiply the two floating point operands
13.0.6.	vmov	R0,S0	@ Move the product into ARM register for display.
13.5.3.	vldm	R6!,{S0,S1}	@*Load next two 32-bit values.
14.0.3.	vldm	R6!,{D0,D1}	@ Load next two 64-bit values.
14.0.4.	vmul.f64	D0,D1	@ Multiply the two floating point operands
14.0.5.	vcvt.f32.f64	S0,D0	@ Convert product back into single precision
14.2.3.	veor.64	D2,D2	@ Initialize running total to zero.
14.2.5.	vmla.f64	D2,D0,D1	@ Multiply 2 operands and add to total.
15.2.9.	vmrs	R0,FPSCR	@* Copy floating point FPSCR to update it.
15.2.12.	vmsr	FPSCR,R0	@* Restore the floating point PSCR.
15.3.6.	vand	D0,D1,D2	@ Perform the logical operation
15.5.22.	vadd.s8	D0,D0,D2	@ Perform 8 additions in parallel
15.11.22.	vaddw.s8	Q0,Q0,D2	@ Perform 8 additions in parallel
15.13.6.	vstm	R1,{D0-D2}	@ Save exact copy for display.
15.13.8.	vst1.8	{D0-D2},[R1]	@ Save as 1 "element" of 8 bits each
15.13.10.	vst3.8	{D0-D2},[R1]	@ Save as 3 "elements" of 8 bits each
15.13.12.	vst3.16	{D0-D2},[R1]	@ Save as 3 "elements" of 16 bits each
18.1.45.	vcvt.f32.s32	S0,S3	@ Convert from integer to floating point format.
18.1.48.	vldr	S1,point1	@ Load from memory into floating point register.
18.3.77.	vldrgt	S1,point1	@*Copy decimal 0.1 into floating point register.
18.3.78.	vldrlt	S1,dec10	@*Copy decimal 10.0 into floating point

19.2.14.	vclz.U32	Q1,Q0	@ Count the leading zeroes
19.2.15.	vshl.U32	Q0,Q1	@ Left justify each 32-bit word.

Table G.0: Locations of first use of vector and floating point instructions in this book

Appendix H
Assembler Directives

Assembler source files contain more than text lines that are simply translated on a one-to-one basis to machine language instructions. These additional instructions are for the assembler, rather than being instructions for the CPU, and are commonly referred to as pseudo-instructions or assembler directives. Almost all assemblers have had these directives, but their exact syntax varies considerably, even for assemblers targeted at the same CPU architecture. The GNU "as" assembler provides several of these directives that assist the compilation of assembler source code into machine language in the following ways:

- Declare which labels will be externalized for the linker: .global
- Separate instructions from data areas: .text, .data
- Allocate data types: .ascii, .asciz, .byte, .ds, .float, .word, .double
- Conditional assembly of code to be included or omitted: .ifc, .ifnc, .endif
- Define substitute variables and parameters: .equ, .set
- Generate repeated sections of code: .irp
- Generate custom sequences of instructions or data: .macro, .endm
- Specify address boundaries for generated code: .align

Access the GNU documentation for the "as" assembler for a complete list of assembler directives and their descriptions. The following table contains just a short description of each directive appearing in this book, so please see the index for the several locations where these directives are described and used in examples.

The second column in Table H.0 contains a three number field indicating the lab number, listing number, and text line number where the assembler directive can be found. For example, 9.1.59 indicates the directive is on line 59 of Listing 1 in Lab 9.

.align	9.1.59	Set memory address to half-word, word, double-word, ... border
.byte	15.15.39	Generate an 8-bit value
.arm	17.6.24	Generate 32-bit ARM machine code (instead of Thumb code)
.ascii	3.0.11	Put string of ASCII characters into memory
.asciz	8.1.19	Put string of ASCII characters into memory with null at end
.data	3.0.10	Inform linker to group following code with other data in memory
.double	14.0.16	Generate double precision floating point number
.ds	15.13.22	Reserve block of memory for data storage
.end	1.0.5	Identifies last line of text file (not required, but comforting to know nothing is missing)
.endif	9.13.24	Marks end of .ifnc and ifc conditional assembly
.endm	9.13.13	Marks end of .macro
.equ	16.1.2	Define assembly time constant
.float	12.4.15	Generate floating point number
.global	1.0.1	Inform the assembler of label to be passed on to the linker.
.ifc	9.13.22	Assemble the following code if two string arguments have same value
.ifnc	9.13.50	Assemble the following code if two string arguments have different values
.macro	9.13.10	Define assembly-time function
.set	9.13.59	Assign value to assembly time variable
.thumb	17.6.8	Start generating Thumb machine language (most are 16-bit)
.text	3.1.4	Inform linker to group following code with other instructions in memory
.word	7.0.64	Generate a 32-bit binary integer

Table H.0: Short descriptions of assembler directives

Appendix I
Linux Service Calls

One of the main responsibilities of an operating system, such as Linux, is to provide services for application programs. A large portion of these services involves reading and writing peripheral devices (display monitor, keyboard, mouse, network, etc.) and disk files (real spinning disks as well as solid-state memory devices). The calling program must provide Linux with the details of what is to be performed:

1. What is to be done (register R7)
2. Which device is to written or read (register R0)
3. Where the data buffer is in the program's memory (register R1)
4. How much data is to be written or read (register R2)

	List of Linux service calls introduced in Labs
1	Terminate the program.
3	Read bytes from device into memory buffer
4	Write array of bytes to device from memory buffer

Service 1: Terminate program

Application programs start when Linux gives them control at the "_start" label, and when a program chooses to quit, it will return control back to Linux using a service call (SVC 0 instruction).

```
mov     R7,#1       @ Service command code 1 terminates this program.
svc     0           @ Issue Linux command to terminate program
```

Listing I.0: Example from Lab 1 to quit program

Service 3: Read data from I/O device into memory buffer

Many custom devices and disk files can be supported, but there are a few device names that have become standard and appear in all Linux and Unix systems. The device *stdin* refers to the standard character input stream that by default is the keyboard, but can be redirected to an alternate device or file. This service call returns register R0 with the number of bytes that were read into the buffer. This byte count may be less than the value provided in register R2, but will not be more.

```
mov    R7,#3         @ Linux service command code to read a string.
mov    R0,#0         @ Code for stdin (standard input, i.e., keyboard)
ldr    R1,=msgtxt    @ Set R1 pointing to input buffer
mov    R2,#25        @ Maximum number of bytes to be entered.
svc    0             @ Issue command to read string from stdin.
```

Listing I.1: Example from Lab 15 to read characters from keyboard

Service 4: Write data from memory buffer to I/O device

A second standard device name is *stdout* which by default is the display monitor, but can be redirected to an alternate device or file.

```
mov    R7,#4         @ Linux service command code to write string.
mov    R0,#1         @ Code for stdout (standard output, i.e., monitor)
ldr    R1,=msgtxt    @ Set R1 pointing to message to be displayed
mov    R2,#10        @ Number of bytes in message
svc    0             @ Issue command to display string on stdout
```

Listing I.2: Example from Lab 3 to write to display monitor

Linux obviously provides many more services than what is listed above. Tables containing over one hundred service codes can easily be found on the Internet. Note: Some of the codes and their responses vary depending on the version of Linux being used.

Appendix J
GDB Debugger Commands

The GNU debugger software ("gdb" command) helps programmers examine what's occurring inside a running machine code program by providing the following features:

1. Set breakpoints to pause program execution
2. Examine register contents
3. Change register contents
4. Examine memory contents
5. Single step through a program
6. Trace program execution

The "as" assembler, "ld" linker, and "gdb" debugger are from the Free Software Foundation, Inc. and are included in the Raspian Linux distribution associated with the Raspberry Pi.

Figure J.0 shows command lines for the creation and testing of a program named "model." For an assembly language program to work with GDB, it should be compiled with the "-g" option. The "gdb model" command doesn't start the "model" program, but starts the debugger which then prompts for a command by outputting the "(gdb)" prompt.

```
pi@raspberrypi ~ $ as -g -o model.o model.s
pi@raspberrypi ~ $ ld -o model model.o
pi@raspberrypi ~ $ gdb model

GNU gdb (GDB) 7.4.1-debian
Copyright (C) 2012 Free Software Foundation, Inc.
License GPLv3+: GNU GPL version 3 or later <http://gnu.org/licenses/gpl.html>
This is free software: you are free to change and redistribute it.
There is NO WARRANTY, to the extent permitted by law. Type "show copying"
and "show warranty" for details.
This GDB was configured as "arm-linux-gnueabihf".
For bug reporting instructions, please see:
<http://www.gnu.org/software/gdb/bugs/>...
Reading symbols from /home/pi/model...done.
(gdb)
```

Listing J.0: Command lines to use GNU gdb debugger

Commands

The most important GDB command is the "help" (or "h" for short) command which provides the documentation on all the GDB commands. Examples using GDB commands appear in this book in the following lab listings:

- **b — break:** Define source code line numbers to suspend execution
  2.7, 3.3, 4.10, 4.18, 5.2, 7.1, 8.6, 12.1
- **c — continue:** Continue program execution after breakpoint
  3.4, 4.12, 4.16, 4.20, 5.4, 5.5, 7.1, 7.2
- **d — delete:** Remove active breakpoint set with **b** command
  5.5
- **i — info:** Dump all registers (r) or breakpoints (b)
  0.4, 2.8, 2.9, 2.10, 3.3, 3.4, 4.11,4.12, 4.14, 4.15, 4.19, 4.20, 5.3, 5.4, 5.5, 5.6, 8.6, 12.2
- **l — list:** Display source code lines
  2.7, 3.2, 4.10, 4.18
- **q — quit:** Exit GDB and return to Linux prompt
  2.10
- **r — run:** Start program execution and pause at next breakpoint
  0.4, 2.8, 2.10, 3.3, 4.11, 4.14, 4.19, 5.3, 7.1, 7.2, 8.6, 12.1
- **s — step:** Single step (execute) the next instruction(s)
  2.9, 4.15, 5.6
- **set — set:** Load new value into a register
  2.10, 4.14, 7.1, 7.2
- **x — dump:** Display section of memory
  3.5, 4.13, 8.1, 8.3, 8.6, 8.8, 8.11, 12.1, 13.5, 17.9

Examples of commands that have parameters:

- **b n:** Set breakpoint at source code line "n"
  b 5 — Set breakpoint at source code line 5
  b +5 — Set breakpoint 5 source code lines from current location
  b hloop — Set breakpoint at label "hloop:"
  b *0x3270 — Set breakpoint at instruction address 0x3270
- **d n:** Remove breakpoint number "n" (see "i b command")
  d 2— Disable breakpoint number 2
- **i n:** Info on "b" breakpoints or "r" registers
  i b — List current breakpoints and tracepoints
  i r — Display contents of all registers
  i t — List tracepoint contents
- **l n:** List 10 lines centered around line number "n"
  l 8 — List lines 3 through 12
- **s n:** Single step "n" source code lines
  s 3 — Continue execution through 3 more source lines
  s 1 — This is the same as no parameter (1 instruction)
- **set $rn = val:** Load new value into a register
  $r0 = 0x3A — Load register R0 with hex value 3A
- **x/nfu a:** Dump "n" units of size "u" in format "f" starting at address "a"
  n = (1 to ...); **f** = (a:address, c:character, d:decimal, f:float, i:instruction, t:binary, x:hexadecimal); **u** = (b:bytes, h:half words, w:words, g:giant double-word); **a** = (hex or decimal)
  x/10cb 0x100A0 — Display 10 characters starting at address 100a0
  x/10wx 0x100A0 — Display 10 words in hexadecimal starting at address 100a0
  x/3wi 0x8088 — Dump three instructions (disassemble)
  x/1cs 0x100a0 — Display memory at hex address 100a0 as characters 1 string long (ends with byte of zero)

Appendix K
Command Lines

The following commands are useful for running the examples in this book. I have provided only summary information here, but complete descriptions can be obtained by entering "man" followed by the command name. The Internet, of course, provides a wealth of information regarding each of the following.

ar rvs view.a v_ascz.o v_hex.o
The archive command builds static object libraries. See Lab 9 for more information and examples.

as -g -o model.o model.s
The "as" command compiles assembler source files to make object files.

cat model.s
The "cat" command is typically used to list the contents of text files.

cd RPi_Asm_Bridge
The "cd" command is used to change to a different directory.

cp RPi_Asm_Bridge/Listing_1_2.txt model.s
The "cp" command copies the first file to the second.

echo $?
Display the contents of the status variable from a previously executed program or command with the echo command. See Lab 1 for an example.

gcc -o model model.c sum.s
The "gcc" command will compile both C and assembler source code as well as link the resulting object files. See Appendix L for more information and an example.

gdb model
The "gdb" debugger helps diagnose assembly language programs. See Appendix J and Lab 2 for details and examples.

git clone https://github.com/robertdunne/RPi_Asm_Bridge.git

See Appendix A for information on using the git utility to download source code used in this book which is stored on the GitHub Internet site.

ld -o model model.o view.a

Link together object files to make an executable program. See Lab 1 for more information and examples.

ls RPi_Asm_Bridge

The "ls" command lists the names of the files in a directory. If no directory name follows the ls command, then the contents of the current directory are listed..

make

The "make" utility is generally used to build a sequence of command lines to compile source code files and then link the resulting object files. The names of all the files and how they should be processed are contained in another file named Makefile..

man cat

Display the manual for a command. This example will provide the documentation for the "cat" command, while "man as" will provide documentation for the "as" assembler.

nano model.s

Edit a file named model.s located in the current directory. See Appendix E and Lab 1 for details and examples.

rm model.s

Remove (i.e., delete) a file.

sudo apt install git

Enter superuser mode for one command line to download and install the *git* application

./model

Execute the program in a file named "model" located in the current directory.

Appendix L
C Programming

Assembly language programs target specific CPU architectures. The C programming language was developed at AT&T Bell Laboratories as a "portable" alternative to assembly language for developing software such as operating systems that needed low-level access to the hardware.

Although almost all of the embedded systems that I have programmed consist of a combination of assembly language with a higher level language like C, I wanted this book to be focused 100% on assembly language. However, I do want to provide a brief clue as to how to call an assembly language subroutine from a C program.

There are basically two techniques used to pass a variable's data in arguments to a subroutine:

- Pass by value: The value is passed in a register or on the stack to the subroutine, and the subroutine has no access to the source variable itself.
- Pass by reference: The memory address of a variable is passed, and the subroutine can actually update the variable in the calling routine's data area.

Listing L.0 contains a very short C main program that calls two assembly language routines that calculate the sum of an array of 32-bit integers:

- thesum: Subroutine that returns the sum to a reference argument
- fcnsum: Function that returns the sum as the return value of the function

```
1.   #include <stdio.h>
2.   #include <stdlib.h>
3.
4.   int main() {
5.       int count = 3, totalA, totalB, tstdat[] = {11, 45, 70};
6.       thesum (&totalA, tstdat, 3);
7.       totalB = fcnsum (tstdat, count);
8.       printf ("Sum from subroutine = %d\n", totalA);
9.       printf ("Sum from function = %d\n", totalB);
10.      return 0;
11.  }
```

Listing L.0: Main C program calling a subroutine and a function

Listing L.1 provides a file that contains both the function and subroutine called by the C main program. Notice that the arguments appear very similar to what we have been using in all the labs.

```
1.              .global    thesum       @ Subroutine entry address to linker
2.              .global    fcnsum       @ Function entry address to linker
3.
4. @            Subroutine thesum adds a variable number of integers.
5. @                       R0: Memory address of variable to receive the sum.
6. @                       R1: Memory address of array of integer values
7. @                       R2: Number of integers in the array
8. @                       LR: Contains the return address
9. @                       Registers R1 through R3 will be not saved.
10.
11. thesum:     ldr        R3,[R1],#4   @ Load first value.
12.             subs       R2,#1        @ Decrement number of integers.
13.             ble        retsub       @ Return with just one value.
14.             push       {R4}         @ R4 contents must be preserved.
15. thelp:      ldr        R4,[R1],#4   @ Load next interger in list.
16.             add        R3,R4        @ Add it to the running total.
17.             subs       R2,#1        @ Number of integers still to add.
18.             bne        thelp        @ Continue with next integer
19.             pop        {R4}
20. retsub:     str        R3,[R0]      @ Return sum to calling program
21.             bx         LR           @ Return to calling program
22.
23. @           Function fcnsum adds a variable number of integers.
24. @                      R0: Memory address of array of integer values
25. @                      R1: Number of integers in the array
26. @                      LR: Contains the return address
27. @                      R0: Return calculated sum to calling program.
28. @                      Registers R0 through R3 will be not saved.
29.
30. fcnsum:     ldr        R3,[R0],#4   @ Load first value.
31.             subs       R1,#1        @ Decrement number of integers.
32.             ble        retfcn       @ Return with just one value.
33. fcnlp:      ldr        R2,[R0],#4   @ Load next interger in list.
34.             add        R3,R2        @ Add it to the running total.
35.             subs       R1,#1        @ Number of integers still to add.
36.             bne        fcnlp        @ Continue with next integer
37. retfcn:     mov        R0,R3        @ Return sum to calling program
38.             bx         LR           @ Return to calling program
39.             .end
```

Listing L.1: Assembly language subroutine and function called by C main program

Use the "gcc" command to compile the C main program and assembly language subprograms as seen on the first line in Listing L.2. This one command line will also link this simple program. The second line of Listing L.2 obviously runs the program, and the final two lines are the output from the program.

```
~$ gcc -o model model.c sum.s
~$ ./model
Sum from subroutine = 126
Sum from function = 126
```

Listing L.2: Compile/link, execute, and two lines of output

Generally speaking, the difference between a function and a subroutine is a function returns a value and a subroutine does not. However, as seen in this example, a subroutine can return one or more values through arguments "passed by reference." In situations where only one value is returned, it is advisable to only use a function because 0) it hides the location of the actual data, 1) it is more efficient, and 2) it is expected to be done (self documenting).

- Arguments are passed in registers R0 through R3. If there are more than four arguments, those are pushed onto the stack.
- The calling program does not expect the contents in registers R0 through R3 to be preserved.
- A function returns its value in register R0.
- The Link Register (LR) contains the return address.
- Arrays are passed by reference.
- Constants and single variables are passed by value.
- A single variable can be passed by reference if preceded by an ampersand.

Arguments	Pass by __	Location
thesum (&totalA, tstdat, 3);	Subroutine	
&totalA	Reference	[R0]
tstdat	Reference	[R1]
3	Value	R2
totalB = fcnsum (tstdat, count);	Function	
tstdat	Reference	[R0]
count	Value	R1
totalB	Value	R0

Table L.0: Subroutine and function arguments in example program

AAPCS Subroutine Interface

The *Procedure Call Standard for the ARM Architecture (AAPCS)* properly describes what I have just briefly introduced. This standard, which is part of the *Application Binary Interface (ABI) for the ARM Architecture*, is not only used for C, but can be used for other languages as well. Very similar techniques have been available on other architectures for decades.

Java, C++, C#

The C++ language is an enhancement to C to incorporate some object oriented programming features. C++ is not a pure object oriented language like Smalltalk, but a hybrid. Calling assembly language subroutines from C++ is possible, but there are more types of arguments possible. I won't describe C++ here as most embedded systems use "straight" C rather than C++.

Although the original incentive for developing the Java language was to develop embedded systems, it has expanded to encompass much of the computer industry. Java source code, which looks very similar to that of C++, is compiled into an intermediate "byte code" object code rather than true CPU-specific machine code instructions. This byte code is then run interpretively on top of a "Java virtual machine." It is interesting to note that there is another coprocessor within the Raspberry Pi that facilitates the running of the "Java virtual machine."

C# is one of Microsoft's *.net* programming languages and has a syntax similar to C++, but compiles into an "assembly" intermediate object code that runs on top of Microsoft's "framework" which is similar to a virtual machine. Neither Java nor any of the .net languages interfaces as easily with native assembly language as does C.

Appendix M
Electronic Interfaces

Input/Output (I/O) programming is somewhat like catching a passenger train. If the train is stopped and at the platform at the station, it's very easy to get on and off. However, catching a moving train from grade level is more complicated. Proper handling of sophisticated I/O devices is a fine balance of synchronization and throttling which is beyond the scope of this book. Assembly language is an excellent environment for handling I/O devices, so I'll use this appendix to whet your appetite for further study. I do recommend that students first use a program like Python to get some experience with devices and I/O ports before embarking on serious assembler I/O.

I/O Processor

In addition to the ARM processor and the IEEE 754 compatible floating point processor, the Raspberry Pi includes other processors, one of which is a specialized I/O processor. Technical details of the GPIO (General Purpose Input Output) processor can be found by downloading the BCM2835 documentation over the Internet.

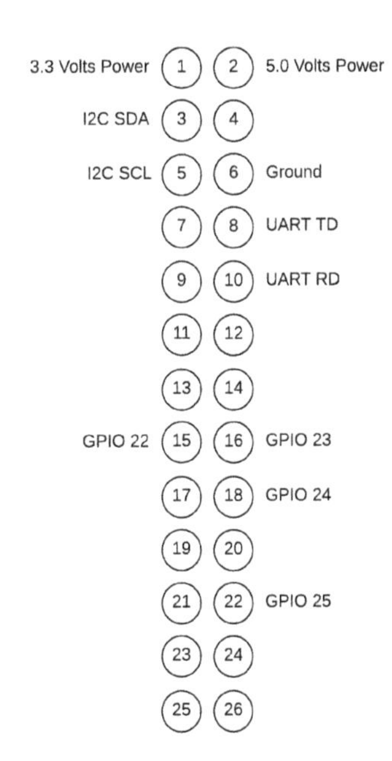

Figure M.0: GPIO pins

The Raspberry Pi includes a 40-pin connector containing a variety of I/O lines, a sampling of which are labeled in Figure M.0. Please note: The original Raspberry Pi had only 26 pins and with the exception of pin 13 has the same pin definitions as the first 26 pins of later editions.

The first thing to notice is that three of the pins are labeled as 3.3 volts, 5.0 volts, and ground. This is a simple warning that shorting the wrong I/O pins together will electrically destroy the Raspberry Pi, so be careful. A second observation is that some of the pins have been specialized for serial I/O, either I^2C or RS232 format, but not with RS232 voltage levels.

I/O processors are not a new idea, but were common on mainframes during the 1960s. Even early microprocessors had separate I/O processors such as the Intel 8089 associated with the very popular Intel 8086.

Virtual Files

If you really want to try out a little I/O, I recommend starting with Linux's virtual file system where Linux acts as an intermediary between the user program and the I/O port. The first line in Listing M.0 tells Linux to construct a virtual file for GPIO port 24 which according to Figure M.0 is pin 18 on the 40 pin connector.

Command lines 2 and 3 of Listing M.0 change the default directory to that associated with the newly created virtual files and list them. The direction file will be either "in" or "out." The first echo command configures this pin for output by writing "out" to the direction file. The next echo command will actually set pin 18 to 3.3 volts (a logical "high" value). The final line of Listing M.0 sets pin 18 to ground level (a logical "low").

```
~$ echo 24 > /sys/class/gpio/export
~$ cd /sys/class/gpio/gpio24
~$ ls
active_low device direction edge power subsystem uevent value
~$ echo out > direction
~$ echo 1 > value
~$ echo 0 > value
```

Listing M.0: Generate virtual I/O file and outputting to GPIO 24 (pin 18).

To read pin 18, the direction must be configured as "in," and then the cat command displays the "value" file. In this example, 3.3 volts will provide a high value of 1.

```
~$ echo in > direction
~$ cat value
1
```

Listing M.1: Read value on pin 18 (GPIO 24 port).

Appendix N
Thumb Code

Thumb is an alternate "instruction set" for the ARM CPU. It uses the same registers, but has a different instruction format. There are three variations of Thumb:

1. Thumb: The original only had instructions that were 16-bits in size.
2. Thumb-2: This enhancement added some 32-bit instructions to the original 16-bit instructions, providing almost all the capability as code written using the regular ARM instruction set. Note: Thumb is really not its own instruction set, but an alternate view of the regular ARM instructions.
3. ThumbEE: This Execution Environment modification of Thumb-2 is optimized for programming languages like Java that use an intermediate level language.

What is the advantage of using the Thumb set of instructions? Performance. Because Thumb instructions are only 16-bits wide, they take up less room in memory and will load faster in hardware configurations having a 16-bit data bus or smaller. Today's configurations, having 32-bit data buses and large memories, could actually run slower in Thumb-coded programs due to the Thumb's lack of features present in the 32-bit ARM format as well as the overhead due to switching between ARM and Thumb modes.

How does the ARM CPU know the difference between regular ARM instructions and Thumb instructions? If the T-bit (CPSR bit 5) is set to 1, the CPU assumes the PC register (Program Counter) is pointing to a Thumb format instruction, while if the T-bit is zero, the instruction is assumed to be a 32-bit ARM instruction. Note: Do not switch between Thumb state and full ARM instruction format by directly setting this bit. The BX (Branch Exchange) instruction is used to change states.

Are the Thumb instructions a subset of the set of ARM instructions? This is close to being true, but not exactly true. Most of the Thumb instructions are a limited version of corresponding 32-bit ARM instructions. Since Thumb instructions are converted to 32-bit ARM instructions before being executed, isn't performance degraded by this extra step of conversion? No, there isn't an extra step. The ARM CPU substitutes the Thumb instruction decoding for the 32-bit ARM instruction decoding when the T-bit is set.

How is the Thumb instruction "set" different from the instruction sets for NEON and VFPv3 floating point coprocessors that are also present in the Raspberry Pi?

- Thumb is not a different processor, but more like a different "face" for many of the most common ARM instructions. Although they are very tightly coupled with the ARM CPU, NEON and VFPv3 are actually separate processors with their own set of data registers.

- The NEON and VFPv3 instructions are coprocessor instructions that exist within the regular 32-bit ARM instruction format. The Thumb instructions do not fit within the ARM format, and the CPU must have the CPSR T-bit set for it to recognize Thumb-format instructions.
- NEON, and to a limited extent VFPv3, can run in parallel with instructions running in the ARM CPU. The Thumb instruction format uses the ARM CPU, and therefore cannot run in parallel with itself.

Lab 17 contains a programming example using Thumb code. Why didn't I provide more information and practical examples using Thumb instructions? If this were 20 years ago, I would have covered it more extensively, but today, memories are large and data buses are wide. Covering Thumb to a greater extent would have added extra complexity without much gain in understanding.

Since thumb instructions are really repackaged ARM instructions, how can a 32-bit ARM instruction be stuffed into a 16-bit package? Obviously, some things have to be eliminated and other things assumed. Figure N.0 provides a sample Thumb instruction being mapped into machine code, and Lab 18 shows several Thumb instructions in 16-bit format.

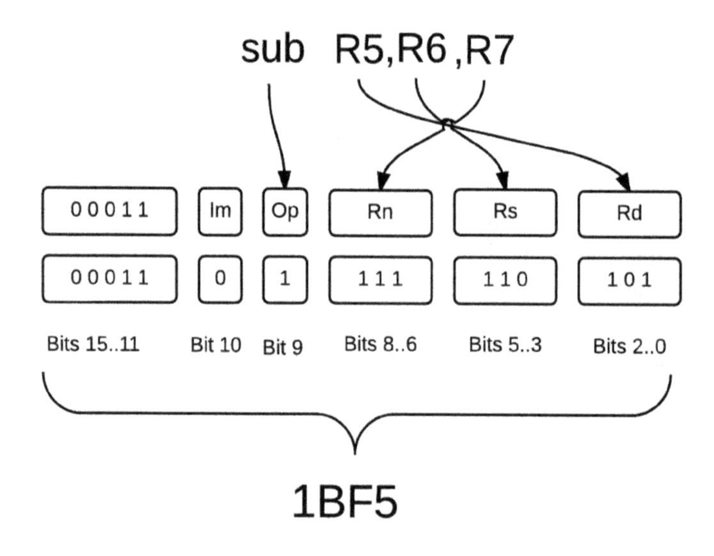

Figure N.0: Thumb add/subtract machine code example

Answers to Selected Questions

Questions marked with an asterisk (*) in the Maintenance section of each lab have their answers, or at least hints, provided below.

0.4 "By hand, without a calculator or computer," convert the following numbers expressed in decimal to binary format. See Appendix B if you need some background in binary.

a. $21_{10} = 10101_2$
b. $63_{10} = 111111_2$
c. $16_{10} = 10000_2$
d. $129_{10} = 10000001_2$

0.5 "By hand, without a calculator or computer," convert the following numbers expressed in binary to decimal format. See Appendix B if you need some background in binary.

a. $1011_2 = 11_{10}$
b. $1100101_2 = 101_{10}$
c. $10110_2 = 22_{10}$
d. $100001_2 = 33_{10}$
e. $1111011_2 = 123_{10}$

1.3 What is the main difference between a procedural computer program and a non-procedural one?

> Non-procedural AI programs must "learn" how to work by being given a large set of sample data along with the desired results. For procedural programs, such as all programs in this book as well as the vast majority of all current computer applications, the programmer "tells" the program how to work. Between these two extremes, there is also a variety of non-procedural "techniques" such as database and data flow manipulation languages.

1.5 When updating a line of source code, should the comment on the line be updated as well?

Yes, usually. However, sometimes the comment is right, but the code was wrong. For example, the comment said why the line of code was present, but the code did not work. The worst case is when someone changed what the code was supposed to be doing, but left the old comment which is now irrelevant and much worse than no comment at all.

2.8 Which register will be changed after executing each of the following instructions, and what is its new value? Assume that each instruction begins with the following contents: R0 = 0, R1 = 1, R2 = 2, R3 = 3, R4 = 4, and R5 = 5.

a. AND R2,R3 changes nothing (R2 is "changed" to 2, but it already was 2)
b. ADD R4, R5 @ changes contents of R4 to 5
c. LSL R1, #4 @ changes value of R1 to 16
d. EOR R3, #1 @ changes value of R3 to 2
e. MUL R4, R5 @ changes value of R4 to 20
f. SVC 0 @ changes nothing or anything (depends on the service call)

3.5 Which register(s) (including Z-flag in CPSR) will be changed after executing each of the following instructions, and what are their new values? Assume that each instruction begins with the following contents: R0 = 0, R1 = 1, R2 = 2, SP = 36010, PC = 32000, and condition code Z=1.

a. ORR R0, R3 @ changes [R0] = 3 and [PC] = 32896
b. ASR R3, #1 @ changes [R3] = 1 and [PC] = 32896
c. BL 34000 @ changes [LR] = 32896 and [PC] = 34000
d. BX LR @ changes [PC] = 36000
e. ADD PC, #4 @ changes [PC] = 32904 Note: If you thought the [PC] should be 32896, then you didn't include 8 bytes for "pipe lining."

4.2. The CMP instruction sets the NZCV status bits. Why do you think its mnemonic isn't cmps and would cmps work also?

The CMP instruction is like a SUBS instruction, except the difference of two operands is not placed into a result register. In other words, the only thing the CMP instruction does is set the NZCV flags (and update the PC, of course). The "S" suffix is always assumed, and therefore not necessary. The assembler even flags an error if CMPS is given. Could the assembler have accepted CMPS? Sure, but it doesn't. The Thumb arithmetic instructions are the same way. All set the NZCV status bits, so ADDS, SUBS, etc. are not allowed.

5.4 The instruction mov R1,#5120 should not be possible (due to the limitations noted in the previous question). What machine code instruction does the assembler generate to make it work? Clue: 5120 = 4096+1024.

MOV is one of the arithmetic/logic instructions that has its immediate value represented by an 8-bit base M rotated to the right by a 4-bit shift count S. The decimal value 4096 is 1 2^{12}, while1024 is 1 2^{10}. Adding the two in binary is 0b1010000000000, which can be represented as 0b101 shifted left 10 bit position or rotated right 22 bit positions. Therefore the lower 12 bits of the MOV instruction which contains the shift count 11 and base 0b101 is 0xB05. Other possibilities that would work are 0xC14 and 0xD50.

6.2. Compare the merits of a byte-addressable computer architecture to one that is word addressable.

Word-addressable:

- Don't have to be concerned about big and little endian.
- Don't have to worry about alignment issues. Many CPUs will either degrade performance or not load/store "word" instructions on addresses that are not multiples of four bytes.
- For CISC architectures that have the memory address inside the instruction, much more memory can be addressed. For example, the Univac 1108 had an address space of 18 bits, which is 256K words. This equated to 1.5M of 6-bit Fieldata characters (bytes). Some instructions even had a 16-bit limitation which is 65K words. Although these sizes seem minuscule today, in the 1960s, they were impressive.

Byte-addressable:

- No complicated sub-word instruction options Many word-addressable CPUs had a special field within their instruction format to select a particular sixth or quarter word. Some didn't even have that, so bytes had to be masked and shifted into places.
- A variety of integer sizes were available: bytes, words, half-words, and double words.
- Packed-decimal arithmetic was generally available.

7.4. In Listing 7.3, the CMP R3,#0 instruction on line 23 can be eliminated if line 16 is modified. What modification is this and why is it probably a bad idea even though it would function perfectly?

The MOVS instruction could set the NZCV status bits. However, there is a potential maintenance problem with having too many intervening instructions between the setting of the status bits and their ultimate use. Consider the possibility of a future maintenance programmer not noticing where the status bits are set and used, and then inserts a new instruction between them that changes the status bits. I do, however, like the fact a "move" instruction can optionally set the status bits. Most CPUs do not have this capability.

8.4. The LDR and STR instructions allow a negative offset. Is this negative direction set with two's complement or sign and magnitude format?

Technically, it's in the sign and magnitude format, but the sign and magnitude are not adjacent to each other. The "sign" is actually the U-bit in the LDR/STR instruction format, where U=1 means positive and U=0 means negative.

9.2. How is a macro different from a subroutine?

- A macro is called while the assembler is running, and a subroutine is called when the application program (being written) is running.
- A macro generates text lines that will later be "assembled," while a subroutine works with numbers and text of the running application.
- Each macro call makes the program physically larger and take up more memory, while subroutines generally reduce memory requirements by eliminating duplicate code.

9.3. Give an example of a useful macro that generates neither any instructions nor any data.

Just a few examples are listed below:

- .align statement to indicate a word or double word boundary will be used next.
- .set or .equ statements to assign values to constants used at assembly time.
- .text or .data to indicate where following instructions and data are to be placed.
- .arm or .thumb to indicate the instruction format for following code.

9.5. An emulator doesn't have to be programmed in assembly language. What would be the advantage of writing one in C or another higher level language?

The C programming language was developed to provide portability to system software. So, an emulator for the ARM could then be run on almost any computer. If the emulator was written in a hardware description language, such as VHDL or Verilog, then it would essentially become a fairly fast substitute for a real ARM CPU or be part of a reconfigurable computing device.

10.1. What are the advantages of fixed point over floating point?

- Fixed point is exact. No error is present.
- Fixed point arithmetic is very fast.
- It is very easy and fast to convert between character representation and fixed point.

11.1 "By hand, without a computer," convert the following decimal fractions into binary and provide the answers in hexadecimal.

a. $0.5_{10} = ½ = 0.1000_2 = 0.8_{16}$
b. $0.625_{10} = 0.1010_2 = 0.A_{16}$
c. $0.25_{10} = 0.0100_2 = 0.4_{16}$
d. $0.03125_{10} = 0.00001_2 = 0.08_{16}$
e. $0.0078125_{10} = 0.0000001_2 = 0.02_{16}$

11.2 "By hand, without a computer," convert the following binary fractions from hexadecimal back into real numbers in base 10.

a. $.C0000000_{16} = 0.11_2 = ½ + ¼ = ¾ = 0.75_{10}$
b. $.E0000000_{16} = 0.111_2 = 0.875_{10}$
c. $.10000000_{16} = 0.0001_2 = 0.0625_{10}$
d. $.50000000_{16} = 0.0101_2 = 0.3125_{10}$

11.3 Using a calculator for division by powers of 2, convert the following binary fractions from hexadecimal into the real number that each is "approaching" in base 10.

a. $.33333333_{16} = 0.00110011001100110011001100110011_2 => 0.2_{10}$
b. $.66666666_{16} => 0.4_{10}$
c. $.CCCCCCCC_{16} => 0.8_{10}$
d. $.E6666666_{16} => 0.9_{10}$

12.1 Convert the following real numbers into single precision IEEE 754 floating point and provide the answers in hexadecimal.

a. 128.0 is 43000000 in floating point
b. 9.25 is 41140000 in floating point
c. -9.25 is C1140000 in floating point
d. 0.03125 is 3D000000 in floating point
e. 128.03125 is 43000800 in floating point
f. 0.0 is 00000000 in floating point
g. -0.0 is 80000000 in floating point

12.2 Convert the following IEEE 754 floating point numbers back into real numbers in base 10.

a. 42a80000 is 84.0
b. C1A80000 is -21.0
c. 424C8000 is 51.125
d. BF100000 is -0.5625
e. 3DCCCCCD is 0.1

13.1 Why does the v_flt program display 0.5 when it should display 0 for a floating point number consisting of all 32 zero bits?

Hint: Is the floating point zero normalized or not? Does subroutine v_flt handle special cases of IEEE 754?

13.2 By examining Figure 13.1, what is the smallest absolute value non-zero normalized number?

Hint: Convert $1 \times 2^{1-127}$ to decimal.

14.3. Is getting that extra 1-bit of precision in the significant more important to the single precision, double precision, or half precision format numbers?

The half-precision floating point format is a 16-bit package containing a sign bit, five bits for the biased exponent, and ten bits for the significant. Note: Half-precision is only used for storage and is not supported for computation within IEEE 754 devices. That extra bit obtained by not taking up a bit-position for the most significant bit improves the resolution of the Half precision format the most.

- Half precision (significant is 10 bits): 1 in 2^{10}
- Single precision (significant is 23 bits): 1 in 2^{23}
- Double precision (significant is 52 bits): 1 in 2^{52}

15.3. If the lane size for integer addition within NEON could be one bit wide, it would be exactly the same as which logical operation?

The exclusive OR operation. Think back to your digital electronics days. How do you build a "half adder"?

16.3. What is a principal danger in using "pass by reference"?

One of the hallmarks of object oriented programming is "information hiding." If a part of a program does not need access to a part of the data, don't give it access. "Pass by reference" provides the location of the data to the subroutine, and if the subroutine makes a mistake, it can write over the original source of the data. In "pass by value," only a copy of the original data is sent as an argument to a subroutine. Of course, if all programs and subroutines worked perfectly, none of this would be a concern.

17.4. We switch between Thumb mode and ARM mode with the BX instruction. Although we could set the T-bit in the CPSR directly, why would this lead to problems if we did?

The quick and easy answer is that the ARM hardware reference manuals say don't do it, and a special way has been set up to perform the switch using the BX instruction. But also remember that the ARM has a "pipelined" architecture where by the time an instruction is actually executed, the following two instructions have already been fetched and are being prepared to execute. The 16-bit/32-bit instruction mode would be switched while instructions are in an intermediate state. The BX, like all branch instructions that execute, will clear the pipeline.

17.5. What simple change can be made to the c_int subroutine so that it works regardless of whether it is called with the T-bit already set or not?

Define a second entry point for a call from a routine already in Thumb mode. Then make sure that bit 0 of the LR is set so that the BX return instruction will leave the CPSR in Thumb mode.

18.1 The loop which multiplies the significant by either 10.0 or 0.1 to accommodate the base ten exponent results in some loss of precision in the conversion of ASCII character format to floating point. What two relatively simple modifications to that technique can greatly improve the resulting precision?

1. Do the multiplication using double precision.
2. Use a table of pre-generated precise powers of ten.

18.2 Why will multiplying by 0.1 always result in a loss of precision in binary computers?

Base ten is not a multiple of base 2, like base 8 and base 16 are multiples. Some numbers like 0.1 cannot exactly be represented as a base two fraction for the same reason 1/3 is $0.333333..._{10}$.

19.2 Name four operations available in NEON that are not available in VFPv3.

Logical, integer, shift, and count leading zeroes

Index

CPSIA information can be obtained
at www.ICGtesting.com
Printed in the USA
LVHW020847290122
709650LV00004B/174